The Classical and Quantum Dynamics of the Multispherical Nanostructures

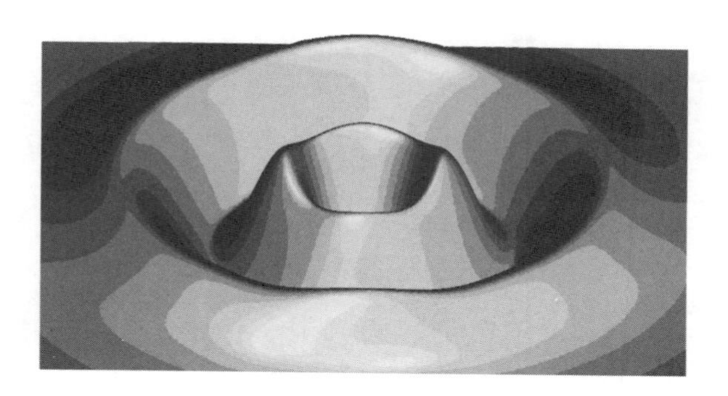

The Classical and Quantum Dynamics of the Multispherical Nanostructures

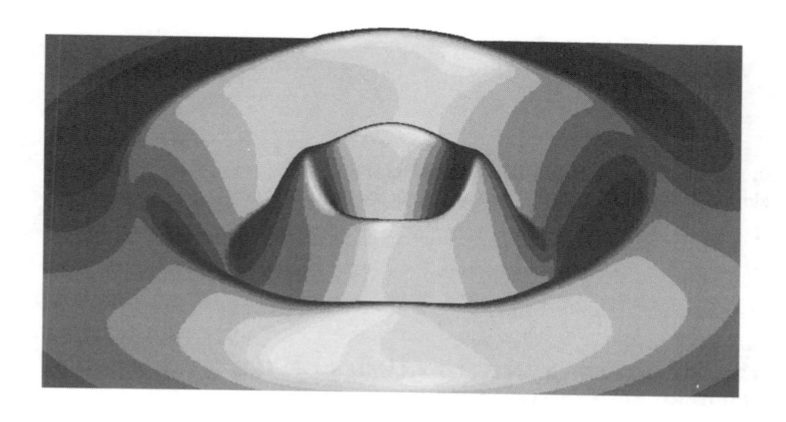

Gennadiy Burlak

Autonomous State University of Morelos, Mexico

ICP Imperial College Press

Published by

Imperial College Press
57 Shelton Street
Covent Garden
London WC2H 9HE

Distributed by

World Scientific Publishing Co. Pte. Ltd.
5 Toh Tuck Link, Singapore 596224
USA office: 27 Warren Street, Suite 401–402, Hackensack, NJ 07601
UK office: 57 Shelton Street, Covent Garden, London WC2H 9HE

British Library Cataloguing-in-Publication Data
A catalogue record for this book is available from the British Library.

THE CLASSICAL AND QUANTUM DYNAMICS OF THE MULTISPHERICAL NANOSTRUCTURES

ISBN 1-86094-444-2

Editor: Tjan Kwang Wei

Printed in Singapore by World Scientific Printers (S) Pte Ltd

Preface

Nowadays there are various emerging possibilities to produce dielectric microspheres with sizes of about 1 micron and less. The number of theoretical and experimental works on the subjects of microspheres increases every year. The most fruitful turns out to be the idea of transition from the passive use of natural volume waves to the active management by the properties of such waves by growing the necessary structures in a surface. Creation of multilayered alternating structures (a dielectric stack) in a surface of microspheres allows one to sharply reduce radiative losses in a necessary frequency range and thus effectively control the parameters of radiation from microspheres. The opportunity of localization of quantum objects (quantum points) in a small working volume of the microsphere allows for the creation of miniature quantum devices. Effects of the thin layers are especially important when the thickness of a layer becomes about a quarter wavelength of radiation. So for wavelengths of about 600 nanometers such thickness becomes 150 nanometers, and for metallo-dielectric layers it is even less. Thus in multilayered microspheres interplay of micro-nano-scales effects occurs, which determines the unique features of the coated microsphere. It has predetermined the theme of this book.

The various spherical micro and nano-structures are now heightened interest with experimenters, and theorists. The reason is that a dielectric microsphere possesses a number of unique features based mainly on an opportunity of energy conservation of optical oscillations in a very small working volume. Such microspheres possess natural modes of light oscillation at characteristic frequencies corresponding to the specific size and to the wavelength ratios. Presently only a spectrum of the optical modes having the large spherical quantum numbers (whispering gallery mode — WGM) is in use, and it is possible to observe the interesting phenomena to find the various engineering applications (see [Bishop *et al.*,

2003; Vahala, 2003 and references therein]). WGM oscillations in microspheres were observed in experiments over 15 years ago as oscillations with a huge quality factor Q ($Q = \mathrm{Re}\, w/2\,\mathrm{Im}\, w$) [Braginsky & Ilchenko, 1987; Braginsky *et al.*, 1989]. However all of them still remain an object of intensive researches. As a result it was possible to lower the spatial scales up to the size when interaction of fields with various quantum subsystems becomes rather effective [Artemyev *et al.*, 2001b and references therein]. Such phenomena are already described by quantum electrodynamics. Due to an opportunity of localization of fields in such a small volume (the radius of microspheres makes about $1{-}2\,\mu$m and less) it is possible to observe the nonlinear effects with very low threshold [Spillane *et al.*, 2002 and references therein]. A variety of interesting nonlinear phenomena in micro-droplets have been reported [Braunstein *et al.*, 1996 and references therein], and finally, the creation of the ensembles of such particles allows for the creation of structures with unusual wave properties [Furukawa & Tenjimbayashi, 2001]. Very often articles on microsphere application sounds the development on quantum computing [Corya *et al.*, 1998; Bouwmeester *et al.*, 2001; Kane, 1998; Khodjasteh & Lidar, 2003; Ozawa, 2002; Pachos & Knight, 2003; Raussendorf *et al.*, 2003; Vrijen *et al.*, 2000; Sorensen & Molmer, 2003 and references therein].

Despite of high cost of such microspheres, many important and interesting features of wave and quantum effects are already discovered. Nevertheless the results of the theorists and the ingenuity of the experimenters have made the microspherical topics far from being exhausted, having many effects still being expected.

From the point of view of the author the situation here reminds us of one earlier described in optics before the development of thin-film coverings. The development of thin-film technologies has led to the creation of new important directions which features are based on various new interference effects in films with thicknesses of about the wavelength of a radiation beam [Born & Wolf, 1980].

Similarly in a microsphere topic only the natural high-quality oscillations (WGM) with large spherical quantum numbers (or orbital angular momentum) in bare microspheres without coating are well investigated. Other oscillations with small spherical numbers (SSN) in such microspheres are not used, for they seem unpromising because of the low quality factor Q due to the leakage of energy in surrounding space. The microsphere is the so-called open system.

However in a number of works [Brady *et al.*, 1993; Sullivan & Hall, 1994; Burlak *et al.*, 2000] the deposition of alternating thin-film structures on a surface of microsphere is shown, which allows one to reduce the energy's leakage to a surrounding medium sharply. As a result the quality factor Q of such modes can increase to values typical for WGM modes. The oscillations having small spherical numbers in such structures are no longer undergoing discrimination and become involved in operation again. Thus the optical mode's spectrum in layered microspheres is used by more fruitful way.

A variety of geometries and a choice of the layers materials make such a coated system richer and it provides new opportunities which were absent in a pure microsphere case. We mention, for example, the occurrence of narrow peaks of the transparency in microspheres with metallo-dielectric layers below the metal plasma frequency [Burlak *et al.*, 2002 and references therein], or an opportunity of control of the threshold of field's generation by a change of number of layers in a spherical stack [Burlak *et al.*, 2002].

Various opportunities of microspheres have caused large interest in various international groups which study both classical and quantum aspects. Some theoretical models and methods become more complicated and not simply comprehensible for the beginner researcher in this theme.

This book is written to cover some classical and quantum aspects of the electromagnetic wave's processes in layered microspheres. Certainly, there are a number of excellent books and textbooks on basics of each of the mentioned aspects [Landau & Lifshits, 1975; Landau *et al.*, 1984; Jackson, 1975; Cohen-Tannoudji *et al.*, 1998; Scully & Zubairy, 1996, etc.]. We have tried to illuminate both aspects and provide references to new works. At the same time we do not discuss the nonlinear aspects which will be covered in another book.

However this book is not the review of new works. Some of such reviews already exist [Bishop *et al.*, 2003; Vahala, 2003; Gulyaev, 1998], and a number of them, apparently, are still in preparation. Even the linear part of a problem appears rather complicated because of the complex structure of a system, and also due to the fact that it is an open system. For example, to calculate the frequency dependences of the reflection or transmittance coefficients or a spectrum of eigenfrequencies and the Q factor of oscillations, it is necessary to use rather complex models and calculations. Due to a large number of relevant factors, the level of the organization of a program code, acquires the same importance as pure computing aspects. Though corresponding computing technologies are known for a long time (object-oriented programming — OOP), the use of this approach has become completely necessary in discussed problems. As OOP has yet to become the conventional technology in the medium of physicists, I have considered a necessity to illustrate in the book the details of such technology with reference to C++ language.

For these reasons the book consists of three parts: classical dynamics, quantum dynamics as well as numerical methods and object-oriented approach.

What does this book cover?

In this book some questions of the theory of classical optics and the quantum optics of the spherical multilayer systems are studied. In such systems the spatial scales have order magnitudes of the wavelength of radiation. This circumstance essentially complicates the analysis of such important electromagnetic properties such as reflectivity, transmission, and the quality factors, etc. Often such quantities cannot be calculated analytically and one has to use numerical calculations. The essential part in such research and development is occupied with computer calculations and modeling. The details of calculating electromagnetic properties of multilayered microspheres are written down comprehensively so that a university student can follow freely.

For skill-oriented point of view, the book covers the following:

1. Electrodynamics of multilayered environments in the spherical geometry.
2. Methods of calculation of both reflection and transmission coefficients for an alternating stack.
3. Calculations of eigenfrequencies and quality factors of electromagnetic oscillations.
4. The radial distribution of the electromagnetic fields.
5. Properties of a quantized electromagnetic field in a spherical cavity.
6. Computer methods of calculation with C++ as a basic language and the construction of the graphical user interface (GUI).

On programming technology, this book covers the C++ manipulation of all the following technologies:

1. The object-oriented approach as a basis of the modern methods of calculation.

2. Construction and calculations with complex vectors and matrices.
3. Practical use of classes for the description of the electrodynamics objects.
4. Methods construction the GUI for the full control over the progress of the computer calculations.
5. Application of various access levels in the classes hierarchy of problem.

What is this book for?

This book is designed for various audiences such as researchers specializing in physics and engineers engaged in classical optics and quantum optics of thin layers who write programs and carry out the average complexity searching calculations on modern computers. Often for such researchers the formulation of a problem and the search of methods for its solution have become inseparable. The details of the philosophy of the problem are crystallized while working under its solution. On the other hand the elegant solution comes easily when the problem is deeply understood. It is difficult sometimes for such people to explain to the support services what they expect from the professional programmers.

This book is also designed for programmers who would like to descend from the theoretical transcendental heights and to look into how their abstract images can find application in concrete, in this case, the electrodynamics calculations. If you are into designing the effective software for real applications using the thin means of the object-oriented technology, then this book is for you.

This book is also written for university students of natural faculties, to physicists who doubt whether it is necessary to study modern programming and to the programmers' students who want to understand why it is necessary to use computers, except that to write the compilers. Whereas *C++* represents the base of the modern programming languages sometimes one can find the solutions suggested in the book useful for experts in Java and C# languages.

The modern opportunities of programming do not allow one to passively watch a stream of white figures on the black screen, but to actively interact with a progress of calculations. For this purpose it is necessary to create the GUI containing a set of parameters, which you can operate. As a case in this point one can recollect any dialog window from MS WinWord. It is easier to construct such GUI, despite what many people think. Then your time spent will be more than compensated by the full control over your problem.

The author's experience shows that the resources and speed of well-known packages often become unacceptably low for even some medium-level problems. What can we do? The answer is single: study the necessary material from *C++*, spend some time creating your own library of classes, and then carry out the engineering research in parallel with numerical simulations on the basis of your model. Such efforts and "any travelling costs" will be generously repaid.

The program code from this book can be reached at gburlak@uaem.mx

This book is organized as follows:

In the first part the questions related to the classical approaches to propagation of electromagnetic radiation through multilayered spherical systems is examined.

In the second part the quantum aspects of this problem are discussed.

The third part is devoted to some modern approaches organizing the computing calculations for the numerical analyzing of the main properties of multilayered spherical systems.

Acknowledgments

I would like to thank Vitaly Datsyuk who has written Secs. 2.2, 2.3 and 2.4 in Chapter 2, Part 1. This work is supported by Consejo National de Ciencia y Tecnologia, project #35455-A.

Contents

Preface v

Introduction xvii

I Classical Dynamics

1. Maxwell Equations 3

1.1 Basic Equations . . . 3
1.1.1 Wave equation . . . 5
1.1.2 Three-dimensional case . . . 5
1.1.3 Electromagnetic waves . . . 6
1.1.4 Potentials of field . . . 7
1.1.5 TM and TE waves . . . 10
1.1.6 Debye potentials . . . 13
1.1.7 Energy of field . . . 22
1.1.8 Metallized sphere . . . 28
1.1.9 Frequency dispersion . . . 30
1.2 The Variational Principle . . . 32
1.2.1 The Whitham's average variational principle . . . 34
1.2.2 Energy in a layered microsphere . . . 36
1.3 Multilayered Microsphere . . . 37
1.4 The Transfer Matrix Method (Solving Equations for a System of Spherical Layers) . . . 37
1.5 Reflection Coefficient and Impedance of a Spherical Stack . 43
1.6 Conclusion . . . 48

2. Electromagnetic Field in Homogeneous Microspheres Without Surface Structures 50

2.1 Experiments with Microspheres 50
2.2 Lorentz–Mie theory and its extensions 57
2.2.1 Lorentz–Mie theory of elastic scattering 58
2.2.2 Theory of spontaneous emission 62
2.2.3 Mie scattering by concentrically stratified spheres . 65
2.3 Peculiarities of the modes of an open spherical cavity . . . 65
2.3.1 Indexes and order of a whispering-gallery mode . . 65
2.3.2 The problem of normalization of the whispering-gallery modes 66
2.4 Quality factor of a whispering-gallery mode 69
2.4.1 Radiative quality factor of an ideal dielectric sphere 69
2.4.2 Effect of light absorption on the quality factor . . . 72
2.4.3 Light scattering on inhomogeneities of the refractive index 72
2.4.4 Effect of a spherical submicrometer-size inclusion . 76
2.4.5 Comparison of different WGM-scattering models . 77
2.4.6 Q factor of a loaded cavity 79

3. Electromagnetic Eigen Oscillations and Fields in a Dielectric Microsphere with Multilayer Spherical Stack 81

3.1 Introduction 81
3.2 Geometry and Basic Equations 84
3.3 Eigenfrequencies of the Spherical Resonator Coated by the Stack 86
3.4 Radial Distribution of Fields 91
3.5 Discussions 95
3.6 Conclusion 99

4. Transmittance and Resonance Tunneling of the Optical Fields in the Microspherical Metal-Dielectric Structures 101

4.1 Introduction 101
4.2 Geometry and Basic Equations 103
4.3 Results and Discussions 106
4.4 Conclusion 112

5. Confinement of Electromagnetic Oscillations in a Dielectric Microsphere Coated by the Frequency Dispersive Multilayers 113

5.1 Introduction 113
5.2 Basic Equations 114
5.3 Results and Discussions 116
5.4 Conclusion 120

6. Oscillations in Microspheres with an Active Kernel 121

6.1 Basic Equations 122
6.2 Results and Discussions 123
6.3 Conclusion 129

7. Transfer Matrix Approach in a Non-Uniform Case 130

7.1 Approach to a Non-Uniform Case 131
7.2 Example. Non-Uniform Electron's Concentration 134

II The Quantum Phenomena in Microspheres

8. Coupling of Two-Level Atom with Electromagnetic Field 145

8.1 Transitions under the Action of the Electromagnetic Field 147
8.2 The Equations for Probability Amplitudes 148
8.3 Derivation of the Equation for Polarization of TLA: Dielectric Permittivity 150
8.4 Temporal Dynamics of Polarization and the Probability Amplitudes 153

9. Classical Field 157

9.1 Schrödinger Equation 157
9.2 Matrix Form for Two-Level Atom 158

10. Quantization of Electromagnetic Field 161

10.1 Energy of Field 162
10.2 Structure of Vacuum Field 166

11. Schrödinger and Interaction Pictures 168

11.1 Equations for the State Vectors 168
11.2 Equations for Operators 171
11.2.1 Operator's calculations 172

12. Two-Level Atom (The Matrix Approach, a Quantized Field) 173

12.1 Equations for Probability Amplitudes in Spherical Coordinates 178

13. Dynamics of Spontaneous Emission of Two-Level Atom in Microspheres: Direct Calculation 182

13.1 Introduction 182
13.2 Basic Equations 185
13.3 Results and Discussions 190
13.4 Triple photon state 200
13.4.1 Basic equations 200
13.4.2 Wigner function 207
13.5 Conclusions 211

III Numerical Methods and Object-Oriented Approach to the Problems of Multilayered Microsystems

14. Use of Numerical Experiment 217

14.1 Introduction 221
14.2 The Brief Review of C++ Operators 222
14.2.1 Data 222
14.2.2 Operators 225
14.2.3 Functions 230
14.2.4 Interacting the data of class with member functions 237
14.2.5 Classes 239
14.2.6 Access to members 240
14.2.7 Virtual functions 243
14.2.8 Overloading the mathematical operator 247

15. Exception Handling 250

15.1 Code 252

16. Visual Programming: Controls, Events and Handlers 254

16.1 DOS and Visual Programming 254
16.2 Controls, Events and Handlers 255
16.3 Graphical User Interface 259

17. Quantum Electromagnetic Field 260

17.1 Introduction 260
17.2 Code 261
17.3 Classes 264

18. Root Finding for Nonlinear and Complex Equations 270

18.1 Introduction 270
18.2 Code 271
18.3 Classes 275

19. Evaluation of Complex ODE 284

19.1 Introduction 284
19.2 Code 285
19.3 Classes 290

20. The Complex Vectorial and Matrix Operations 293

20.1 Introduction 293
20.2 Code 294
20.3 Classes 305

21. Spontaneous Emission of Atom in Microsphere 324

21.1 Introduction 324
21.2 Code 327
21.3 Classes 327

22. Electromagnetic Oscillations in Layered Microsphere 340

22.1 Introduction 340

Appendix A: Calculation of Field's Energy in a Sphere 349

Appendix B: Calculation of Surface Integral 352

Appendix C: Continuity of Tangential Fields 353

Appendix D: Integral on Bessel Functions 354

Appendix E: Surface Integrals for Dipole 355

Appendix F: Some Mathematical Formulas 357

Appendix G: Various Head *.h Files 360

Bibliography 364
Index 377

Introduction

Various structures of periodic layers in a planar geometry are used as elements of optical filters in resonators, and as quasi-optical reflection systems of integrated optics [Kogelnik & Li, 1966; Hodgson & Weber, 1997; Ramo *et al.*, 1994; Hummel & Guenther, 1995].

Nowadays different optical systems with the use of microspheres as one of the important elements are under wide investigation. The use of microspheres provides a possibility to achieve very narrow resonant lines [Vassiliev *et al.*, 1998]. The basic regime for the operation of these microspheres is a whispering gallery mode (WGM) for a microsphere with a radius of order 20–100 μm. Extreme great values of a Q factor in a narrow frequency range are observed [Vassiliev *et al.*, 1998; Ilchenko *et al.*, 1998], but in a general case it is desirable to get the spherical resonators to possess high quality Q factors beyond WGM. Such a regime may be reached in a microsphere of a submicron size in a low spherical mode regime. An application of the well-known idea on coating by quarter-wave layers gives a possibility of a sharp increase of the Q factor of such a system by up to $Q \sim 10^9$ as in a WGM.

The layered systems with a spherical geometry are quite complex and electromagnetic oscillations have been investigated rather completely only for an asymptotically great number of lossless layers [Brady *et al.*, 1993]. In [Li *et al.*, 2001], the dyadic Green's function was constructed and it was applied for multi-layer media. However the intermediate case with tens of layers was investigated insufficiently. Spherical geometry is rather attractive for two reasons. First, both amplitudes and phases of spherical waves depend on a radius. This provides additional possibilities with respect to a plane case, since the local properties of oscillations in a stack depend not only on the number and the thickness of layers, but also on the place of the layer in a stack. This causes a large variety of properties based not only on a choice of a material but also on geometrical properties of such an object. In such a case, an additional possibility emerges, namely, to operate with spherical modes of the lowest orders. In the presence of a dielectric stack, those modes possess high Q factors in a rather wide frequency range. This frequency range is determined by a spatial period of the stack and it can be controlled by either design or means of external influences like a pressure under operation [Ilchenko *et al.*, 1998]. Second, successes in a modern technology based, for instance, on the ultrasonic levitation [Ueha, 1998] or other technologies (see, for example, Internet site [Laboratories & Inc., 2001 and references therein]) allows one to design the spherical samples of submicron sizes with a multilayer structure.

Realization of high-quality multilayer spherical resonator provides a possibility to insert active elements of small sizes into the central cavity inside the dielectric stack.

It is known that in systems with a spherical geometry only two cases can be easily studied analytically: with $r \gg \lambda$ (far zone) or $r \ll \lambda$ (near zone), where r is a distance from the centre of a sphere to the given point. However, in a spherical multilayered microsphere, the intermediate case can be realized when r is close to the wavelength. In such a situation, the general theory of quasi-optical systems becomes invalid and the problem should be solved more exactly. Therefore, for a spherical stack, a more general method should be developed, similarly to a plane stack case [Born & Wolf, 1980; Solimeno *et al.*, 1986]. This method must provide the possibility of taking into account material absorption (or amplification in the case of active materials) for any ratio between r, λ and for arbitrary thicknesses of layers, and also for random deviations of thicknesses from optimal quarter-wave length value.

We note similar problems in acoustics, namely, acoustic wave scattering by a sphere in water and the scattering by multilayered spherical structures was studied in [Gaunaurd & Uberall, 1983; Gerard, 1983]. In [Ewing *et al.*, 1957, Chap. 5], an influence of a curvature on propagating waves was analyzed for the geophysical problems. More references can be found in Chapter 7.

In this book, we investigate microspheres of a submicron size coated by a system of contacting concentric spherical dielectric micro and nano layers (spherical stack) in optical frequency range.

PART I

Classical Dynamics

CHAPTER 1

Maxwell Equations

This chapter serves as an introduction to the subject on classical electro dynamics. More details and information can be found in [Born & Wolf, 1980; Ginzburg, 1989; Jackson, 1975; Landau & Lifshits, 1975; Landau *et al.*, 1984; Panofsky & Phillips, 1962; Solimeno *et al.*, 1986; Stratton, 1941; Vainstein, 1969; Vainstein, 1988]. Some basic knowledge from the theory of electromagnetic fields is necessary for deeper understanding of the subsequent materials. It is also intended for references from the subsequent chapters.

1.1 Basic Equations

Maxwell equations are the basis for the theory of electromagnetic field. They have been written by English physicist G. Maxwell in 1873, and they were the generalization of the experimental facts available then. The state of field is described by the vectors of electric $\vec{E}$ and magnetic $\vec{H}$ fields accordingly, which can variate both in space $\vec{r}$, and in time t. The Maxwell equations have the forms

$$\nabla \times \vec{H} = \vec{j} + \frac{\partial}{\partial t}\vec{D}, \tag{1.1}$$

$$\nabla \times \vec{E} = -\frac{\partial}{\partial t}\vec{B}, \tag{1.2}$$

$$\nabla \cdot \vec{D} = \rho, \tag{1.3}$$

$$\nabla \cdot \vec{B} = 0, \tag{1.4}$$

where $\vec{D} = \bar{\varepsilon}\vec{E}$, $\vec{B} = \bar{\mu}\vec{H}$ are the vectors of the electric and magnetic induction respectively, $\bar{\varepsilon} = \varepsilon_0\varepsilon$ and $\bar{\mu} = \mu_0\mu$ are the dielectric and magnetic permittivities, ε and μ are the relative dielectric and magnetic permittivities, and ε_0 and μ_0 are the dielectric and the magnetic permitivities of the vacuum, connected by a relation $\varepsilon_0\mu_0 = c^{-2}$ (where $c \approx 3\cdot 10^8$ m/sec is the light velocity in vacuum). ρ is the density of the electric charge and $\vec{j}$ is the vector of density of the electric current. From Eqs. (1.1)–(1.4) one can derive the important equation binding the electrical charge and current in the form:

$$\frac{\partial \rho}{\partial t} + \nabla \cdot \vec{j} = 0. \tag{1.5}$$

In the simplest case of homogeneous, isotropic and linear environment ε and μ are constant scalar quantities. In the case of non-magnetic materials, which is considered in this book, $\mu = 1$, so $\bar{\mu} = \mu_0$. In the anisotropic linear environment (crystals) both ε and μ have dependence on the direction, i.e. become tensors [Born & Wolf, 1980; Stratton, 1941]. In non-uniform environment these variables also depend on the spatial variable $\vec{r}$, but in nonlinear materials ε depends on the amplitude of field $\vec{E}$. In (1.1)–(1.4) we can exclude the field $\vec{B}$ by

$$\begin{aligned} \nabla \times [\nabla \times \vec{E}] &= -\frac{\partial}{\partial t}[\nabla \times \vec{B}] = -\mu_0 \frac{\partial}{\partial t}[\nabla \times \vec{H}] = -\mu_0 \frac{\partial}{\partial t}\left(\frac{\partial \vec{D}}{\partial t} + \vec{j}\right) \\ &= -\varepsilon_0 \varepsilon \mu_0 \frac{\partial^2}{\partial t^2}\vec{E} - \mu_0 \frac{\partial}{\partial t}\vec{j} = -\frac{\varepsilon}{c^2}\frac{\partial^2}{\partial t^2}\vec{E} - \mu_0 \frac{\partial}{\partial t}\vec{j}. \end{aligned} \tag{1.6}$$

But $\nabla \times [\nabla \times \vec{E}] = \nabla \cdot [\nabla \cdot \vec{E}] - \nabla^2 \vec{E} = \nabla \bar{\varepsilon}^{-1} \nabla \cdot \vec{D} = \bar{\varepsilon}^{-1}\nabla\rho - \nabla^2 \vec{E}$. Substituting this in Eq. (1.6) we get

$$\nabla^2 \vec{E} - \bar{c}^{-2}\frac{\partial^2}{\partial t^2}\vec{E} = \mu_0 \frac{\partial}{\partial t}\vec{j} + \varepsilon^{-1}\nabla\rho, \tag{1.7}$$

where $\bar{c} = c/\varepsilon^{1/2}$ is the light velocity in a material. In a vacuum case $\varepsilon = 1$, $\rho = 0$, $\vec{j} = 0$ and Eq. (1.7) becomes the wave equation in form

$$\nabla^2 \vec{E} - \frac{1}{c^2}\frac{\partial^2}{\partial t^2}\vec{E} = 0. \tag{1.8}$$

1.1.1 *Wave equation*

The equation of the following type

$$\frac{\partial^2}{\partial x^2}U - \frac{1}{v^2}\frac{\partial^2}{\partial t^2}U = 0 \tag{1.9}$$

has general solution in a form of

$$U(x,t) = \psi_1(t - x/v) + \psi_2(t + x/v), \tag{1.10}$$

where $\psi_{1,2}(z)$ are arbitrary functions of the argument $z = t \mp x/v$. The values of $\psi_{1,2}(z)$ remain constant if $z = t \mp x/v = z_0 = const$ for x and t, such that $x = \pm vt + const$. From here one can find that $dx/dt = \pm v$, and so v is a velocity of the wave propagation of $\psi_{1,2}(z)$ for the one-dimensional case. The sign $\pm$ describes waves propagation to the opposite directions of the x axis. From Eq. (1.9) one can see that the solution does not change if one were to multiply both x and t by the same constant factor.

Practically the most important case is the oscillate mode, when $\psi_{1,2} = \cos(\omega t - kx) = \cos\omega(t - xk/\omega)$, with ω and k as the arbitrary constants jointed by the ratio

$$\omega/k = \pm v. \tag{1.11}$$

In Eq. (1.11) one refers to the quantities ω and k as the frequency and waves number respectively, and v is the phase velocity of wave. Similar result can be received, if you take $\psi_{1,2} = \sin(\omega t - kx)$. As (1.9) is the linear equation, the sum of its solutions is also the solution. Therefore it is convenient to write down the common oscillating solution as

$$\psi(x,t) = A_0\{\cos(\omega t - kx) + i\sin(\omega t - kx)\} = A_0 e^{i(\omega t - kx)}, \tag{1.12}$$

where A_0 is a constant amplitude of wave, which in general can be a complex number. Until the initial or boundary conditions are applied to (1.9) the amplitude A_0 has an arbitrary value.

1.1.2 *Three-dimensional case*

Generally the wave equation describing the wave propagation in space (3D case) reads

$$\left(\frac{\partial^2}{\partial x^2} + \frac{\partial^2}{\partial y^2} + \frac{\partial^2}{\partial z^2}\right)U - \frac{1}{v^2}\frac{\partial^2}{\partial t^2}U = 0. \tag{1.13}$$

By analogy with (1.9), (1.12) one can search for the solution of (1.13) in the form

$$U(x, y, z, t) = U_0 e^{i(\omega t - \vec{k}\vec{r})}, \tag{1.14}$$

where $\vec{k} = \{k_x, k_y, k_z\}$ is a wave vector, and U_0, ω, k_x, k_y, k_z are some of the complex numbers. To find these numbers, we substitute (1.14) in (1.13), giving

$$\left(-k^2 + \frac{\omega^2}{v^2}\right) U_0 e^{i(\omega t - \vec{k}\vec{r})} = 0, \tag{1.15}$$

where $k^2 = (\vec{k} \cdot \vec{k}) = k_x^2 + k_y^2 + k_z^2$. Since $U_0 \neq 0$, and $e^{i(\omega t - \vec{k}\vec{r})} \neq 0$, from Eq. (1.15) one can receive the following relation

$$\omega\ /k = \pm v, \tag{1.16}$$

which is similar to Eq. (1.11), but is generalized to a three-dimensional case.

Now we will talk about the wave front, i.e. a surface on which $\omega t - \vec{k}\vec{r} = -z_0 = const$. For any moment of time $t = t_1 = const$ one can write $\vec{k}\vec{r} = z_0 + \omega t_1$ or $\vec{n}\vec{r} = p = (z_0 + \omega t_1)/k = const$, $\vec{n} = \vec{k}/k$. From the mathematics textbooks [Korn & Korn, 1961; Riley *et al.*, 1998], it is known that the equation $\vec{n}\vec{r} = p$ is the equation of the plane shifted on distance p from the origin of coordinates, and the vector $\vec{n}$ is normal to this plane (see Fig. 1.1). This allows one to recognize the waves front as the plane, and accordingly we refer to such a wave as a plane wave. One can see that the wave vector $\vec{k}$ is perpendicular to such a plane.

Due to a periodicity $e^{iz} = e^{i(z+2n\pi)}$, $n = 0, 1, \ldots$, similar reasoning can be applied for all other planes parallel to the given one, shifted for a distance of $\lambda = 2\pi/k$. Quantity λ is referred to as the wavelength, and instead of angular frequency ω is often used the frequency $f = \omega/2\pi$. Then in (1.16) for the sign + one can write $\omega/k = f\lambda = v$. Note that in materials the phase velocity is c/n, where $n = \sqrt{\varepsilon}$ is the refraction index of this material.

1.1.3 *Electromagnetic waves*

Thus, for the solution of the free field equation without sources (1.8) $\rho = 0$ and $\vec{j} = 0$ has the form of a plane wave

$$\vec{E}(x, y, z, t) = \vec{E}_0 e^{i(\omega t - \vec{k}\vec{r})} + \text{c.c.}, \tag{1.17}$$

where c.c. refers to a complex conjugate value. The formula $2\text{Re}(z) = z + z^*$ allows one to separate a real part in any complex expression. Often the multiplier 2 is dropped, making it equivalent to renormalize the amplitude of a field $\vec{E}_0 \to 2\vec{E}_0$. Such renormalization is not essential in problems of calculating a spectrum of field eigenvalues. However it is necessary to take it into account while solving the initial or boundary problem.

Generally the field is the superposition of waves with various wave vectors $\vec{k}$. Thus it is necessary to write expression (1.17) as

$$\vec{E}(x, y, z, t) = \sum_{\vec{k}} \vec{E}_{0\vec{k}} e^{i(\omega t - \vec{k}\vec{r})} + \text{c.c.} \tag{1.18}$$

In free space the wave vector $\vec{k}$ takes the continuous values, causing the sum in (1.18) to become the Fourier integral.

The solution for free fields $\vec{E}$ and $\vec{B}$ can then be written as the sum of plane waves with arbitrary amplitudes. It is also the solution of the Eq. (1.8) and the similar equation for $\vec{B}$. However generally it is necessary to take into account the presence of currents $\vec{j}$ which can radiate a field, i.e. to be the sources of fields. In contrast to the vacuum case the problem of calculations of such field remains difficult.

1.1.4 *Potentials of field*

Often the problems with sources become simpler if one uses the auxiliary variables referred to as the potentials of field. In the equation $\nabla \cdot \vec{B} = 0$

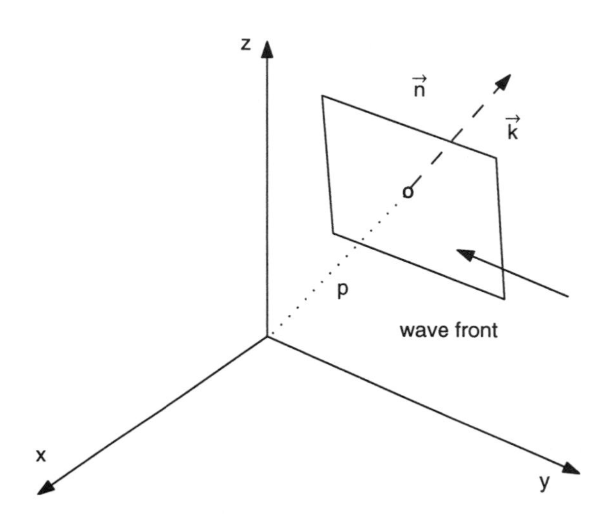

Fig. 1.1 Propagation of plane wave.

the quantity $\vec{B}$ in view of identity $\nabla \cdot (\nabla \times \vec{A}) = 0$ can be written as $\vec{B} = \nabla \times \vec{A}$, where $\vec{A}$ is an arbitrary vector. Substituting this expression in the equation $\nabla \times \vec{E} = -\frac{\partial}{\partial t}\vec{B}$, we obtain $\nabla \times \left(\vec{E} + \frac{\partial}{\partial t}\vec{A}\right) = 0$. In view of identity $\nabla \times (\nabla\Phi) = 0$, where Φ is an arbitrary scalar, we get the next presentation for fields as

$$\vec{E} = -\frac{\partial}{\partial t}\vec{A} - \nabla\Phi, \quad \vec{B} = \nabla \times \vec{A}, \tag{1.19}$$

where $\vec{A}$ and Φ are called the vectorial and scalar potentials accordingly. Now we rewrite the remaining Maxwell equations using (1.19). For the simple case of a homogeneous environment we have $\nabla \cdot \vec{D} = \bar{\varepsilon}\nabla \cdot \vec{E} = \rho$ and $\nabla \cdot (\partial\vec{A}/\partial t + \nabla\Phi) = -\rho/\bar{\varepsilon}$ and finally

$$\frac{\partial \nabla \vec{A}}{\partial t} + \nabla^2\Phi = -\frac{\rho}{\bar{\varepsilon}}. \tag{1.20}$$

Furthermore, from Maxwell equation $\nabla \times \vec{H} = \vec{j} + \frac{\partial}{\partial t}\vec{D}$ one can obtain $\frac{1}{\bar{\mu}}\nabla \times \nabla \times \vec{A} = -\bar{\varepsilon}\frac{\partial}{\partial t}\left(\frac{\partial}{\partial t}\vec{A} + \nabla\Phi\right) + \vec{j}$ or

$$\nabla \cdot (\nabla \cdot \vec{A}) - \nabla^2\vec{A} = -\bar{\varepsilon}\bar{\mu}\frac{\partial^2}{\partial t^2}\vec{A} + \bar{\varepsilon}\bar{\mu}\frac{\partial}{\partial t}\nabla\Phi + \bar{\mu}\vec{j}, \tag{1.21}$$

where the formula $\nabla \times (\nabla \times \vec{A}) = \nabla \cdot (\nabla \cdot \vec{A}) - \nabla^2\vec{A}$ was applied [Korn & Korn, 1961].

To remove the remained arbitrariness in the definition of potentials $\vec{A}$ and Φ the following additional conditions are often used.

(i) Coulomb gauge $\nabla \cdot \vec{A} = 0$. Thus (1.20) and (1.21) become

$$\nabla^2\Phi = -\frac{\rho}{\bar{\varepsilon}}, \quad \nabla^2\vec{A} - \frac{\varepsilon}{c^2}\frac{\partial^2}{\partial t^2}\vec{A} = \frac{\varepsilon}{c^2}\frac{\partial}{\partial t}\nabla\Phi - \bar{\mu}\vec{j}. \tag{1.22}$$

(ii) Lorenz gauge $\nabla \cdot \vec{A} + \frac{\varepsilon}{c^2}\frac{\partial}{\partial t}\Phi = 0$. In this case (1.20) and (1.21) become

$$\nabla^2\Phi - \frac{\varepsilon}{c^2}\frac{\partial^2}{\partial t^2}\Phi = -\frac{\rho}{\bar{\varepsilon}}, \tag{1.23}$$

$$\nabla^2\vec{A} - \frac{\varepsilon}{c^2}\frac{\partial^2}{\partial t^2}\vec{A} = -\bar{\mu}\vec{j}. \tag{1.24}$$

In the case of a Lorenz gauge both equations for Φ and $\vec{A}$ are independent and they have a form of inhomogeneous equations with sources. For scalar potential (1.23) the source is the electrical charge ρ whereas in vector potential equation (1.24) the source is the electrical current $\vec{j}$.

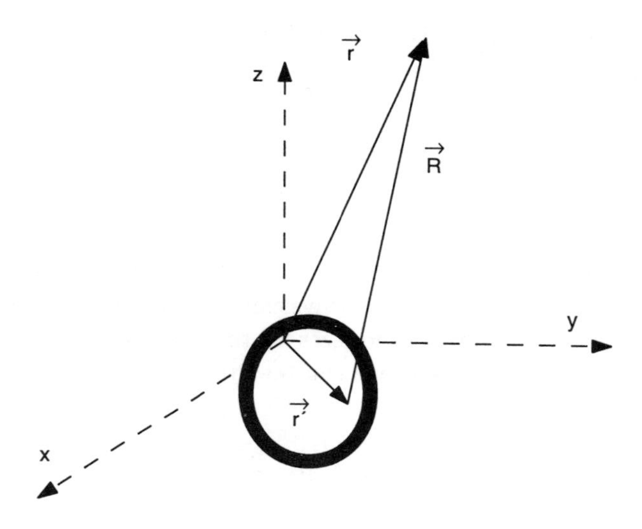

Fig. 1.2 Geometry of problem.

An important circumstance is the relativistic covariant form of the Lorenz gauge [Landau & Lifshits, 1975; Panofsky & Phillips, 1962].

The solution of the Eq. (1.24) for $\vec{A}$, is determined by a current $\vec{j}(\vec{r}, t)$ and can be written down as [Stratton, 1941]

$$\vec{A}(\vec{r}, t) = \frac{\bar{\mu}}{4\pi} \int_V \frac{\vec{j}(\vec{r'}, t - R/c)}{R} d\vec{r'}, \tag{1.25}$$

where V is the volume occupied with current $\vec{j}$, $\vec{R} = \vec{r} - \vec{r'}$, and $\vec{r}$ is vector of a point of observations, whereas $\vec{r'}$ is the position of current $\vec{j}$ (see Fig. 1.2). One can see from (1.25) that for a retarded case the value of potential $\vec{A}(\vec{r}, t)$ at moment t is determined by the value of the current $\vec{j}$ in the previous moment $t - R/c$, where the shift R/c is the time of propagation of a signal from a point $\vec{r'}$ to a point observation $\vec{r}$.

For the evaluation of (1.25) the following approximations are normally used

(i) If $r \gg r'$ then $\vec{R} = \vec{r} - (\vec{r}\vec{r'})/r$, in this case (1.25) becomes

$$\vec{A}(\vec{r}, t) = \frac{\bar{\mu}}{4\pi r} \int_V \vec{j}(\vec{r'}, t - r/c + (\vec{r}\vec{r'})/rc) d\vec{r'}. \tag{1.26}$$

(ii) The second assumption is connected to the wavelength of radiation λ. In (1.26) we write $\vec{j}(r,t) = \vec{j}(r)e^{i\omega t}$, then

$$\vec{A}(\vec{r},t) = \frac{\mu}{4\pi r} e^{i\omega(t-r/c)} \int_V j_0(\vec{r}) \cdot e^{i\omega(\vec{r}\vec{r'})/rc} d\vec{r'}. \quad (1.27)$$

If $\omega r'/c = kr' \ll a/\lambda \ll 1$, where $a = \max(r')$, one can neglect the last term in the exponent for Eq. (1.27). Physically it means that the wavelength of radiation λ is larger than the sizes of the area of radiation. If both mentioned conditions are valid one can rewrite (1.25) as

$$\vec{A}(\vec{r},t) = \frac{\bar{\mu}}{4\pi r} \int_V \vec{j}(\vec{r'}, t - r/c) d\vec{r'}. \quad (1.28)$$

Usually they refer to Eq. (1.28) as an approximation of the electromagnetic dipole radiations or dipole approximation. Now $\vec{A}$ is determined by the spatial structure of $\vec{j}$, therefore the analysis of wave effects is simplified.

However in the case of spherical multilayered systems both mentioned assumptions often appear unusable. If the layers becomes $a \simeq \lambda$, which physically reflects the impossibility of neglect the pattern effects in such systems. Further we shall see that such interference effects form the strong reflection zones of electromagnetic radiation in multilayered microspherical structures.

1.1.5 *TM and TE waves*

In homogeneous space can propagate transverse electromagnetic waves with two independent polarizations. There are no differences between them in a homogeneous space. However at the presence of a boundary such degeneracy is no longer valid and both polarizations become unequal. We shall consider a simple plane case when the boundary plane coincides with plane XY, and the normal to it parallel to axis Z (see Fig. 1.3).

We first consider, that the orientation of axes X, Y and Z is chosen such that there is no dependence on coordinate y ($d/dy = 0$). Then one vectorial equation $\nabla \times \vec{E} = -\frac{\partial}{\partial t}\vec{B}$ can be written down as the following

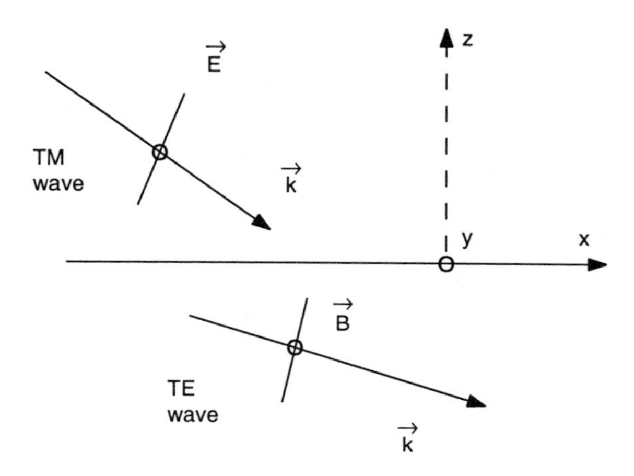

Fig. 1.3 TE and TM waves.

three scalar equations

$$
\begin{aligned}
-\frac{\partial E_y}{\partial z} &= -\frac{\partial}{\partial t} B_x, \\
-\frac{\partial E_z}{\partial x} + \frac{\partial E_x}{\partial z} &= -\frac{\partial}{\partial t} B_y, \\
\frac{\partial E_y}{\partial x} &= -\frac{\partial}{\partial t} B_z.
\end{aligned}
$$

Similarly the equation $\nabla \times \vec{H} = \frac{\partial}{\partial t}\vec{D} + \vec{j}$ results in the following

$$
\begin{aligned}
-\frac{\partial B_y}{\partial z} &= \frac{\varepsilon}{c^2}\frac{\partial}{\partial t} E_x + \mu_0 j_x, \\
-\frac{\partial B_z}{\partial x} + \frac{\partial B_x}{\partial z} &= \frac{\varepsilon}{c^2}\frac{\partial}{\partial t} E_y + \mu_0 j_y, \\
\frac{\partial B_y}{\partial x} &= \frac{\varepsilon}{c^2}\frac{\partial}{\partial t} E_z + \mu_0 j_z.
\end{aligned}
$$

One can separate the above six equations into two independent subsystems. The first subsystem only E_x, B_y, E_z contains and the second subsystem consist of B_x, E_y, B_z. Corresponding waves subsystems are called TM and TE waves accordingly. The equations for TM waves are given by

$$-\frac{\partial B_y}{\partial z} = \frac{\varepsilon}{c^2}\frac{\partial}{\partial t}E_x + \mu_0 j_x, \quad -\frac{\partial E_z}{\partial x} + \frac{\partial E_x}{\partial z} = -\frac{\partial}{\partial t}B_y,$$
$$\frac{\partial B_y}{\partial x} = \frac{\varepsilon}{c^2}\frac{\partial}{\partial t}E_z + \mu_0 j_z, \tag{1.29}$$

while the equations for TE waves read

$$-\frac{\partial E_y}{\partial z} = -\frac{\partial}{\partial t}B_x, \quad -\frac{\partial B_z}{\partial x} + \frac{\partial B_x}{\partial z} = \frac{\varepsilon}{c^2}\frac{\partial}{\partial t}E_y + \mu_0 j_y, \quad \frac{\partial E_y}{\partial x} = -\frac{\partial}{\partial t}B_z. \tag{1.30}$$

It is important that the wave processes for TM and TE modes can be studied independently.

Let's consider more general case. We assume that the waves propagate along the z axis, and all amplitudes can depend arbitrarily from x and y. However the dependence on time t and z coordinate is given by $e^{i(\omega t - \beta z)}$. In this case a field can also be separated in the TM and TE waves. First we consider a TM wave is case, where $\vec{E} = \{E_x, E_y, E_z\}$, $\vec{B} = \{B_x, B_y, 0\}$. From equation $\nabla\vec{D} = 0$, we have

$$\varepsilon_0\varepsilon\nabla\vec{E} = \varepsilon_0\varepsilon\left(\frac{\partial E_x}{\partial x} + \frac{\partial E_y}{\partial y} + \frac{\partial E_z}{\partial z}\right) = \varepsilon_0\varepsilon\left(\frac{\partial E_x}{\partial x} + \frac{\partial E_y}{\partial y} - i\beta E_z\right) = 0. \tag{1.31}$$

From the equation $\nabla\times\vec{E} = i\omega\vec{B}$ we obtain

$$-i\omega B_x = \frac{\partial E_z}{\partial y} + i\beta E_y, \quad i\omega B_y = \frac{\partial E_z}{\partial x} + i\beta E_x, \quad \frac{\partial E_y}{\partial x} - \frac{\partial E_x}{\partial y} = 0, \tag{1.32}$$

and from the equation $\nabla\times B = \frac{i\omega\varepsilon}{c^2}\vec{E}$, $\vec{j} = 0$ we obtain

$$\beta B_y = \frac{\omega\varepsilon}{c^2}E_x, \quad \beta B_x = -\frac{\omega\varepsilon}{c^2}E_y, \quad \frac{\partial B_y}{\partial x} - \frac{\partial B_x}{\partial y} = i\frac{\omega\varepsilon}{c^2}E_z. \tag{1.33}$$

Combining (1.32) and (1.33) we can write

$$B_x = -i\frac{\omega\varepsilon}{\beta^2 c^2}\frac{\partial E_z}{\partial y}, \quad B_y = i\frac{\omega\varepsilon}{\beta^2 c^2}\frac{\partial E_z}{\partial x}, \quad \kappa^2 = \frac{\omega^2\varepsilon}{c^2} - \beta^2. \tag{1.34}$$

Now we can calculate

$$\frac{\partial^2 E_z}{\partial x^2} + \frac{\partial^2 E_z}{\partial y^2} = i\frac{\kappa^2}{\beta^2}\left(\frac{\partial E_x}{\partial x} + \frac{\partial E_y}{\partial y}\right) = -\kappa^2 E_z. \tag{1.35}$$

One can see that the transverse field components $E_{x,y}$ are written through the longitudinal component of electric field E_z, which satisfies the

Helmholtz equation

$$\frac{\partial^2 E_z}{\partial x^2} + \frac{\partial^2 E_z}{\partial y^2} + \kappa^2 E_z = 0. \tag{1.36}$$

Similar formulas can be obtained and for TE waves. In this case all transverse field amplitudes are expressed through B_z. Similar formulas hold for spherical geometry. In this case the transverse field components are also expressed through a longitudinal component in a spherical case through the radial field.

1.1.6 *Debye potentials*

The method of studying electromagnetic waves processes in spheres, in which the Debye potentials where used, has been developed in the works of Debye and Mie (see references in [Stratton, 1941]).

In the case of $\rho = 0$ the next Maxwell equations become the identical forms $\nabla \cdot \vec{B} = 0$ and $\nabla \cdot \vec{D} = 0$. As we have seen above, from equation $\nabla \cdot \vec{B} = 0$ follows the possibility in representing $\vec{B} = \nabla \times \vec{A}_M$. Now such a symmetry allows one to write the same for $\vec{D} = \nabla \times \vec{A}_E$. Ultimately this leads to the separation of the field to TM and TE waves. Such approach is very useful for the investigation of problems in cylindrical or spherical coordinates. So far the spherical geometry plays the central role in this book, we describe below the details of corresponding calculations for spherical system.

(i) *TM case*

First we study TM case. Introducing the vectorial $\vec{A}_M$ and scalars f_M potentials accordingly, we can write down

$$\vec{B} = \nabla \times \vec{A}_M, \quad \vec{E} = -\frac{\partial \vec{A}_M}{\partial t} - \nabla f_M, \tag{1.37}$$

where $c = (\mu_0 \varepsilon_0)^{-0.5}$ is the light velocity, and a dielectric permittivity ε can depend on radius of sphere $\varepsilon = \varepsilon(r)$. From $\nabla \times \vec{H} = \frac{\partial}{\partial t}\vec{D} + \vec{j}$ the equation for $\vec{A}$ follows

$$\nabla \times \nabla \times \vec{A}_M + \frac{\varepsilon}{c^2}\frac{\partial^2 \vec{A}_M}{\partial t^2} = -\frac{\varepsilon}{c^2}\frac{\partial \nabla f_M}{\partial t} + \mu_0 \vec{j}. \tag{1.38}$$

In spherical system we search for a solution of (1.38) in the form

$$\vec{A}_M = \{\Pi_M, 0, 0\}\ , \tag{1.39}$$

where $\Pi_M = \Pi_M(r, \theta, \varphi)$ is the Debye potential. Calculation of $\vec{B} = \nabla \times \vec{A}_M$ gives

$$\vec{B} = \widehat{i}_\theta \frac{1}{r \sin\theta} \frac{\partial \Pi_M}{\partial \varphi} + \widehat{i}_\varphi \frac{1}{r} \left(-\frac{\partial \Pi_M}{\partial \theta} \right), \tag{1.40}$$

whence one can see that in TM waves $B_r = 0$. Now we project (1.38) on the spherical axes

$$r: \quad \left(-\frac{1}{r^2} \Delta_T + \frac{\varepsilon}{c^2} \frac{\partial^2}{\partial t^2} \right) \Pi_M = -\frac{\varepsilon}{c^2} \frac{\partial}{\partial t} \frac{\partial f_M}{\partial r} + \mu_0 j_r, \tag{1.41}$$

$$\theta: \quad \frac{\partial}{\partial \theta} \left(\frac{\partial \Pi_M}{\partial r} + \frac{\varepsilon}{c^2} \frac{\partial f_M}{\partial t} \right) = \mu_0 j_\theta r, \tag{1.42}$$

$$\varphi: \quad \frac{\partial}{\partial \varphi} \left(\frac{\partial \Pi_M}{\partial r} + \frac{\varepsilon}{c^2} \frac{\partial f_M}{\partial t} \right) = \mu_0 j_\varphi r \sin\theta, \tag{1.43}$$

where $\Delta_T = \frac{1}{\sin\theta} \frac{\partial}{\partial \theta} \sin\theta \frac{\partial}{\partial \theta} + \frac{1}{\sin^2\theta} \frac{\partial^2}{\partial \varphi^2}$. As follows from (1.42) and (1.43) the compatibility condition of the two equations is given by

$$\frac{\partial j_\theta}{\partial \varphi} - \frac{\partial j_\varphi \sin\theta}{\partial \theta} = 0, \tag{1.44}$$

whence follows, that $\left[\nabla \times \vec{j}\right]_r = 0$. In the simplest case when $j_\theta = j_\varphi = 0$ the condition (1.44) becomes valid then for a current $\vec{j}$ and we have $\vec{j} = \{j_r, 0, 0\}$. In a general case of $\vec{j}$ one should apply more complicate Green's-function technique [see Li *et al.*, 1994, 2001, Sullivan & Hall, 1994].

After integration of equations (1.42) and (1.43) on θ or φ accordingly they will become the next identical form

$$\frac{\partial \Pi_M}{\partial r} + \frac{\varepsilon}{c^2} \frac{\partial f_M}{\partial t} = \widetilde{\Phi}_M(r, t), \tag{1.45}$$

where $\widetilde{\Phi}_M(r, t)$ it is an arbitrary function. Further it is convenient to write down $\widetilde{\Phi}_M$ in a slightly different form $\widetilde{\Phi}_M(r, t) = \varepsilon(r) \int \frac{\Phi_M(r, t)}{\varepsilon(r)} dr$. Then after differentiation of (1.45) one can obtain

$$\frac{1}{c^2}\frac{\partial}{\partial t}\frac{\partial f_M}{\partial r} = -\frac{\partial}{\partial r}\left(\frac{1}{\varepsilon(r)}\frac{\partial \Pi_M}{\partial r}\right) + \frac{\Phi_M(r,t)}{\varepsilon(r)}. \qquad (1.46)$$

Substituting (1.46) in (1.41) we receive the equations for Π_M in the next closed form

$$\frac{1}{r^2}\Delta_T \Pi_M + \varepsilon\frac{\partial}{\partial r}\left(\frac{1}{\varepsilon}\frac{\partial \Pi_M}{\partial r}\right) - \frac{\varepsilon}{c^2}\frac{\partial^2 \Pi_M}{\partial t^2} = \mu_0 j_r - \Phi_M(r,t). \qquad (1.47)$$

Since $\Phi_M(r,t)$ is an arbitrary function we have opportunity to choose it by most convenient way. One can see that equation (1.47) can be simplified by choosing $\Phi_M(r,t) = \mu_0 j_r$, after that (1.47) becomes the homogeneous equation (without source). From (1.47) we can conclude that such approach is valid when both Φ_M and j_r depend on radius r and t only. Now the equation (1.47) for Π_M is given by

$$\frac{1}{r^2}\Delta_T \Pi_M + \varepsilon\frac{\partial}{\partial r}\left(\frac{1}{\varepsilon}\frac{\partial \Pi_M}{\partial r}\right) - \frac{\varepsilon}{c^2}\frac{\partial^2 \Pi_M}{\partial t^2} = 0. \qquad (1.48)$$

Using (1.37) and (1.46) the expressions for electric and magnetic fields for TM waves can be written down as follows:

$$\frac{\partial E_r}{\partial t} = \frac{c^2}{\varepsilon}\left[\varepsilon\frac{\partial}{\partial r}\frac{1}{\varepsilon}\frac{\partial \Pi_M}{\partial r} - \frac{\varepsilon}{c^2}\frac{\partial^2 \Pi_M}{\partial t^2} - \mu_0 j_r\right], \qquad (1.49)$$

$$\frac{\partial E_\theta}{\partial t} = \frac{c^2}{r\varepsilon}\frac{\partial^2 \Pi_M}{\partial r \partial \theta}, \quad \frac{\partial E_\varphi}{\partial t} = \frac{c^2}{r\varepsilon\sin\theta}\frac{\partial^2 \Pi_M}{\partial r \partial \varphi}, \qquad (1.50)$$

$$H_r = 0, \quad H_\theta = \frac{1}{\mu_0 r\sin\theta}\frac{\partial \Pi_M}{\partial \varphi}, \quad H_\varphi = -\frac{1}{\mu_0 r}\frac{\partial \Pi_M}{\partial \theta}. \qquad (1.51)$$

Further we use the temporal dependence of fields as $e^{i\omega t}$. After that Eq.(1.48) becomes

$$\frac{1}{r^2}\Delta_T \Pi_M + \varepsilon\frac{\partial}{\partial r}\left(\frac{1}{\varepsilon}\frac{\partial \Pi_M}{\partial r}\right) + \frac{\omega^2\varepsilon}{c^2}\Pi_M = 0. \qquad (1.52)$$

We solve (1.52) by the separation variables method [Riley *et al.*, 1998] and we search for a solution (1.52) in the form

$$\Pi_M(r,\theta,\varphi) = R(r)Y(\theta,\varphi). \qquad (1.53)$$

Substituting (1.53) in (1.52) and separating the variables we receive

$$-\frac{\Delta_{\perp}Y}{Y}=+\frac{r^2}{R}\left\{\frac{\omega^2\varepsilon}{c^2}R+\varepsilon\frac{\partial}{\partial r}\left(\frac{1}{\varepsilon}\frac{\partial R}{\partial r}\right)\right\}=K, \tag{1.54}$$

where ε can depend on radius $\varepsilon=\varepsilon(r)$ and K is the constant of separation. Separating variables from (1.54) we obtain two equations. The equation for the angular part $Y=Y(\theta,\varphi)$ one can write in the form

$$\Delta_{\perp}Y+KY=\frac{1}{\sin\theta}\frac{\partial}{\partial\theta}\left(\sin\theta\frac{\partial Y}{\partial\theta}\right)+\frac{1}{\sin^2\theta}\frac{\partial^2 Y}{\partial\varphi^2}+KY=0. \tag{1.55}$$

Equation (1.55) is spherical functions equation if K is an integer $K=m(m+1)$. Further we use such values, therefore $Y(\theta,\varphi)\to Y_m^j(\theta,\varphi)$, where $m=0,1,2,\ldots$ is a spherical quantum number, whereas $j=0,\pm1,\ldots\pm m$ is an azimuthal quantum number. Equation for radial part $R=R(r)$ becomes

$$\varepsilon\frac{\partial}{\partial r}\left(\frac{1}{\varepsilon}\frac{\partial R}{\partial r}\right)+\left[\frac{\omega^2\varepsilon}{c^2}-\frac{m(m+1)}{r^2}\right]R=0. \tag{1.56}$$

It is clear from (1.56) that $R=R_m(r)$ can depend on m and does not depend on j. As we have noted in a general case ε can be a function on the radial variable r and frequency ω i.e. $\varepsilon=\varepsilon(\omega,r)$. In this case we have to use the numerical methods to solve (1.56). We will analyze the TE waves case, in the next subsection.

(ii) *TE case*

For a TE waves case and in a free charges case $\rho=0$ from the continuity equation one obtains $\nabla\vec{j}=0$. The latter allows one to write down $\vec{j}=\nabla\times\vec{M}$ where $\vec{M}$ is a vector of the magnetic moment. Then from the equations $\nabla\cdot\vec{D}=0$ and $\nabla\times\vec{H}=\partial\vec{D}/\partial t$ one can obtain the next potential representation for fields

$$\vec{D}=\nabla\times\vec{A}_E,\quad \vec{H}=\frac{\partial\vec{A}_E}{\partial t}+\vec{M}+\nabla f_E, \tag{1.57}$$

where $\vec{A}_E$ and f_E it is vectorial and scalar potentials accordingly for TE waves. From the equation $\nabla\times\vec{E}=\nabla\times\left(\vec{D}/\varepsilon(r)\varepsilon_0\right)=-\partial\vec{B}/\partial t$ follows

$$\nabla\times\left[\frac{1}{\varepsilon(r)}\nabla\times\vec{A}_E\right]+\frac{1}{c^2}\vec{A}_E=-\frac{1}{c^2}\frac{\partial}{\partial t}\left(\vec{M}+\nabla f_E\right). \tag{1.58}$$

In spherical coordinates we search for a solution of (1.58) in the form

$$\vec{A}_E = \{\varepsilon(r)\Pi_E, 0, 0\} \,, \tag{1.59}$$

where $\Pi_E = \Pi_E(r, \theta, \varphi)$ is the Debye potential for TE waves. Calculating $\vec{D} = \varepsilon_0\varepsilon(r)\vec{E} = \nabla \times \vec{A}_E$ allows us to rewrite the electrical field $\vec{E}$ in the form

$$\vec{E} = \widehat{i}_\theta \frac{1}{\varepsilon_0 r \sin\theta} \frac{\partial \Pi_E}{\partial \varphi} + \widehat{i}_\varphi \frac{1}{\varepsilon_0 r} \left(-\frac{\partial \Pi_E}{\partial \theta} \right), \tag{1.60}$$

whence one can see that for TE waves $E_r = 0$. Projecting (1.58) on the spherical axes we obtain

$$r: \quad \left(-\frac{1}{r^2}\Delta_T + \frac{\varepsilon}{c^2}\frac{\partial^2}{\partial t^2} \right) \Pi_E = -\frac{1}{c^2}\frac{\partial M_r}{\partial t} - \frac{1}{c^2}\frac{\partial}{\partial t}\frac{\partial f_E}{\partial r}, \tag{1.61}$$

$$\theta: \quad \frac{\partial}{\partial \theta} \left(\frac{\partial \Pi_E}{\partial r} + \frac{1}{c^2}\frac{\partial f_E}{\partial t} \right) = -\frac{r}{c^2}\frac{\partial M_\theta}{\partial t}, \tag{1.62}$$

$$\varphi: \quad \frac{\partial}{\partial \theta} \left(\frac{\partial \Pi_E}{\partial r} + \frac{1}{c^2}\frac{\partial f_E}{\partial t} \right) = -\frac{r\sin\theta}{c^2}\frac{\partial M_\varphi}{\partial t}, \tag{1.63}$$

where $\Delta_T = \dfrac{1}{\sin\theta}\dfrac{\partial}{\partial\theta}\sin\theta\dfrac{\partial}{\partial\theta} + \dfrac{1}{\sin^2\theta}\dfrac{\partial^2}{\partial\varphi^2}$. For TE waves the compatibility condition of (1.62) and (1.63) has the form

$$\frac{\partial M_\theta}{\partial \varphi} - \frac{\partial M_\varphi \sin\theta}{\partial \theta} = 0, \tag{1.64}$$

whence follows that $\left[\nabla \times \vec{M}\right]_r = 0$. For the simplest case $M_\theta = M_\varphi = 0$ the condition (1.64) is valid and $\vec{M} = \{M_r, 0, 0\}$. In this case the current is given by

$$\vec{j} = \nabla \times \vec{M} = \widehat{i}_\theta \frac{1}{r\sin\theta}\frac{\partial M_r}{\partial \varphi} + \widehat{i}_\varphi \frac{1}{r}\left(-\frac{\partial M_r}{\partial \theta} \right). \tag{1.65}$$

From (1.62) and (1.63) for TE waves one obtains a scalar potential f_E as follows:

$$\frac{1}{c^2}\frac{\partial f_E}{\partial t} = -\frac{\partial \Pi_E}{\partial r} - \int \Phi_E(r,t)dr, \tag{1.66}$$

where $\Phi_E(r,t)$ is an arbitrary function. Substituting (1.66) in (1.61) we obtain the equations for Π_E in the form

$$\frac{1}{r^2}\Delta_T \Pi_E + \frac{\partial^2 \Pi_E}{\partial r^2} - \frac{\varepsilon}{c^2}\frac{\partial^2 \Pi_E}{\partial t^2} = \frac{1}{c^2}\frac{\partial M_r}{\partial t} - \Phi_E(r,t). \tag{1.67}$$

Since $\Phi_E(r,t)$ is an arbitrary function we again have the opportunity to choose it by the most convenient way. One can see that equation (1.67) can be simplified by choosing

$$\Phi_E(r,t) = \frac{1}{c^2}\frac{\partial M_r}{\partial t}, \tag{1.68}$$

after that (1.67) becomes the homogeneous equation (without source). From (1.66) we can conclude that such approach is valid when both Φ_E and M_r depend on radius r and time t only. Now the homogeneous equation (1.67) reads

$$\frac{1}{r^2}\Delta_T \Pi_E + \frac{\partial^2 \Pi_E}{\partial r^2} - \frac{\varepsilon}{c^2}\frac{\partial^2 \Pi_E}{\partial t^2} = 0. \tag{1.69}$$

This equation will coincide with the equation for the TM case (1.48) if $\dfrac{\partial^2 \Pi}{\partial r^2} \to \varepsilon \dfrac{\partial}{\partial r}\left(\dfrac{1}{\varepsilon}\dfrac{\partial \Pi}{\partial r}\right)$ in (1.69) is to change and therefore it can be resolved by the same way as above.

From (1.57) one obtains the expressions for electric and magnetic fields for TE waves in the form

$$E_r = 0, \quad E_\theta = \frac{1}{\varepsilon_0 r \sin\theta}\frac{\partial \Pi_E}{\partial \varphi}, \quad E_\varphi = -\frac{1}{\varepsilon_0 r}\frac{\partial \Pi_E}{\partial \theta}, \tag{1.70}$$

$$\frac{\partial}{\partial t}H_r = -c^2\left[\frac{\partial^2 \Pi_E}{\partial r^2} - \frac{\varepsilon}{c^2}\frac{\partial^2 \Pi_E}{\partial t^2}\right], \tag{1.71}$$

$$\frac{\partial H_\theta}{\partial t} = -\frac{c^2}{r}\frac{\partial^2 \Pi_E}{\partial r \partial \theta}, \quad \frac{\partial H_\varphi}{\partial t} = -\frac{c^2}{r\sin\theta}\frac{\partial^2 \Pi_E}{\partial r \partial \varphi}. \tag{1.72}$$

(iii) *Uniform dispersive medium*

In the important case of oscillating fields the temporal multiplier we write as $e^{i\omega t}$. In this case formulas (1.49)–(1.51) for TM case read

$$E_r = -i\frac{c^2}{\omega\varepsilon}\left[\frac{\partial^2 \Pi_M}{\partial r^2} + \frac{\varepsilon\omega^2}{c^2}\Pi_M - \mu_0 j_r\right], \tag{1.73}$$

$$E_\theta = -i\frac{c^2}{\omega r \varepsilon}\frac{\partial^2 \Pi_M}{\partial r \partial \theta}, \quad E_\varphi = -i\frac{c^2}{\omega r \varepsilon \sin\theta}\frac{\partial^2 \Pi_M}{\partial r \partial \varphi}, \tag{1.74}$$

$$H_r = 0, \quad H_\theta = \frac{1}{\mu_0 r \sin\theta}\frac{\partial \Pi_M}{\partial \varphi}, \quad H_\varphi = -\frac{1}{\mu_0 r}\frac{\partial \Pi_M}{\partial \theta}. \tag{1.75}$$

Corresponding expressions (1.70)–(1.72) for TE case have the form

$$E_r = 0, \quad E_\theta = \frac{1}{\varepsilon_0 r \sin\theta}\frac{\partial \Pi_E}{\partial \varphi}, \quad E_\varphi = -\frac{1}{\varepsilon_0 r}\frac{\partial \Pi_E}{\partial \theta}, \tag{1.76}$$

$$H_r = i\frac{c^2}{\omega}\left[\frac{\partial^2 \Pi_E}{\partial r^2} + \frac{\varepsilon\omega^2}{c^2}\Pi_E\right], \tag{1.77}$$

$$H_\theta = i\frac{c^2}{\omega r}\frac{\partial^2 \Pi_E}{\partial r \partial \theta}, \quad H_\varphi = i\frac{c^2}{\omega r \sin\theta}\frac{\partial^2 \Pi_E}{\partial r \partial \varphi}. \tag{1.78}$$

For spatial homogeneous case $\varepsilon = \varepsilon(\omega)$, Eq. (1.56) for the radial part $R = R(r)$ has the same form for both TM and TE cases and it reduces to the following:

$$\frac{\partial^2 R}{\partial r^2} + \left[\frac{\omega^2\varepsilon}{c^2} - \frac{m(m+1)}{r^2}\right] R = 0. \tag{1.79}$$

In (1.79) ε can have the frequency dependence $\varepsilon = \varepsilon(\omega)$. From (1.79) one can obtain

$$\frac{\partial^2 \Pi_M}{\partial r^2} + \frac{\omega^2\varepsilon}{c^2}\Pi_M = \frac{m(m+1)}{r^2}\Pi_M, \tag{1.80}$$

and to further simplify the equation for E_r we substitute (1.80) in (1.73). Further we also use $\Pi_M \to -i\Pi_M$. After that equations (1.73)–(1.75) acquire the final form

$$E_r = \frac{c^2}{\omega\varepsilon}\left[\frac{m(m+1)}{r^2}\Pi_M + i\mu_0 j_r\right], \tag{1.81}$$

$$E_\theta = \frac{c^2}{\omega r\varepsilon}\frac{\partial^2 \Pi_M}{\partial r \partial \theta}, \quad E_\varphi = \frac{c^2}{\omega r\varepsilon \sin\theta}\frac{\partial^2 \Pi_M}{\partial r \partial \varphi}, \tag{1.82}$$

$$H_r = 0, \quad H_\theta = \frac{i}{\mu_0 r \sin\theta}\frac{\partial \Pi_M}{\partial \varphi}, \quad H_\varphi = -\frac{i}{\mu_0 r}\frac{\partial \Pi_M}{\partial \theta}. \tag{1.83}$$

In a simplest case $\varepsilon = \varepsilon_{00} = const$ one can use a new variable $y = \omega\sqrt{\varepsilon_{00}}/c$, after that Eq. (1.79) can be rewritten as

$$\frac{\partial^2 R}{\partial y^2} + \left[1 - \frac{m(m+1)}{y^2}\right] R = 0, \tag{1.84}$$

and has a general form. Solutions of Eqs. (1.79) and (1.84) can be written throw the various spherical Bessel functions [Korn & Korn, 1961, Riley *et al.*, 1998]. We use Hankel functions $h_m^{(1,2)}$, where m is an index of function, whereas the upper index $(1, 2)$ means that the Hankel function is of first or second order. Such functions are similar to the forward and backward plane waves accordingly, only it is generalized on the spherical geometry case. We write such functions in the next equation to underline such similarity explicitly:

$$h_m^{(1,2)}(y) = P_m^{(1,2)}(y)e^{\pm iy}. \tag{1.85}$$

We also use the derivative of the Hankel function in the form:

$$\partial/\partial y)h_m^{(1,2)}(y) = G_m^{(1,2)}(y)e^{\pm iy}, \tag{1.86}$$

where $P_m^{(1,2)}(y)$, $G_m^{(1,2)}(y)$ are rational pre-exponential factors of the Hankel spherical functions. The separation of the exponential part of the Hankel functions allows us to make the calculations in the simplest way. Also in this way, the final results can be easily compared with the well-known expressions for a plane case. In general case, using the standard properties of Hankel functions [Korn & Korn, 1961], one can write down the useful recursive relations for $P_m^{(1,2)}(y)$ and $G_m^{(1,2)}(y)$ in the next form

$$\begin{aligned} P_m^{(1,2)}(y) &= \frac{2m-1}{y}P_{m-1}^{(1,2)}(y) - P_{m-2}^{(1,2)}(y), \\ G_m^{(1,2)}(y) &= -\frac{m}{y}P_m^{(1,2)}(y) + P_{m-1}^{(1,2)}(y). \end{aligned} \tag{1.87}$$

For the cases of $m = 0, 1$ one can easily calculate

$$\begin{aligned} &P_0^{(1,2)}(y) = \pm\frac{i}{y}, \quad G_0^{(1,2)}(y) = \frac{1}{y}, \quad P_1^{(1,2)}(y) = -\frac{\pm i + y}{y^2}, \\ &G_1^{(1,2)}(y) = \frac{\pm i(1-y^2) + y}{y^3}. \end{aligned} \tag{1.88}$$

Together with (1.87) this allows the calculation of $P_m^{(1,2)}(y)$ and $G_m^{(1,2)}(y)$ for any m. For instance

$$\begin{aligned} &P_2^{(1)}(y) = (-3i + (-3 + iy)y)/y^3, \\ &\quad P_3^{(1)} = (-15i + (-15 + (6i + y)y)y)/y4), \quad P_2^{(2)}(y) = P_2^{(1)*}(y^*). \end{aligned} \tag{1.89}$$

So the solution of homogeneous equation for Debye potential (1.52) reads

$$\Pi(r,\theta,\varphi) = \sum_{m} \sum_{,j=-m}^{m} R_m(r) Y_m^j(\theta,\varphi), \tag{1.90}$$

where $Y_m^j(\theta,\varphi)$ are the spherical functions of order m with a positive integer m and $j = m, m-1, \ldots -m$. So the radial part $R_m(r)$ in (1.90) can be written through the Hankel spherical functions [Panofsky & Phillips, 1962].

One can see that in the case of $m = 0$ and $Y_0^0(\theta,\varphi) = 1$ fields H_φ, E_θ are zero and current $j_r(r)$ in (1.49) contributes in the spherical symmetric part in E_r only. Thus $j_r(r)$ can be responsible for the processes of inverting of population in the active core of microsphere. But $j_r(r)$ does not contribute to the resonance multipole radiation described by (1.49). Further the $m = 0$ case is excluded from our analysis while this case does not contribute to the resonance wave's solutions.

Omitting in (1.90) the angular part $Y_m^j(\theta,\varphi)$, the general solution for components of magnetic and electric fields for TM case can be written as following

$$H_\varphi = \frac{in}{c\mu_0}\left\{aP_m^{(2)}(y)e^{-iy} + bP_m^{(1)}(y)e^{iy}\right\}, \tag{1.91}$$

$$E_\theta = aG_m^{(2)}(y)e^{-iy} + bG_m^{(1)}(y)e^{iy} \ , \ y = k_0 nr, \tag{1.92}$$

where a and b are arbitrary constants, $k_0 = \omega/c$ and n is a refraction index of material. The expressions (1.91) and (1.92) describe a superposition of forward and backward spherical waves for magnetic and electric fields that is due to the reflections and transmissions in each spherical layer. Amplitudes of waves are proportional to functions $P_m^{(1,2)}(y)$ and $G_m^{(1,2)}(y)$ and, accordingly, they are inhomogeneous waves depending on a radial coordinate $y = k_0 nr$.

The above equations allow us to find the amplitudes of the electric and magnetic fields for arbitrary m. Now one can calculate the ratio E_θ/H_φ in every layer. For a forward wave, it becomes

$$\frac{E_\theta}{H_\varphi} = -i\frac{c\mu_0}{n}\left[-\frac{m}{y} + \frac{P_{m-1}^{(2)}(y)}{P_m^{(2)}(y)}\right]. \tag{1.93}$$

This formula is important for understanding the essential difference between spherical and plane stacks. From (1.93), it is apparent that for a spherical stack the structure of fields depends on a distance of given layer to a center of the cavity y and, in general, such a ratio cannot be reduced to a well-known plane case value $(-1/n)$. Therefore, the waveguide properties of a spherical stack are determined by an additional parameter of a distance to the center. Such a parameter may be chosen as a distance from the center to the boundary of deepest layer. Only for the far-zone layers with $y >> 1$ one obtains the plane case expression $E_\theta/H_\varphi = -1/n$, when using the asymptotic values of Hankel spherical functions [Korn & Korn, 1961, Riley *et al.*, 1998]. In this a case a dependence on $y = k_0 nr$ already does not take place. But generally in a spherical stack, an intermediate case with $y \sim 1$ may be valid for some deep layers.

Similar formulas can be written for TE waves case. To do that one has only to change in (1.91)–(1.93) $G_m^{(1,2)} \leftrightarrow P_m^{(1,2)}$. So in (1.90) omitting the angular part $Y_m^j(\theta, \varphi)$, the general solution for components of magnetic and electric fields for the TE case reads

$$H_\varphi = \frac{in}{c\mu_0} \left\{ aG_m^{(2)}(y)e^{-iy} + bG_m^{(1)}(y)e^{iy} \right\}, \tag{1.94}$$

$$E_\theta = aP_m^{(2)}(y)e^{-iy} + bP_m^{(1)}(y)e^{iy} \ , \ y = k_0 nr, \tag{1.95}$$

where a and b are arbitrary constants. The relation E_θ/H_φ in every layer for a TE forward wave reads

$$\frac{E_\theta}{H_\varphi} = -i\frac{c\mu_0}{n}\left[-\frac{m}{y} + \frac{P_{m-1}^{(2)}(y)}{P_m^{(2)}(y)}\right]^{-1}. \tag{1.96}$$

1.1.7 *Energy of field*

From Maxwell's equations (1.1)–(1.4) we can obtain the expression for the energy of electromagnetic field. We start from the identity law, which is valid for any vectors $\vec{E}$ and $\vec{H}$

$$\vec{H} \cdot [\nabla \times \vec{E}] - \vec{E} \cdot [\nabla \times \vec{H}] = \nabla \cdot [\vec{E} \times \vec{H}]. \tag{1.97}$$

If we substitute expressions for $\nabla \times \vec{E}$ and $\nabla \times \vec{H}$ from (1.1) and (1.2) in (1.97), and after some algebra we get

$$\begin{aligned}\nabla \cdot [\vec{E} \times \vec{H}] &= -\vec{H} \cdot \frac{\partial \vec{B}}{\partial t} - \vec{E} \cdot \frac{\partial \vec{D}}{\partial t} - \vec{E} \cdot \vec{j} \\ &= -\frac{1}{2}\frac{\partial}{\partial t}(\varepsilon_0 \varepsilon E^2 + \mu_0 \mu H^2) - \vec{E} \cdot \vec{j}.\end{aligned}$$

This equation can be rewritten in the next simple form

$$\frac{\partial w}{\partial t} + \nabla \cdot \vec{p} = -\vec{E} \cdot \vec{j}, \tag{1.98}$$

where

$$w = \tfrac{1}{2}(\varepsilon_0 \varepsilon E^2 + \mu_0 \mu H^2) \tag{1.99}$$

is the density of energy of electromagnetic field and

$$\vec{p} = \vec{E} \times \vec{H} \tag{1.100}$$

is the Poynting vector. Formula (1.99) is valid for non-dispersive materials $\varepsilon \neq \varepsilon(\omega)$ whereas (1.100) has a general form (about a dispersive case see Sec. 1.2). One can rewrite (1.98) in other useful form. If to integrate (1.98) with respect to volume V (occupied with field) we obtain

$$\frac{dW}{dt} = -P - \int_V \vec{E} \cdot \vec{j} dV, \tag{1.101}$$

where

$$W = \frac{1}{2} \int_V (\varepsilon_0 \varepsilon E^2 + \mu_0 H^2) dV, \tag{1.102}$$

and $P = \int_V (\nabla \cdot \vec{p}) dV = \iint_S (\vec{p} \cdot \vec{n}) dS$ is a flux of energy (flux of Poynting vector) over the whole surface S of which encloses the volume V. One can see from (1.101) that decreasing energy W occurs due to radiation (the energy flux) through surface S as well as due to the heating caused by the currents $\vec{j} = \sigma \vec{E}$, $\vec{E} \cdot \vec{j} = \sigma \vec{E}^2$, (where $\sigma > 0$ is a conductivity of material).

Due to the spherical geometry of problem we will characterize the state of electromagnetic field in microsphere with indexes ν, m and j (radial, spherical and azimuthal quantum numbers) accordingly, where index ν numerates the eigenfrequencies of a boundary problem. Let us calculate the field energy for a sphere with radius a_0.

The expressions for electric and magnetic fields $\vec{E}$, $\vec{H}$ in spherical polar coordinates (r, θ, φ) can be calculated by means of scalar functions called the Debye potentials, and in general it can be written as

$$\vec{E}(r,\theta,\varphi) = \sum_{\nu,m,j} \vec{E}_{\nu mj}(r,\theta,\varphi),$$
$$\vec{H}(r,\theta,\varphi) = \sum_{\nu,m,j} \vec{H}_{\nu mj}(r,\theta,\varphi), \tag{1.103}$$

$$\vec{E}_{\nu mj}(r,\theta,\varphi) = A_{\nu m}[\varepsilon(r,\theta,\varphi)\widehat{e}_r + \varepsilon_\theta(r,\theta,\varphi)\widehat{e}_\theta + \varepsilon_\varphi(r,\theta,\varphi)\widehat{e}_\varphi], \tag{1.104}$$

$$\vec{H}_{\nu mj}(r,\theta,\varphi) = \sqrt{\frac{\varepsilon_0}{\mu_0}} A_{\nu m}[h_r(r,\theta,\varphi)\widehat{e}_r + h_\theta(r,\theta,\varphi)\widehat{e}_\theta + h_\varphi(r,\theta,\varphi)\widehat{e}_\varphi], \tag{1.105}$$

where $A_{\nu m} = A_{0\nu m}e^{i\omega t}$, $A_{0\nu m}$ is a constant complex amplitude, $\varepsilon = \varepsilon(r,\omega)$ is the relative dielectric permittivity and $\widehat{e}_{r,\theta,\varphi}$ is the basis set for spherical coordinates. Due to the eigenmode orthogonality we can study the mode oscillations separately. In this stage we omit the indexes ν, m, j for notational simplicity.

Two different polarization, TM and TE, are allowed. For the TM case $h_r = 0$, while for the TE case $e_r = 0$. We now start from the TM wave case. Using the Debye potential approach the field's components in (1.104), (1.105) can be written in following form

$$\varepsilon_s(r,\theta,\varphi) = AF_s, \quad h_s(r,\theta,\varphi) = iAf_s, \tag{1.106}$$

where $F_s = F_s(r,\theta,\varphi)$, $f_s = f_s(r,\theta,\varphi)$, $s = r,\theta,\varphi$ and

$$F_r = \frac{m(m+1)}{k_0^2 r^2 \varepsilon}\Pi, \quad F_\theta = \frac{1}{k_0^2 r\varepsilon}\frac{\partial^2 \Pi}{\partial r \partial \theta}, \quad F_\varphi = \frac{1}{k_0^2 r\varepsilon \sin\theta}\frac{\partial^2 \Pi}{\partial r \partial \varphi}, \tag{1.107}$$

$$f_r = 0, \quad f_\theta = +\frac{1}{k_0 r \sin\theta}\frac{\partial \Pi}{\partial \varphi}, \quad f_\varphi = -\frac{1}{k_0 r}\frac{\partial \Pi}{\partial \theta}. \tag{1.108}$$

In Eqs. (1.107) and (1.108), $\Pi(r,\theta,\varphi)$ is the Debye potential, which is given by

$$\Pi(r,\theta,\varphi) = R_m(r)Y_m^j(\theta,\varphi), \tag{1.109}$$

where $Y_m^j(\theta,\varphi)$ are the spherical functions, $k_0 = \omega/c$ and the radial part $R_m(r)$ for TM waves obey the next equation

$$\varepsilon\frac{d}{dr}\left[\frac{1}{\varepsilon}\frac{dR_m}{dr}\right] + \left[\varepsilon - \frac{m(m+1)}{r^2}\right]R_m = 0. \tag{1.110}$$

In inhomogeneous microspheres the dielectric permittivity has a radial structure $\varepsilon = \varepsilon(r)$. Since the equations for TE fields can be obtained by a similar way, we do not write down the corresponding formulas separately. Equations (1.103)–(1.110) contain all the necessary physical information pertaining to the field energy calculation.

To calculate the energy of the spherical system we rewrite the fields (1.104) and (1.105) in the next real-value form

$$E_s \to \frac{1}{2}(E_s + E_s^*) = \frac{1}{2}(A + A^*)F_s, \tag{1.111}$$

$$H_s \to \frac{1}{2}(H_s + H_s^*) = \frac{i}{2}(A - A^*)f_s, \tag{1.112}$$

where $s = r, \theta, \varphi$. The asterisk in A^* represents the complex conjugate of A. Here we restrict ourselves with dispersiveless materials. The energy of dispersive medium will be derived in Sec. 2. The fields energy (1.99) in sphere with volume V can be written as follows

$$W = \frac{1}{2}\int_V (\varepsilon_0 \varepsilon E^2 + \mu_0 H^2) dV = \beta_e^2 + \beta_h^2, \tag{1.113}$$

where β_e^2 and β_h^2 are the electric and magnetic portions of the total energy, which with the help of (1.113) can be rewritten as

$$\beta_e^2 = \frac{\varepsilon_0}{8}(A + A^*)^2 \bar{\beta}_e^2, \tag{1.114}$$

$$\bar{\beta}_e^2 = \int_V \varepsilon(r)(F_r^2 + F_\theta^2 + F_\varphi^2) dV, \tag{1.115}$$

and

$$\beta_h^2 = -\frac{\varepsilon_0}{8}(A - A^*)^2 \bar{\beta}_h^2, \tag{1.116}$$

$$\bar{\beta}_h^2 = \int_V (f_\theta^2 + f_\varphi^2) dV. \tag{1.117}$$

In (1.114) and (1.116) A is an arbitrary complex constant. Substituting (1.107) and (1.108) in (1.114)–(1.116), and after some algebra (see

Appendix A), the quantities $\bar{\beta}_e^2$ and $\bar{\beta}_h^2$ acquire rather simple forms:

$$\bar{\beta}_e^2 = a_0^3 G_m^2 \left\{ I_m(\alpha) + \frac{1}{\varepsilon(\alpha)} R_m(\alpha) R_m'(\alpha) - \frac{1}{\varepsilon(0)} R_m(0) R_m'(0) \right\}, \quad (1.118)$$

$$\bar{\beta}_h = a_0^3 G_m^2 I_m(\alpha), \quad (1.119)$$

where $\alpha = k_0 a_0 = \omega a_0 / c$ and the integral

$$I_m(\alpha) = \frac{1}{\alpha^3} \int_0^\alpha R_m^2(y) dy \quad (1.120)$$

is determined only by the radial structure of field. The angular part G_m^2 in (1.118) and (1.119) can be calculated in the general form (see Appendix B)

$$G_m^2 = \int_0^{2\pi} d\varphi \int_0^\pi d\theta \sin\theta \left[\frac{1}{\sin^2\theta} \left(\frac{\partial Y_m^j}{\partial \varphi} \right)^2 + \left(\frac{\partial Y_m^j}{\partial \theta} \right)^2 \right] = m(m+1). \quad (1.121)$$

Note for a sphere with radius a_0 the boundary conditions $R_m(0) = 0$, $R_m'(0) = 0$ and $R_m(\alpha) = 0$, $R_m'(\alpha) = 0$ completely determine a spectrum of eigenfrequencies $\alpha_{\nu m} = \omega_{\nu m} a_0 / c$. Index ν numerates the eigenfrequencies (radial or principal quantum numbers). First conditions correspond to the field limitation in the center of sphere $r = 0$, whereas the second reflects the vanishing of transverse electrical and magnetic fields at $r = a_0$. This is true for metallized microspheres, but on the other hand such a condition is often used in a quantum theory in the field's quantization procedures (see Part II). With Eqs. (1.107) and (1.108) the latter indicates that in the boundary of sphere $r = a_0$ the Poynting vector flux is zero and full field's energy in volume V is preserved ($W = const$). One can see from (1.118)–(1.119) that when such boundary conditions apply, the important rigorous equality of electric and magnetic energies of field in a sphere follows

$$\bar{\beta}_e^2 = \bar{\beta}_h^2 = a_0^3 m(m+1) I_m(\alpha) \equiv \bar{\beta}^2. \quad (1.122)$$

Equation (1.122) generalizes the formula of equality of the energy of electric and magnetic fields of the plane-waves case [Loudon, 1994] to a spherical geometry case. Note that the form of equality $\bar{\beta}_e^2 = \bar{\beta}_h^2$ does not change for more complex multilayered structures (see Appendix C). Now we can write

$$\beta_e^2 = \frac{\varepsilon_0}{8}(A + A^*)^2\bar{\beta}^2, \quad \beta_h^2 = -\frac{\varepsilon_0}{8}(A - A^*)^2\bar{\beta}^2,$$

and energy (1.113) has the form

$$W = \frac{\varepsilon_0}{8}\bar{\beta}^2[(A + A^*)^2 - (A - A^*)^2] = \frac{\varepsilon_0}{4}\bar{\beta}^2(AA^* + A^*A). \qquad (1.123)$$

Since A is a number, we have $AA^* + A^*A = 2AA^*$. But in the quantum mechanics case A and A^* become operators (operator annihilation and operator creation, see Part II) and do not commutate.

Let's write $A = \sqrt{2}\mathcal{E}_0 e^{i\varsigma}$, where $\mathcal{E}_0$ and ς are real numbers, $\mathcal{E}_0 > 0$. Then

$$\beta_e^2 = \varepsilon_0\mathcal{E}_0^2\bar{\beta}^2\cos^2\varsigma, \quad \beta_h^2 = \varepsilon_0\mathcal{E}_0^2\bar{\beta}^2\sin^2\varsigma, \qquad (1.124)$$

and W becomes

$$W = \beta_e^2 + \beta_h^2 = \varepsilon_0\mathcal{E}_0^2\bar{\beta}^2.$$

If we were to introduce a new variable $\mathcal{E}(\alpha)$ which has the field dimension as

$$\mathcal{E}(\alpha)^2 = \beta_e^2/a_0^3 = \beta_h^2/a_0^3 = m(m+1)I_m(\alpha), \qquad (1.125)$$

then the energy W in (1.113) can be rewritten as

$$W = \varepsilon_0 a_0^3\mathcal{E}(\alpha)^2. \qquad (1.126)$$

In the quantum electrodynamics of microsphere the quantity $\mathcal{E}(\alpha)$ is related to the field per photon (see Part II).

Formula (1.125) has a general form and it is valid for any confocal spherical structures deposited on the surface of microsphere. Now with the help of (1.107) one can write down the final expression for the tangential electric field in microsphere as follows:

$$E_\theta = \mathcal{E}_0 \cdot \left[\frac{16\pi\alpha_{\nu m}^3}{3m(m+1)I_m(\alpha_{\nu m})}\right]^{1/2} \frac{1}{\alpha_{\nu m}^2 z\varepsilon(z)}\frac{\partial R_m(\alpha_{\nu m}z)}{\partial z}\frac{\partial Y_m^j(\theta,\varphi)}{\partial\theta}, \qquad (1.127)$$

where $z = r/a_0$, $\alpha_{\nu m} = \omega_{\nu m}a_0/c = k_{\nu m}a_0$, $\omega_{\nu m}$ are eigenfrequencies of the corresponding spherical structure.

1.1.8 *Metallized sphere*

As it was already mentioned, further concreteness in the definition of $I_m(\alpha)$ requires detail knowledge of the radial structure of the electromagnetic oscillations. We apply derived formulas to the simplest case of a hollow metallized microsphere with $\varepsilon = 1$ (see Fig. 1.4). For such a microsphere the solution of (1.110) has the form $R_m(y) = \left(\frac{2y}{\pi}\right)^{1/2} J_{m+1/2}(y)$, where $J_m(y)$ is the Bessel function, $y = k_0 r$. Due to the perfectly conducting walls the transverse components of the electromagnetic field vanish in the boundary of sphere and the field's volume fits the volume of the microsphere.

The boundary conditions $E_\theta = 0$ and $E_\varphi = 0$ at $r = a_0$ for TE waves result in the eigenfrequency equation $R_m(\alpha) = 0$ and can be written as

$$J_{m+1/2}(\alpha) = 0. \tag{1.128}$$

The solutions of this equation are $\alpha_{\nu m}$, where ν is a radial quantum number. Now for the calculation of $I_m(\alpha)$ in (1.127) we will use (1.128) (see

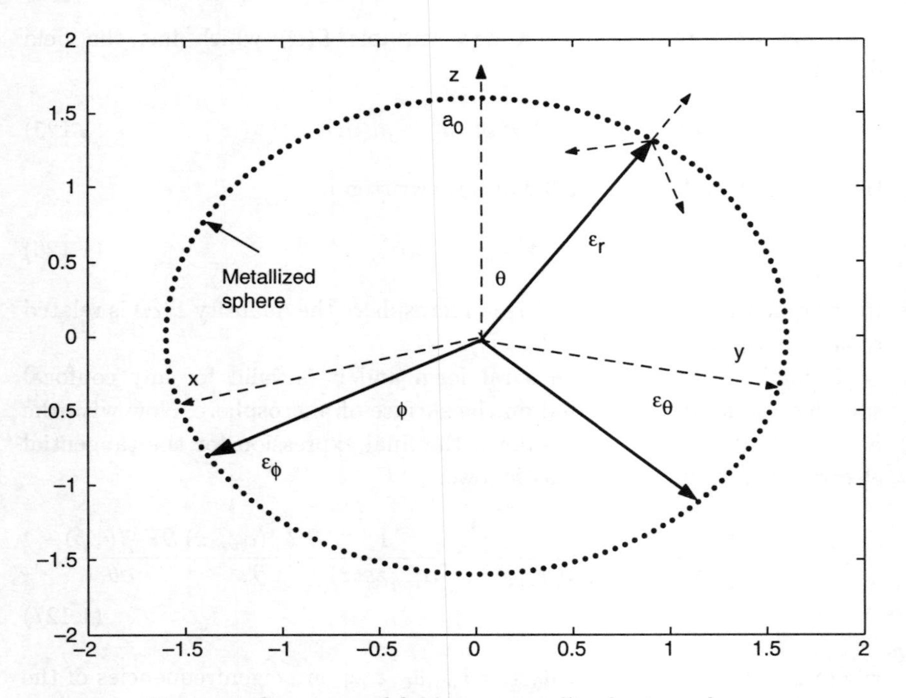

Fig. 1.4 Geometry of fields in a metallized microsphere.

Appendix D). Thereafter $\mathcal{E}(\alpha)$ in (1.125) for TE waves becomes

$$\mathcal{E}(\alpha) = \mathcal{E}_0 \left[m(m+1) \frac{\pi\alpha}{16} J^2_{m+3/2}(\alpha) \right]^{-1/2}, \quad \alpha = \alpha_{\nu m}. \tag{1.129}$$

Similar calculations for TM waves result in the eigenfrequencies equation $R'_m(\alpha) = 0$ or

$$J_{m+1/2}(\alpha) + 2\alpha J'_{m+1/2}(\alpha) = 0. \tag{1.130}$$

In this case $\mathcal{E}(\alpha)$ in (1.125) has the next form

$$\mathcal{E}(\alpha) = \mathcal{E}_0 \left[m(m+1) \frac{\pi\alpha}{4} \left(1 - m\frac{m+1}{\alpha^2} \right) J_{m+1/2}(\alpha) \right]^{-1/2}, \quad \alpha = \alpha_{\nu m}. \tag{1.131}$$

If $\alpha_{\nu m} \gg 1$ and $\alpha_{\nu m} \gg m$ from Eqs. (1.129) and (1.131) follows, that $\mathcal{E} \sim \mathcal{E}_0/\sqrt{m(m+1)}$, i.e. quantity $\mathcal{E}$ decreases with an increase of m spherical number.

Now we examine the asymptotic of the received formulas for a case where $\alpha_{\nu m} \gg m$. For $\alpha_{\nu m} z \gg 1$ and $\alpha_{\nu m} \gg m$ we use the asymptotic of Bessel functions [Korn & Korn, 1961; Riley *et al.*, 1998]: $J_{m+1/2}(\alpha) \sim \left(\frac{2}{\pi\alpha}\right)^{1/2} \sin\left(\alpha - \frac{m\pi}{2}\right)$. In this case Eq. (1.130) has the form $\cos\left(\alpha_{\nu m} - \frac{m\pi}{2}\right) = 0$ and (1.127) becomes

$$E_\theta = \mathcal{E}_0 \left[\frac{32\pi}{m(m+1)3} \right]^{1/2} \frac{\cos\left(\alpha_{\nu m} z - \frac{m\pi}{2}\right)}{z} \frac{\partial Y^j_m(\theta,\varphi)}{\partial\theta}, \tag{1.132}$$

which is similar to the regular spherical wave. From the above formulas one can derive the approximate values for the eigenfrequencies: for the TM case $\alpha_{\nu m} = (\pi/2)(m+2\nu-1)$, and for the TE case $\alpha_{\nu m} = (\pi/2)(m+2\nu)$, where $m = 1, 2, \ldots$, $\nu = 1, 2, \ldots$, which is in agreement with [Vainstein, 1988]. However the direct solution of (1.128) and (1.130) shows that for large $m \sim \alpha_{\nu m}$ these formulas are no longer valid.

To use the obtained formulas in theory we analyze the details of radial fields distribution, numerically. In Fig. 1.5 the radial distribution of the tangential field $E_\theta(\alpha_{\nu m} r/a_0)/\mathcal{E}_0$ in (1.127) for various m is shown at corresponding values of eigenfrequencies $\alpha_{\nu m}$ (calculated numerically). One can see from Fig. 1.5 is that the first maximum of the field is placed in the center of the microsphere only at $m = 1$. For larger m the maximums migrate to periphery, and the field in a vicinity of the center is exponentially small. For large $m(= 100)$ practically all fields are concentrated on the periphery

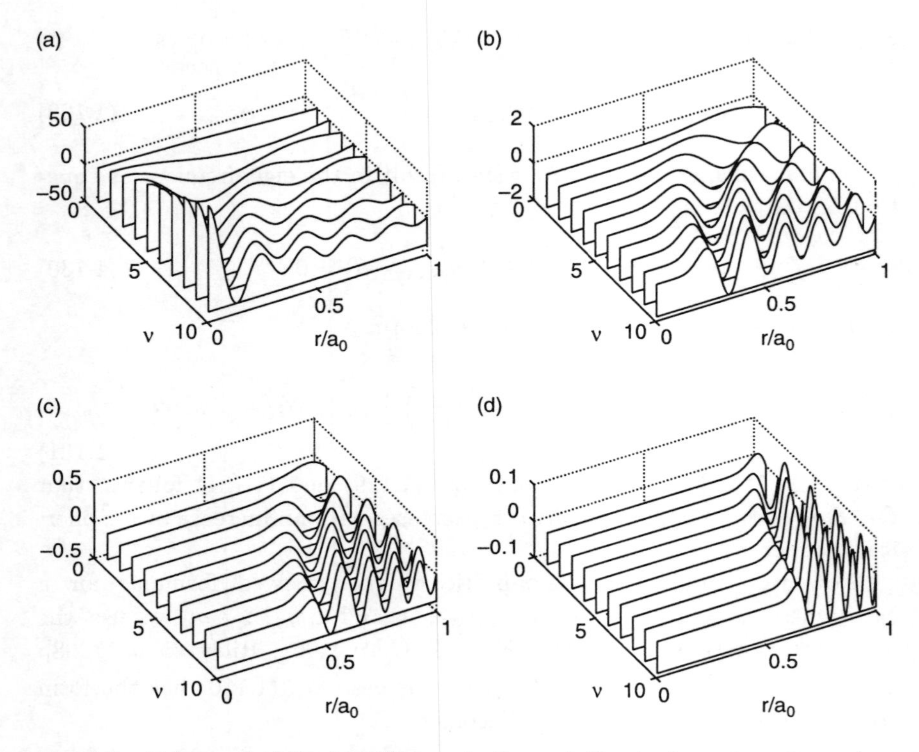

Fig. 1.5 $E_\theta(\alpha_{\nu m} r/a_0)/\mathcal{E}_0$ for TM case in the metallized microsphere for various radial $\nu = 1, \ldots, 10$ and spherical m quantum numbers. (a) $m = 1$; (b) $m = 10$; (c) $m = 30$; (d) $m = 100$; $\alpha_{\nu m}$ are dimensionless eigenfrequencies (see text), a_0 is a radius of microsphere.

(see Fig. 1.5d). Due to strong oscillations the radial distribution of the fields nodes at large $m \gg 1$ becomes quite intricate.

1.1.9 *Frequency dispersion*

Now we study the influence of a medium on the propagation of electromagnetic waves. We start with a simple model (see Fig. 1.6), which however results in reasonable expressions for dielectric permittivity. In this model the negative charged particles can move to the vicinities of positive charged node lattices which occupy the fixed position with shift d. It is considered that negative particles are linked to nodes by elastic forces, which are modeled by a spring with rigidity κ. The mass of the particles is m, and their electric charge is q. Two forces from the external electrical field qE, and the

Fig. 1.6 Simple model of dispersive medium.

dissipative force dependent on velocity $2\widetilde{\gamma}dx/dt$ act on such particles. If x is a displacement of negative charged particles from the equilibrium points, the motion equation is given by

$$m\frac{d^2x}{dt^2} = qE - \kappa x - 2\widetilde{\gamma}\frac{dx}{dt}, \tag{1.133}$$

or

$$\frac{d^2x}{dt^2} + 2\gamma\frac{dx}{dt} - \omega_0^2 x = \frac{q}{m}E, \tag{1.134}$$

where $\gamma = \widetilde{\gamma}/m$, $\omega_0 = (\kappa/m)^{1/2}$ are the eigenfrequencies, $E = E_0 e^{i(\omega t - kx)}$ is an external electric field of electromagnetic wave. We suppose that $d \ll \lambda = 2\pi/k$ and that one can neglect the spatial nonuniformity of the field.

We search for a solution of (1.134) in form

$$x = x_0 e^{i\omega t}. \tag{1.135}$$

Substituting (1.135) in (1.134) gives

$$x_0 = \frac{qE_0/m}{-\omega^2 + \omega_0^2 + 2i\gamma\omega}. \tag{1.136}$$

Electrical induction is given by

$$D = \varepsilon_0 E + P, \tag{1.137}$$

where $P = qxN$ is a polarization of medium and N is number of particles. Substituting (1.136) in (1.137) we obtain

$$D = \varepsilon_0 \left(1 + \frac{q^2 N/m\varepsilon_0}{-\omega^2 + \omega_0^2 + 2i\gamma\omega}\right) E, \tag{1.138}$$

and therefore the relative dielectric permittivity $\varepsilon = D/(\varepsilon_0 E)$ can be written in the form

$$\varepsilon = 1 - \frac{\omega_p^2}{\omega^2 - \omega_0^2 - 2i\gamma\omega}, \tag{1.139}$$

where $\omega_p^2 = q^2 N/m\varepsilon_0$ does not depend on the sign of the particle's charge q, and ω_p is referred to as the plasma frequency. In metals $\omega_0 = 0$, $N \sim 10^{22}\,\mathrm{cm}^{-3}$, $\omega_p \sim 6 \cdot 10^{15}\,\mathrm{sec}^{-1}$. We note in a quantum case that the quantity N can become negative (the inverse populations). In this case such a medium acquires the amplifier properties (see Part II). The expression for dielectric permittivity in the quantum case is derived in Sec. 8.3.

1.2 The Variational Principle

The variational principle provides us with a nice procedure which allows one to derive the equations of electromagnetic field from the compact integral expression for an action and the Lagrangian of the field [Landau & Lifshits, 1975]. Moreover the variational approach supplies a powerful base for various approaches in the perturbations theory. Whitham was the first to develop the important method in the perturbations theory [Whitham, 1974] on the basis of the variational approach, which is now referred to as Whitham's averaged Lagrangian method. First we consider how one can derive the Maxwell equations, starting from a standard variational principle [Jackson, 1975; Landau & Lifshits, 1975]. To make the idea clear, we consider a boundless linear dispersiveless environment with no electrical charges and currents, i.e. $\rho = 0$ and $\vec{j} = 0$. The Lagrangian of the electromagnetic field L then reads

$$L = L(E, B) = \frac{1}{2}\varepsilon\varepsilon_0 \vec{E}^2 - \frac{1}{2\mu_0}\vec{B}^2, \tag{1.140}$$

where $\vec{E}$ and $\vec{B}$ are the electric and magnetic fields accordingly, ε_0 and μ_0 are the dielectric and the magnetic permeability of vacuum, ε is a relative dielectric permittivity of medium. Now to obtain the action function we have to integrate L with respect to time and over volume V occupied by the field:

$$\mathcal{S}(E, B) = \int_{t_1}^{t_2} dt \int_V dV \cdot L(E, B), \tag{1.141}$$

where t_1 and t_2 are the initial and final time respectively. The variational principle sounds as: the equations of a field can be obtained from the conditions of the stationary of the action function (1.141) $\mathcal{S}(E, B)$ (i.e. $\delta\mathcal{S} = 0$) while small variations $\delta\vec{E}$ and $\delta\vec{B}$ are done. Note that till now we did not use $\vec{E}$ and $\vec{B}$. However to obtain the correct equations, quantity $\vec{E}$ and

$\vec{B}$ have to possess some a potential representation. Usually one uses the following form

$$\vec{E} = -\frac{\partial \vec{A}}{\partial t}, \quad \vec{B} = \text{curl}\vec{A} = \nabla \times \vec{A}, \tag{1.142}$$

where $\vec{A}$ is the vectorial potential, and we assume a scalar potential Φ (for simplicity) to be zero, i.e. $\Phi = 0$. The condition that $\delta\mathcal{S}$ in (1.141) vanishes $\delta\mathcal{S} = 0$ at small variations of fields (or potential $\delta\vec{A}$) is well-known in a calculus of variations and reduces to the Euler-Lagrange equation [Riley *et al.*, 1998]. In the case of (1.140) the Euler-Lagrange equation is given by

$$\frac{\partial}{\partial t}\left(\frac{\partial L}{\partial \frac{\partial \vec{A}}{\partial t}}\right) + \frac{\partial}{\partial x_i}\left(\frac{\partial L}{\partial \frac{\partial \vec{A}}{\partial x_i}}\right) = \frac{\partial L}{\partial \vec{A}}, \tag{1.143}$$

where x_j are the spatial coordinates inside a volume V. Here and in the following we adopt the convention of summation over repeated vector-component indices (i in this case). It is convenient to write down the equation (1.143) in a slightly different form, taking into account that Lagrangian L in (1.140) explicitly depends only on fields $\vec{E}$ and $\vec{B}$ and does not explicitly contain potential $\vec{A}$. From (1.143) we obtain

$$\frac{\partial}{\partial t}\left(\frac{\partial L}{\partial E_k}\frac{\partial E_k}{\partial \frac{\partial A_j}{\partial t}}\right) + \frac{\partial}{\partial x_i}\left(\frac{\partial L}{\partial B_k}\frac{\partial B_k}{\partial \frac{\partial A_j}{\partial x_i}}\right) = 0. \tag{1.144}$$

The calculations we have performed from (1.142) result in

$$E_k = -\frac{\partial A_k}{\partial t} \quad \text{and} \quad B_k = \text{curl}_{km}Am = \varepsilon_{kjm}\frac{\partial Am}{\partial x_j}, \tag{1.145}$$

where ε_{kjm} is a three-subscript antisymmetric tensor Levi-Civita, the value of which is given by

$$\varepsilon_{kjm} = \begin{cases} 1 & \text{if } k \neq j \neq m \text{ is an even permutation of } 1,2,3 \\ -1 & \text{if } k \neq j \neq m \text{ is an even permutation of } 2,1,3 \\ 0 & \text{if any of } k,m,j \text{ are equal} \end{cases}. \tag{1.146}$$

Following that, the consecutive steps of calculations can be done easily. From (1.142) one can obtain

$$\frac{\partial E_k}{\partial \frac{\partial A_j}{\partial t}} = -\delta_{kj} \quad \text{and} \quad \frac{\partial B_k}{\partial \frac{\partial A_j}{\partial x_i}} = \varepsilon_{kij}. \tag{1.147}$$

The result of substitution of these expressions in (1.144) reads

$$\frac{\partial}{\partial t}\left(-\frac{\partial L}{\partial E_j}\right)+\frac{\partial}{\partial x_i}\left(\frac{\partial L}{\partial B_k}\varepsilon_{jik}\right)=0, \tag{1.148}$$

and furthermore with the help of (1.140) we obtain

$$\frac{\partial}{\partial t}\left(-\varepsilon_0\varepsilon E_j\right)+\frac{1}{\mu_0}\frac{\partial}{\partial x_i}\left(\varepsilon_{jik}B_k\right)=0. \tag{1.149}$$

Equation (1.149) can be rewritten in the vectorial form

$$-\mu_0\varepsilon_0\frac{\partial}{\partial t}(\varepsilon\vec{E})+\nabla\times\vec{B}=0, \tag{1.150}$$

which coincides with the first Maxwell equation. The second Maxwell equation results through the elimination of potential $\vec{A}$ in the equation (1.142), namely

$$\nabla\times\vec{E}=-\nabla\times\frac{\partial\vec{A}}{\partial t}=-\frac{\partial}{\partial t}\nabla\times\vec{A}=-\frac{\partial}{\partial t}\vec{B}. \tag{1.151}$$

In addition we can derive the Hamiltonian of field H (which in this case coincides with the energy). We take into account the equality $\vec{E}=-\frac{\partial\vec{A}}{\partial t}$ and that the Lagrangian L in (1.140) does not depend on time explicitly (i.e. $\partial L/\partial t=0$). Hamiltonian H can be written as [Landau & Lifshits, 1975]

$$H=\frac{\partial A_k}{\partial t}\frac{\partial L}{\partial\frac{\partial A_k}{\partial t}}-L=E_k\frac{\partial L}{\partial E_k}-L=\frac{1}{2}\varepsilon\varepsilon_0\vec{E}^2+\frac{1}{2\mu_0}\vec{B}^2=const. \tag{1.152}$$

This formula gives us the energy of a free electromagnetic field in a dispersiveless medium.

1.2.1 *The Whitham's average variational principle*

Now we use the complex expressions for the fields and we assume that fields depend on time as $\sim e^{i\omega t}$. In this case from (1.142), for electric and magnetic fields we can obtain

$$\vec{E}=\vec{E}_0(r)e^{i\omega t}=-i\omega\vec{A}_0(r)e^{i\omega t}, \tag{1.153}$$

$$\vec{B}=\vec{B}_0(r)e^{i\omega t}=e^{i\omega t}[\nabla\times\vec{A}_0(r)], \tag{1.154}$$

where $\vec{E}_0(r)$, $\vec{B}_0(r)$ and $\vec{A}_0(r)$ can depend on cooditates and frequency ω. In contrast to the precedent part, we consider that ε has a weak frequency dispersion. In this case the electric induction can be rewritten as

$\vec{D}(t) = \varepsilon_0 \int_0^\infty \varepsilon(\tau)\vec{E}(t-\tau)d\tau$, i.e. the contribution of a field in $\vec{D}(t)$ occurs only from the previous moments of time t. This means that at time t only values of the electric field prior to the time in determining the displacement are in accordance with the fundamental ideas of causality in physical phenomena. Thus $\vec{D}(\omega) = \varepsilon_0\varepsilon(\omega)\vec{E}(\omega)$, where $\varepsilon(\omega) = \int_0^\infty \varepsilon(\tau)e^{-i\omega\tau}d\tau$. We substitute the expression (1.153) and (1.154) into (1.140) and then average on the period of oscillations $T = 2\pi/\omega$. As a result the rapidly oscillating terms like $e^{\pm 2i\omega t}$ vanish, and the expression for the averaged Lagrangian L becomes

$$\begin{aligned}\bar{L} = \bar{L}(E,B) &= \frac{1}{T}\int_0^T \left(\frac{1}{2}\varepsilon\varepsilon_0\vec{E}^2 - \frac{1}{2\mu_0}\vec{B}^2\right)dt \\ &= \frac{1}{2}\varepsilon\varepsilon_0\vec{E}_0(r)\vec{E}_0^*(r) - \frac{1}{2\mu_0}\vec{B}_0(r)\vec{B}_0^*(r). \qquad (1.155)\end{aligned}$$

In general the dielectric permittivity ε is a complex function $\varepsilon = \varepsilon_r + i\varepsilon_i$, where the imaginary part ε_i is determined by an absorption in material. In the case when ε depends on frequency $\varepsilon = \varepsilon(\omega)$, we consider (in (1.155)) such a dependence very slow, and it is possible to neglect the imaginary part ε_i in comparison with the real part ε_r. Such a material is referred to as a transparent material. In this case both $\bar{L}$ and the energy are real quantities. With the help of (1.153), (1.154) the expression for $\bar{L}$ in (1.155) becomes

$$\bar{L}(\omega,\vec{A}_0) = \frac{\varepsilon_0}{2}\varepsilon(\omega)\omega^2\vec{A}_0\vec{A}_0^* - \frac{1}{2\mu_0}[\nabla\times\vec{A}_0(r)]\cdot[\nabla\times\vec{A}_0^*(r)]. \qquad (1.156)$$

Now, following Whitham, we consider, that the temporal dependence can be rewritten in the equivalent form as $e^{i\Theta(t)}$, where $\Theta(t) = \omega t$ is the phase of oscillations, which depends on time t only. One can see from (1.156) that now the varied quantities have to be Θ and $\vec{A}_0$. The averaged Lagrangian principle states that the dynamical equations (Euler-Lagrange equations) have to derive from the vanish of the variations $\delta\bar{\mathcal{S}} = 0$, where

$$\bar{\mathcal{S}} = \bar{\mathcal{S}}(E,B) = \int_{t_1}^{t_2} dt \int_V dV \cdot \bar{L}(E,B). \qquad (1.157)$$

The variation $\delta\Theta$ results in the Euler-Lagrange equation takes the form

$$\frac{\partial}{\partial t}\frac{\partial\bar{L}}{\partial\frac{\partial\Theta}{\partial t}} = \frac{\varepsilon_0}{2}\frac{\partial}{\partial\omega}[\varepsilon(\omega)\omega^2]\cdot\frac{\partial}{\partial t}[\vec{A}_0\vec{A}_0^*] = 0, \qquad (1.158)$$

whence follows, that at a weak frequency dispersion $\varepsilon(\omega)$ the amplitude $\vec{A}_0$ can slowly vary with time, so the product of such derivatives is a second order smallness. The variation $\delta\vec{A}_0^*$ gives the Euler-Lagrange equation $\nabla\times\vec{B}_0=\varepsilon_0\varepsilon(\omega)\omega^2\vec{A}_0$, which in view of (1.153) and (1.154) coincides with the Maxwell equation (1.150). One can now calculate the average Hamiltonian in the form

$$\begin{aligned}\bar{H}&=\frac{\partial\theta}{\partial t}\frac{\partial\bar{L}}{\partial\frac{\partial\theta}{\partial t}}-\bar{L}=\omega\frac{\partial\bar{L}}{\partial\omega}-\bar{L}\\&=\frac{\varepsilon_0}{2}\frac{\partial\left[\varepsilon(\omega)\omega\right]}{\partial\omega}\omega^2\vec{A}_0\vec{A}_0^*+\frac{1}{2\mu_0}\vec{B}_0\vec{B}_0^*=\frac{\varepsilon_0}{2}\frac{\partial\left[\varepsilon(\omega)\omega\right]}{\partial\omega}\vec{E}_0\vec{E}_0^*+\frac{1}{2\mu_0}\vec{B}_0\vec{B}_0^*.\end{aligned}\tag{1.159}$$

As a result the average energy becomes

$$\bar{w}=\frac{\varepsilon_0}{2}\frac{\partial\left[\varepsilon(\omega)\omega\right]}{\partial\omega}\vec{E}_0\vec{E}_0^*+\frac{1}{2\mu_0}\vec{B}_0\vec{B}_0^*,\tag{1.160}$$

where we again have used (1.153). The expression for field's energy $\bar{w}$ has a characteristic form for a transparent dispersive environment with a multiplier $\partial[\varepsilon(\omega)\omega]/\partial\omega$. Such a formula can be derived also by different ways [Jackson, 1975; Landau *et al.*, 1984]. Nevertheless the requirement of the weakness frequency dispersion remains basic.

The correct expression on such a quantity in dispersive mediums is discussed in [see Barash & Ginzburg, 1976 and references therein]. Other approach is applied to problems of the energy transport velocity (see [Ruppin, 2002] and references therein).

1.2.2 *Energy in a layered microsphere*

Now we apply Eq. (1.159) for a case of layered microsphere. The energy of field in such a microsphere was considered in Sec. 1.1.7. For fields we can use equations (1.103)–(1.110) which contain all the necessary physical information pertaining to the field energy calculation. Such formulas can be summarized as

$$E_{s_0}(r,\theta,\varphi)=AF_s,\quad B_{s_0}(r,\theta,\varphi)=i\mu_0Af_s,\quad s=r,\theta,\varphi,\tag{1.161}$$

where A is a constant complex amplitude. If substitute Eq. (1.161) into (1.159) and integrate over a volume of microsphere for TM mode, we obtain

$$\begin{aligned}
&\int_V \left(\frac{\varepsilon_0}{2} \frac{\partial \left[\varepsilon(\omega)\omega\right]}{\partial\omega} \vec{E}_0 \vec{E}_0^* + \frac{1}{2\mu_0} \vec{B}_0 \vec{B}_0^* \right) dV \\
&= \frac{\varepsilon_0}{2} \frac{\partial \left[\varepsilon(\omega)\omega\right]}{\partial\omega} \int_V \vec{E}_0 \vec{E}_0^* dV + \frac{\mu_0}{2} \int_V \vec{H}_0 \vec{H}_0^* dV \\
&= \frac{\varepsilon_0}{2} \frac{\partial \left[\varepsilon(\omega)\omega\right]}{\partial\omega} |A|^2 \int_V (F_r^2 + F_\theta^2 + F_\varphi^2) dV + \frac{\varepsilon_0}{2} |A|^2 \int_V (f_\theta^2 + f_\varphi^2) dV \\
&= \frac{|A|^2}{2} \varepsilon_0 \left(\frac{\partial \left[\varepsilon(\omega)\omega\right]}{\varepsilon(\omega)\partial\omega} + 1 \right) \bar{\beta}^2, \qquad (1.162)
\end{aligned}$$

where $\bar{\beta}_e^2 = \bar{\beta}_h^2 = \bar{\beta}^2 = a_0^3 m(m+1) I_m(\alpha)$, see (1.122), a_0 is the radius of a layered microsphere, m is a spherical quantum number, and quantity $I_m(\alpha)$ is defined by the field's structure and is written in (1.120). Equation (1.162) can be a starting point for the field's quantization in weak-dispersive transparent materials (see Part II).

1.3 Multilayered Microsphere

Now we consider the multilayered spherical structures. Consider a dielectric sphere of radius r_0 and a concentric system of spherical layers contacting with the sphere. The layers are localized at the distances r_k from the center (see Fig. 1.7) where $r_k - r_{k-1} = d_k$ is the thickness of the kth layer; $k = 1, \ldots, N-1$.

1.4 The Transfer Matrix Method (Solving Equations for a System of Spherical Layers)

Now we discuss how to satisfy the boundary conditions of continuity for the fields H_φ and E_θ in the interface of the spherical layers at $r = r_k$. Due to a great number of layers, we develop a general matrix method [Burlak *et al.*, 2000] to calculate the reflectance of such a spherical stack of layers deposited onto a spherical substrate (Fig. 1.7). In order to preserve the generality of our study, we assume that the thickness of each layer can be arbitrary. This method is similar to a well-known matrix method for plane layers [Born & Wolf, 1980; Sakaguchi & Kubo, 1999].

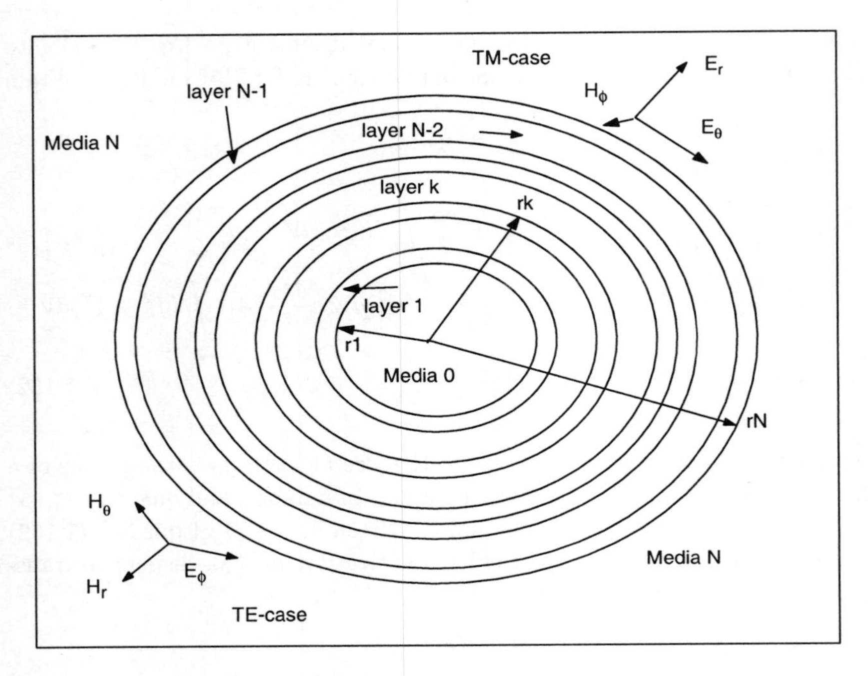

Fig. 1.7 Geometry of layered microsphere.

The general solution for components of magnetic and electric fields for the TM case in every layer of the stack is given by

$$H_\varphi = in\{aP_m^{(2)}(y)e^{-iy} + bP_m^{(1)}(y)e^{iy}\}, \tag{1.163}$$

$$E_\theta = aG_m^{(2)}(y)e^{-iy} + bG_m^{(1)}(y)e^{iy}, \quad y = k_0 nr, \tag{1.164}$$

where a and b are arbitrary constants, $k_0 = \omega/c$ and $n = n(\omega) = \sqrt{\varepsilon(\omega)}$ is the refractive index of a particular layer. Equations (1.163), (1.164) can be represented in the next matrix form

$$\vec{u} = \begin{bmatrix} H_\phi \\ E_\theta \end{bmatrix} = \widehat{D} \cdot \begin{bmatrix} a \\ b \end{bmatrix} = \widehat{D} \cdot \vec{q}, \tag{1.165}$$

where matrix $\widehat{D} = \widehat{D}(y)$ is given by

$$\widehat{D} = \begin{bmatrix} inP_m^{(2)}(y)e^{iy} & inP_m^{(1)}(y)e^{-iy} \\ G_m^{(2)}(y)e^{iy} & G_m^{(1)}(y)e^{-iy} \end{bmatrix}, \tag{1.166}$$

$$\vec{q} = \begin{bmatrix} a \\ b \end{bmatrix}. \tag{1.167}$$

$P_m^{(1,2)}(y)$ is the rational part of the Hankel spherical functions $h_m^{(1,2)}(y) = P_m^{(1,2)}(y)e^{\pm iy}$, $G_m^{(1,2)}(y)$ is the rational part of a derivative of Hankel spherical functions $(\partial/\partial y)h_m^{(1,2)}(y) = G_m^{(1,2)}(y)e^{\pm iy}$, and m is the number of spherical harmonic, $y = \omega n(\omega)r/c$. The recursive relations for the calculations of $P_m^{(1,2)}(y)$ and $G_m^{(1,2)}(y)$ are given in Eq. (1.87).

Let's supply the matrix $\widehat{D}$, and vectors $\vec{q}$ and $\vec{u}$ with indices according to the layer number in the stack (see Fig. 1.7). As the vector $\vec{q}$ in every layer is constant, for any two points r_1 and r_2 of a kth layer from (1.165) we obtain the next formula (see also [Chew, 1996])

$$\vec{q}_k = \widehat{D}_k^{-1}(r_1) \cdot \vec{u}_k(r_1) = \widehat{D}_k^{-1}(r_2) \cdot \vec{u}_k(r_2). \tag{1.168}$$

Assume that the points $r_2 = r_1 + d_1$, d_1 are the thickness of a layer. In the boundary $r = r_2$ between layers k and $k+1$ the continuity of fields gives us $\vec{u}_k(r_2) = \vec{u}_{k+1}(r_2)$. With Eq. (1.168) one can rewrite:

$$\vec{u}_k(r_1) = \widehat{D}_k(r_1) \cdot \widehat{D}_k^{-1}(r_2) \cdot \vec{u}_{k+1}(r_2) \equiv \widehat{M_k} \cdot \vec{u}_{k+1}(r_2), \tag{1.169}$$

where

$$\widehat{M_k} = \widehat{D}_k(r_1) \cdot \widehat{D}_k^{-1}(r_2). \tag{1.170}$$

Equation (1.169) can be extended to a general case as follows. Let's start from the bottom layer of the stack with number $k = 1$, then

$$\vec{u}_1 \equiv \vec{u}_1(r_1) = \widehat{M_1} \cdot \vec{u}_2 = \widehat{M_1} \cdot \widehat{M_2} \cdot \vec{u}_3 = \widehat{M_1} \cdot \widehat{M_2} \cdots \widehat{M}_{N-1} \cdot \vec{u}_N \equiv \widehat{M} \cdot \vec{u}_N(r_N), \tag{1.171}$$

or

$$\vec{u}_1 = \widehat{M} \cdot \vec{u}_N, \tag{1.172}$$

where

$$\widehat{M} = \prod_{k=1}^{N-1} \widehat{M_k} \tag{1.173}$$

is the transfer matrix $\widehat{M}$ between the inner and outer layers in the stack, which is determined by the refractive indices and thicknesses of total layers of the spherical stack. It allows the calculation of the values of the fields in the input of the stack when the output values are known.

In the simplest case of spherical mode $m = 1$, the matrix elements $\widehat{M_k}$ in (1.171) are given by

$$M_{11} = \frac{(y_2 - y_1 + y_1\, y_2{}^2)\cos(\Phi_k)}{y_1{}^2 y_2} + \frac{(-1 + y_2{}^2 - y_1\, y_2)\sin(\Phi_k)}{y_1{}^2 y_2}, \quad (1.174)$$

$$M_{12} = n_k \frac{i\,(y_2 - y_1)\cos(\Phi_k)}{y_1{}^2} - n_k \frac{i\,(1 + y_2 y_1)\sin(\Phi_k)}{y_1{}^2}, \quad (1.175)$$

$$M_{21} = \frac{-i\,(1 + y_2 y_1)\,(-y_2 + y_1)\cos\Phi_1}{y_2 y_1^3 n_k} - i\frac{(1 - y_2^2 + y_2 y_1 - y_1^2 + y_1^2 y_2^2)\sin(\Phi_k)}{y_1^3 y_2 n_k}, \quad (1.176)$$

$$M_{22} = \frac{(-y_2 + y_1 + y_1{}^2 y_2)\cos(\Phi_k)}{y_1{}^3} + \frac{(1 + y_1\, y_2 - y_1{}^2)\sin(\Phi_k)}{y_1{}^3}. \quad (1.177)$$

Here $y_1 \equiv y_{k,k}$, $y_2 \equiv y_{k,k+1}$, $y_{k,l} = k_0 n_k r_l$, $y_2 - y_1 = \Phi_k = k_0 n_k d_k$; d_k is the thickness of the kth layer. Note if $r_k \gg d_k$ or $y_1 \sim y_2 \gg 1$ in (1.174)–(1.177), one gets the well-known plane case expression for $\widehat{M_k}$ [Born & Wolf, 1980; Chew, 1996; Sakaguchi & Kubo, 1999]

$$\widehat{M_k} = \begin{bmatrix} \cos\Phi_k & -in_k \sin\Phi_k \\ -i\frac{1}{n_k}\sin\Phi_k & \cos\Phi_k \end{bmatrix}. \quad (1.178)$$

In this approximation, the matrix elements in (1.178) depend on the thickness of layers d_k through the value $y_2 - y_1 = \Phi_k$ but it do not depend on the absolute values of coordinates y_k. Nevertheless in the general case in (1.174)–(1.177), such a dependence takes place. It means again that the local properties of electromagnetic oscillations depend on the position of a layer with respect to the center of the spherical cavity.

Equation (1.172) can be rewritten through the waves amplitudes in other useful forms such as

$$\vec{q}_1 = \begin{bmatrix} a_1 \\ b_1 \end{bmatrix} = \widehat{D}_1^{-1} \cdot \widehat{M} \cdot \widehat{D}_{N-1} \cdot \vec{q}_N = \widehat{Q} \cdot \begin{bmatrix} a_N \\ b_N \end{bmatrix}, \quad (1.179)$$

where matrix Q is given by

$$\widehat{Q} = \widehat{D}_1^{-1} \cdot \widehat{M} \cdot \widehat{D}_{N-1}. \quad (1.180)$$

So far $\vec{q}$ in (1.179) are $2d$ vectors where we have two equations for four unknowns, namely a_1, b_1, a_N and b_N. To start the calculations we need two more relations. First since (1.179) is a homogeneous equation, we can exclude one of the unknowns equating it to 1, that is equivalent to normalization. The second condition is normally due to the boundary properties. For instance. In outer boundaries very important is the Sommerfeld's radiation conditions: there is only outgoing wave, because there is no reflecting interface to generate a backward wave. For such a condition one has to use $b_N = 0$ in the expression for $\vec{q}_N$. As a result, the waves amplitudes in (1.172) has the form

$$\vec{q}_1 = a_1 \cdot \begin{bmatrix} 1 \\ R \end{bmatrix} = \widehat{D}_1^{-1} \cdot \widehat{M} \cdot \widehat{D}_{N-1} \cdot \vec{q}_N = a_1 \sqrt{\frac{n_1}{n_N}} \cdot \widehat{Q} \cdot \begin{bmatrix} T \\ 0 \end{bmatrix}. \tag{1.181}$$

In (1.181) two important quantities are introduced: R and T. The complex reflection coefficient of the spherical stack in the internal boundary R is calculated as a ratio of the amplitudes of forward (a_1) and backward (b_1) waves, i.e.

$$R = \frac{b_1}{a_1}. \tag{1.182}$$

The transmittance coefficient of the stack T is the ratio of the forward (a_1) and transmitted (a_N) amplitudes and it is given by

$$T = \frac{a_N}{a_1} \cdot \sqrt{\frac{n_1}{n_N}}, \tag{1.183}$$

where $n_1$and n_N are the refraction indices of the inner layer and the outer medium, respectively. a_k and b_k are the constant amplitudes of waves in the kth layer. The complex impedance of the stack Z is defined by the following ratio

$$Z = \frac{E_\theta}{H_\varphi} = \frac{(\vec{u}_1)_2}{(\vec{u}_1)_1} \equiv Z_{\text{stack}}. \tag{1.184}$$

After solving (1.181) one easily obtains R and T in the next form

$$R = \frac{Q_{21}(\omega)}{Q_{11}(\omega)}, \quad T = \frac{1}{Q_{11}(\omega)} \sqrt{\frac{n_1}{n_N}}. \tag{1.185}$$

In Eq. (1.185) two equations relate to three variables: R, T and frequency ω. Defining ω one can calculate the frequency dependence of $R(\omega)$, $T(\omega)$ for the stack.

Other example is the metallized microsphere, where the boundary condition in the external boundary of microsphere is $E_\theta = 0$. In this case $\vec{u}_N = \{H_\varphi, 0\}$.

It is easy to see from (1.166) that the matrix $\widehat{D}_k(y)$ can be factorized into a product of two matrices:

$$\widehat{D}_k(y) = \widehat{F}_k(y) \cdot \widehat{H}_k(y), \tag{1.186}$$

where

$$\widehat{F}_k(y) = \begin{bmatrix} inP_m^{(2)}(y) & inP_m^{(1)}(y) \\ G_m^{(2)}(y) & G_m^{(1)}(y) \end{bmatrix}, \quad \widehat{H}_k(y) = \begin{bmatrix} e^{iy} & 0 \\ 0 & e^{-iy} \end{bmatrix}. \tag{1.187}$$

Using these formulas, after obvious calculations, we can write the expression (1.173) for matrix $\widehat{M_k}$ in the form

$$\begin{aligned} \widehat{M_k} &= \left[\widehat{F}_k(r_{k-1}) \cdot \widehat{H}_k(r_{k-1})\right] \left[\widehat{F}_k(r_k) \cdot \widehat{H}_k(r_k)\right]^{-1} \\ &= \widehat{F}_k(r_{k-1}) \cdot \widehat{H}_k(r_{k-1}) \cdot \left[\widehat{H}_k(r_k)\right]^{-1} \left[\widehat{F}_k(r_k)\right]^{-1}, \end{aligned}$$

or

$$\widehat{M_k} = \widehat{F}_k(r_{k-1}) \cdot \widehat{E}_k(\Phi_k) \cdot (\widehat{F}_k(r_k))^{-1}, \tag{1.188}$$

where matrix $\widehat{E}_k$ is given by

$$\begin{aligned} \widehat{E}_k(\Phi_k) &\equiv \widehat{H}_k(r_{k-1}) \cdot \left[\widehat{H}_k(r_k)\right]^{-1} = \begin{bmatrix} e^{i\Phi_k} & 0 \\ 0 & e^{-i\Phi_k} \end{bmatrix}, \\ \Phi_k &= y_{k-1} - y_k = -k_0 n_k d_k. \end{aligned} \tag{1.189}$$

$d_k = r_{k+1} - r_k$ is the thickness of the kth layer. Such a form of matrix $\widehat{M}$ allows one to separate the exponential e^{iy} into a single matrix $\widehat{E}$. Since y_k may have a complex value, such an approach allows the improvement the accuracy of calculations.

With the help of the expression for the Wronskian of Hankel functions [Korn & Korn, 1961] $W\{H_\nu^{(1)}(z)H_\nu^{(2)}(z)\} = -4i/\pi z$, one can calculate that $\det(\widehat{M_k}) = (y_{k+1}/y_k)^2 = (r_{k+1}/r_k)^2$. Such a value of the determinant corresponds to the conservation of intensity of spherical fields in any solid angle. For $r_k \gg d_k$ and $y_{k-1} \simeq y_k \gg 1$ we have known that the plane case value is $det(\widehat{M_k}) = 1$ [Born & Wolf, 1980; Chew, 1996; Sakaguchi & Kubo, 1999].

Till now, we have considered the thickness for each layer in the stack as an arbitrary number. However, the case with the periodic alternative layers is of the most practical interest. So, when considering $\Phi_k = k_0 n_k d_k = \pi/l$, one obtains $d_k = \pi/l k_0 n_k$ where l is integer. In the analogy with the plane case, we consider a stack with $1/k_0 = \Lambda_0/2\pi$, where Λ_0 defines the periodicity in a stack. We often consider the case $d_k n_k = \Lambda_0/4$ that approximately corresponds to a quarter-wave case.

1.5 Reflection Coefficient and Impedance of a Spherical Stack

Now we use the transfer matrix method for calculations. We numerically study the frequency dependence of reflection coefficient R and impedance Z of the stack with the various number of layers (Fig. 1.7). The following parameters have been used in calculations: the geometry of a system is $ABCBCBC \cdots BD$ where the letters A, B, C, D indicate the materials in the system, $\Lambda_0 = 2\pi/k_0 = 1.75\,\text{mkm}$, $f_0 = 171.5\,\text{THz}$. We have used the

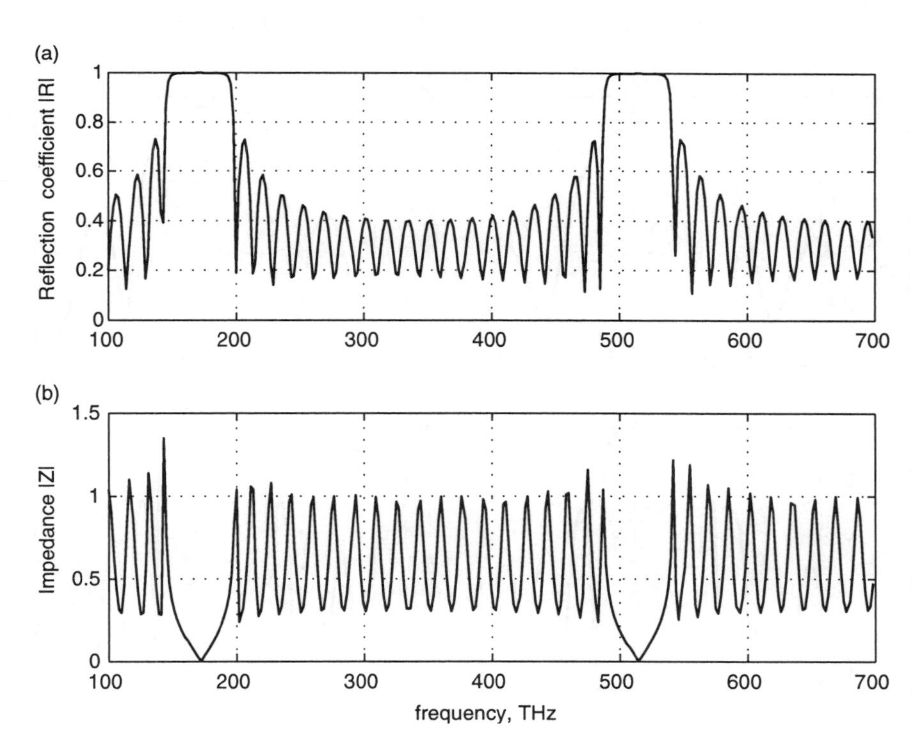

Fig. 1.8 A module of a reflection coefficient and a module of an impedance of spherical stack versus frequency (THz) for a case $\Lambda_0/4$. Designed 20 layers. A lossless case.

parameters of materials [Hodgson & Weber, 1997]

$$\begin{aligned}
&A: \text{LiF}, \quad n_A = 1.4,\\
&B: \text{TiO}_2, n_B = 2.8,\\
&C: \text{SiO}_2, n_C = 1.56,\\
&D: \text{air}, \quad n_D = 1.
\end{aligned}$$

The losses in materials and random deviations of thicknesses of layers will be taken into account. Note that already in the case with layers in a stack $N \leq 5$, the matrix $\widehat{M}$ is too long. Therefore, the calculations of both R and Z can be done only by means of computations.

The computer program performed the calculations of complex R and Z against an electromagnetic radiation frequency f. More details about computation one can be found in Part III. The results are presented in Figs. (1.8)–(1.13). Another task is to find a solution for the eigenfrequency equation for the resonator coated by a spherical stack. However the analysis

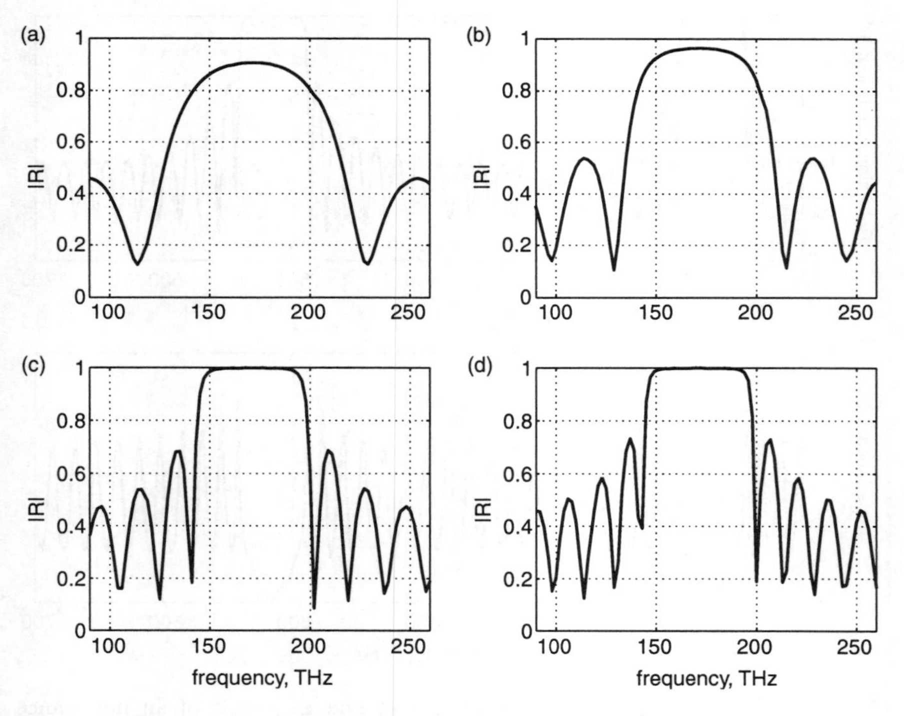

Fig. 1.9 A module of the reflection coefficient for the lossless layers for various number of layers in the stack (a) $N = 5$, (b) $N = 10$, (c) $N = 15$, (d) $N = 20$.

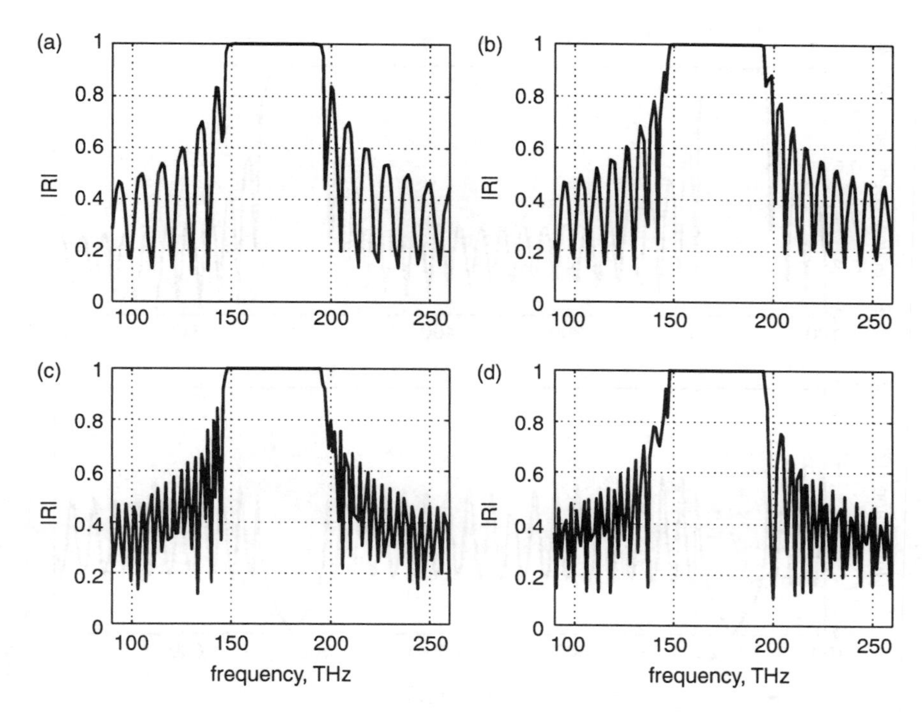

Fig. 1.10 Module of the reflection coefficient for lossless layers for various number of layers in the stack (a) $N = 30$, (b) $N = 40$, (c) $N = 80$, (d) $N = 150$.

of eigenfrequency problem for optical resonator loaded by such a stack is rather complicated and we will discussed later.

We begin from the simplest spherical mode with $m = 1$. In Fig. 1.8, the reflection coefficient R and the impedance Z of a stack versus frequency f (for $d_k = \Lambda_0/4$ and the spherical mode $m = 1$) are presented. One can see that R has a stop band around $f = f_0, 3f_0$. Figures 1.9 and 1.10 demonstrate the reshaping of the stop-band zone versus a number of layers in the stack N. One can see that R has a rather sharp shape at $N = 15$ and it has almost no essential modification for $N > 15$. The steepness of R with respect to the frequency f is quite high, starting from $N = 30$, and at $N = 40$ the zone of stop-band looks like a complete rectangle.

In the real case, one should pay attention to two additional factors. First, one should take into account the material losses. For this, it is necessary to consider the refraction index n as a complex value ($n = n_r + in_i$, where n_r and n_i are the real and imaginary parts of n). In the range of

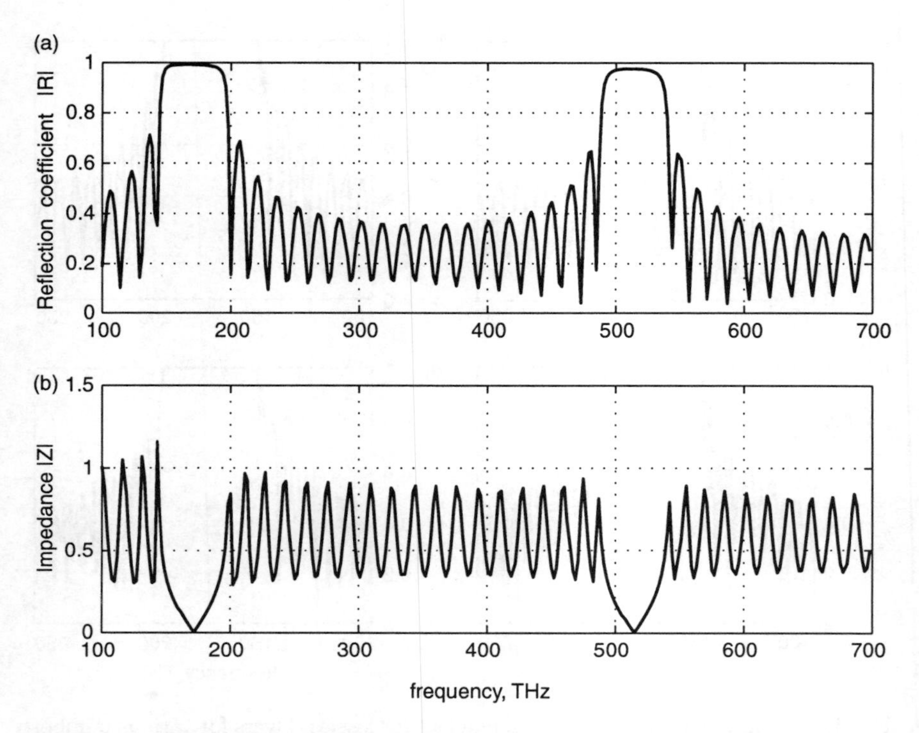

Fig. 1.11 Module of a reflection coefficient and module of an impedance of stack against frequency (THz) in case $\Lambda_0/4$ for lossy layers. Designed 20 layers.

material transparency, the imaginary part is usually small $n_i \simeq 0.005$. The reflectance in the case of the lossy materials (Fig. 1.11) changes insignificantly to compare with a lossless case (Fig. 1.8).

Second, it is impossible to realize exactly the quarter-wave thickness of layers. Due to technological problems on the preparation of samples, the random factor presents in the size of the layers. It is important that the periodicity in the system can be broken due to the random thicknesses of layers. We have simulated the random factor by considering that the thicknesses of layers l have random values distributed in intervals of $l_0+\delta l$, where δl is an additive term randomly distributed in the interval $[-\alpha l_0, +\alpha l_0]$. We have used α up to 20%.

From Fig. 1.12b one can see that in the case of a random deviation of thicknesses, the first stop-band can keep the original profile, but at higher order stop-bands may be destroyed completely.

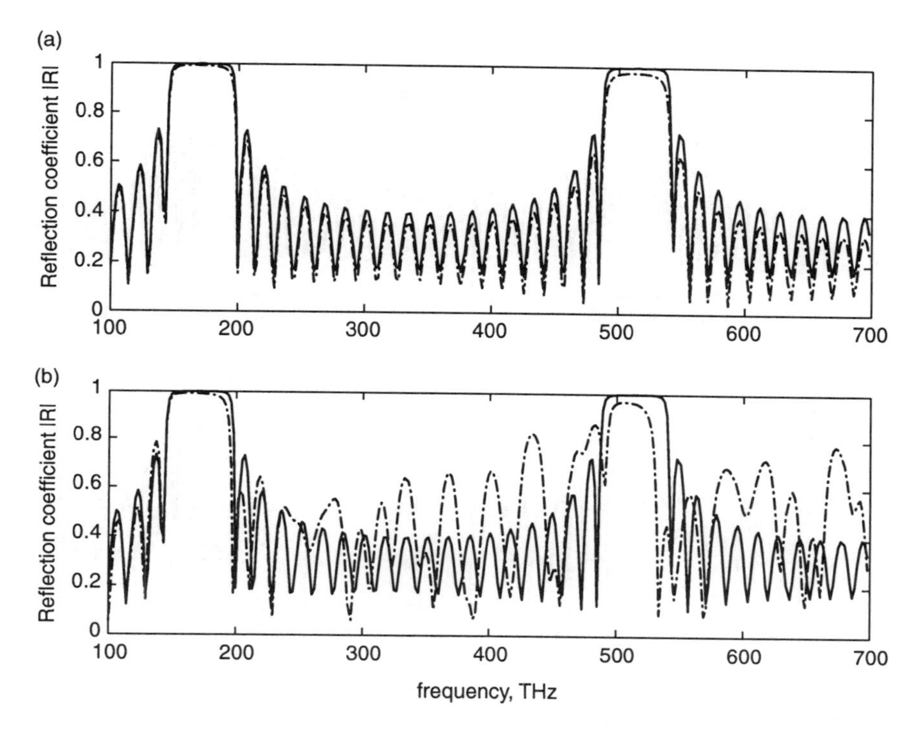

Fig. 1.12 Module of a reflection coefficient of a stack versus frequency (THz) in cases of $\Lambda_0/4$. Designed 20 layers. (a) Lossless case (solid line) and the case of lossy layers (dash line). (b) Lossless case (solid line) and the case when both lossy layers and random factor (20%) (dash line).

Till now, an influence of the sphericity on properties of a stack has been inessential, because even for microspheres with $r_0 = 20\,\mu\text{m}$ and for a radiation with a wavelength $\lambda = 1\,\mu\text{m}$ the far-zone case [Ramo *et al.*, 1994] $r_0/\lambda \gg 1$ is realized. However, when utilizing the submicron substrate microspheres with $r_0 < 1\,\mu\text{m}$, one can expect much more influence of a number of spherical mode on the reflection properties of the stack. In Fig. 1.13 the frequency dependence of the reflection coefficient for the different numbers of spherical modes is shown. The coating microsphere with $r_0 = 0.5\,\mu\text{m}$ has been studied. One can see that for a few lowest modes $m = 1, 2$ the picture (see Fig. 1.13a,b) looks like the plane stack case [Sakaguchi & Kubo, 1999; Convertino *et al.*, 1999] and the influence of sphericity is unessential here. However, for higher modes with $m = 4$ (Fig. 1.13c), the additional zone of strong reflection arises from the lower frequency side. Such a shift of

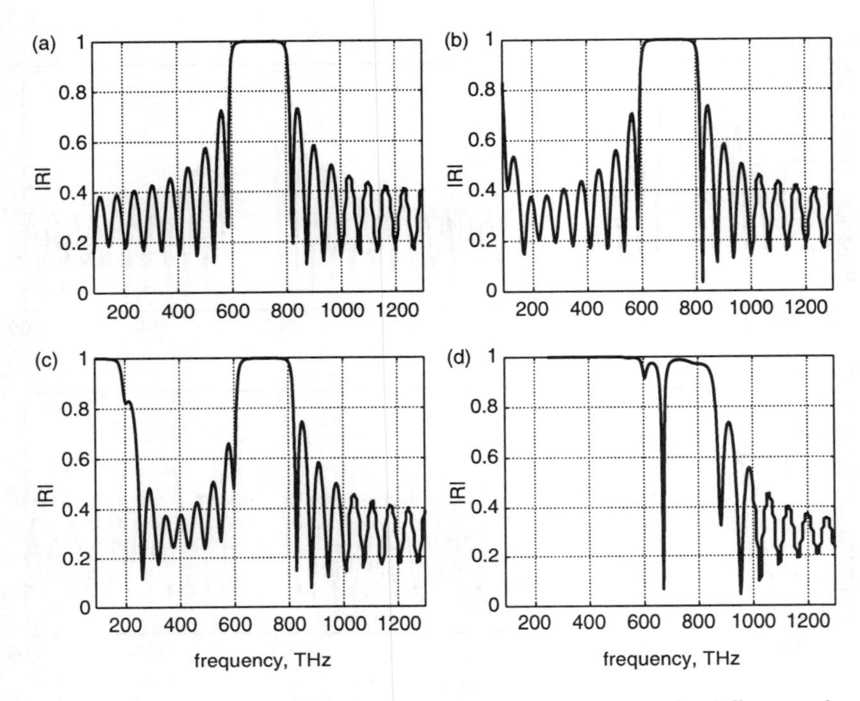

Fig. 1.13 Module of a reflection coefficient for a loss-free layer for different spherical modes with numbers (a) $m = 1$, (b) $m = 2$, (c) $m = 4$, and (d) $m = 11$.

reflectivity to higher frequencies for higher values of m is in an agreement with [Brady *et al.*, 1993]. Figure 1.13c shows that in this case the symmetry on a center of the first stop-band no longer takes place. For $m = 11$ (Fig. 1.13d) the additional reflection zone approaches the first stop-band, and a narrow transparency gap is reshaped in the interspace (Fig. 1.13d). Such gap is encircled by the areas with a high reflection ($R \approx 1$). Therefore, one can expect the essential increase of both frequency selectivity and Q factor, similar to the case of a microsphere resonator with a whispering-gallery mode [Vassiliev *et al.*, 1998]. Figure 1.13d shows that in this case almost full transparency ($R = 0.04$) in the bottom of the gap takes place.

1.6 Conclusion

We have studied in this chapter the submicron size microspheres coated by a system of contacting concentric spherical dielectric layers (spherical stack) in the optical frequency range. The frequency dependence of the reflection coefficient and the impedance of a stack have been investigated. The general matrix formalism has been developed for the spherical geometry layers with an arbitrary ratio between r and λ. The quarter-wave length layers have been studied in details. A case of the random deviated layers has been investigated. Numerical calculations of the frequency dependence of the coefficient reflection show that a suitable reflectance may be designed when a number of quarterwave-length layers is of an order of ten. The appearance of the narrow gap of the transparency in the zone of high reflectance of a stack under the suitable parameters has been revealed.

The sizes of the coating microsphere in an optical frequency range are rather small. Therefore such a system can be designed as a solid one and it can possess, hence, a weak sensitivity on deformations. Thus, it has an advantage in comparison with the long-size quasi-optical systems. A separate problem is the influence of a coating on a spectrum of eigenfrequencies and a Q factor of a coated spherical resonator. For lossless layers, Q grows exponentially against the number of layers in a stack [Brady *et al.*, 1993]. But a real case of lossy materials as well as a random deviation of thickness is of a great practical interest. Another problem is to investigate the dispersive properties of materials in a spherical stack, similar to a plane case [Kumar *et al.*, 1999].

CHAPTER 2

Electromagnetic Field in Homogeneous Microspheres Without Surface Structures

2.1 Experiments with Microspheres

In this section we briefly touch on the experiments with microspheres. The experiments were carried out in various groups and universities.

In [Artemyev & Woggon, 2000] a micrometer-sized spherical microcavity, the photonic dot, is made from semiconductor nanocrystals, which are the quantum dots. The coupling of electronic and photonic states is demonstrated for a single photonic dot by the observation of whispering gallery modes (WGM) in the spectrum of spontaneous emission of the embedded CdSe quantum dots.

In [Artemyev *et al.*, 2001a] the incorporation of *CdSe* quantum dots into a thin ($<1\,\mu$m) surface shell of polymer microspheres ($R \sim 2$–$4\,\mu$m) is achieved. The room-temperature emission spectra of single, hollow microcavities show several spectrally well-separated cavity modes in the red-orange spectral range which have been assigned to high-Q WGMs with radial quantum number $n = 1$ and high angular quantum number l. An enhancement of the cavity finesse Q by a factor of about 10 with respect to *CdSe*-doped bulk polymer microspheres is found.

In one of the first articles of [Braginsky *et al.*, 1989] the properties of optical resonators with quality-factor $Q \sim 10^8$, effective volume of electromagnetic field localization $V_{eff} \sim 10^{-9}\,\text{cm}^3$ and the threshold power of optical bistability $W_{bist} \sim 10^{-5}$ W are described. The prospects of reducing

V_{eff} and W_{bist} are discussed. With possible reduction of the controlling energy of optical switching down to a single quantum and the employment of the monophotonic states of light, the whispering-gallery microresonators can open the way to realize the Feynman's quantum-mechanical computer.

In [Brun & Wang, 2000] coupling nanocrystals (quantum dots) at a high-Q WGM of a silica microsphere can produce a strong coherent interaction between the WGM and the electronic states of the dots. Shifting the resonance frequencies of the dots, for instance, by placing the entire system in an electric potential, allows this interaction to be controlled, permitting entangling interactions between different dots in a way analogous to the ion-trap computer of Cirac and Zoller. Thus, a more advanced system of this type could potentially be used to implement a simple quantum computer.

In [Buck & Kimble, 2003] the WGMs of quartz microspheres are investigated for the purpose of strong coupling between single photons and atoms in cavity quantum electrodynamics (cavity QED). Within current understanding of the loss mechanisms of the WGMs, the saturation photon number n_0 and critical atom number N_0 cannot be minimized simultaneously, so that an "optimal" sphere size is taken to be the radius for which the geometric mean $\sqrt{n_0 N_0}$, is minimized. While a general treatment is given for the dimensionless parameters used to characterize the atom-cavity system, detailed consideration is given to the D_2 transition in atomic cesium at $\lambda_0 = 852\,\text{nm}$ using fused-silica microspheres, for which the maximum coupling coefficient $g_a/(2\pi) \approx 750\,\text{MHz}$ occurs for a sphere radius of $a = 3.63\,\mu\text{m}$ corresponding to the minimum for $n_0 \approx 6.06 \times 10^{-6}$. By contrast, the minimum for $N_0 \approx 9.00 \times 10^{-6}$ occurs for a sphere radius of $a = 8.12\,\mu\text{m}$, while the optimal sphere size for which it is minimized occurs at $a = 7.83\,\mu\text{m}$. On an experimental front, authors have fabricated fused-silica microspheres with radii of $a \sim 10\,\mu\text{m}$ and they have consistently observed quality factors of $Q \geq 0.8 \times 10^7$. These results for the WGMs are compared with corresponding parameters achieved in Fabry-Perot cavities to demonstrate the significant potential of microspheres as a tool for cavity QED with strong coupling.

Authors [Deumie *et al.*, 2002] produced overcoated microspheres and they try to obtain artificial optical powders with specific spectral properties. Electron-beam deposition is used with a particular vibration system. Calibration and characterization results are presented that validate the techniques and procedures for single-layer and quarter-wave mirrors.

In [Gorodetsky & Ilchenko, 1994] high-Q optical whispering-gallery modes $TE(TM)lmq$ in spheroidal dielectric microresonators have shown to be equivalent to precessing circular modes $TE(TM)llq'$ with the rate of precession depending on the value of the residual resonator nonsphericity. This approach gives a natural interpretation to the characteristic two-lobe emission patterns of these modes with prism couplers. Spatial recognition of different modes on the basis of their emission patterns facilitates optimal coupling to selected modes, which is of interest to applications. Experimental results on the observation of two-lobe emission patterns and the imaging of internal mode field distribution are presented. The feasibility of a single-prism transient filter and anoptical delay line based on precessing modes is discussed.

In [Gorodetsky & Ilchenko, 1999] a general model is presented for the coupling of high-Q WGM in optical microsphere resonators with coupler devices that possess a discrete and continuous spectrum of propagating modes. In contrast to conventional high-Q optical cavities, in microspheres the independence of high intrinsic quality-factor and the controllable parameters of coupling via an evanescent field offer a variety of regimes similar to those that are already available in RF devices. The theory is applied to data reported earlier on different types of couplers on the microsphere resonators and is complemented by the experimental demonstration of enhanced coupling efficiency (80) and variable loading regimes with $Q \sim 10^8$ fused-silica microspheres.

In [Fujiwara & Sasaki, 1999] the blue and red upconversion lasing of a $Tm3+$-doped fluorozirconate glass microsphere was demonstrated. Tens-of-micrometer-sized, genuinely spherical glass particles were produced by a melting method with a burner. The single microsphere was pumped by a focused beam of a $Nd:YAG$ laser operating at 1064 nm. Three-photon-excited lasing emission could be observed in 480 and 800 nm regions at room temperature, and their lasing thresholds were determined to be 20 and 5 mW, respectively. Applications of the microspherical upconversion laser to near-field scanning optical microscopy are also discussed.

Authors [Ilchenko *et al.*, 1998] demonstrated the tuning of high-Q optical whispering-gallery modes in a fused silica microsphere by applying mechanical strain.

In [Sasaki *et al.*, 1997] the photon tunneling of lasing emission from a dye-doped microspherical particle to an object was investigated by the use of a microspectroscopy system combined with a laser manipulation technique. An emission spectrum changed drastically when the lasing microsphere approaches a glass plate. The intensity ratio between resonant peaks exhibited exponential dependence on the sphere-object distance, whose decay constant agreed with the penetration depth of an evanescent field just outside the microsphere. The variation in the spectral profile can be explained with the Mie scattering theory. The applicability of the lasing microsphere as a probe for a near-field scanning optical microscope is discussed.

Authors [An & Moon, 2003] report laser oscillations in WGMs of a microsphere with its cavity quality factor Q unchanged for varied optical pumping intensities well above the laser threshold. Laser gain was present only in the evanescent-wave region of the lasing WGM around the microsphere equator, and thus the pumping-induced Q degradation could be minimized, resulting in Q values maintained at $8(2) \cdot 10^9$ regardless of the pumping strength.

In [Cai *et al.*, 2000] the authors presented the observation of critical coupling in a high-Q fused-silica microsphere WGM resonator coupled to a fiber taper. Extremely efficient and controlled power transfer to the high-Q ($\sim 10^7$) resonators has been demonstrated. Off-resonance scattering loss was measured to be less than 0.3%. On-resonance extinction in transmitted optical power through the fiber coupler was measured to be as high as 26 dB at the critical coupling point. This result opens up a range of new applications in fields as diverse as near-field sensing and quantum optics.

In [Yamasaki *et al.*, 2000] the periodic dielectric structures, consisting of hexagonally closed-packed arrays of silica microspheres with diameters of 550 nm, were incorporated into organic light-emitting devices with a conventional two-layer structure produced through vacuum-sublimation. The arrays acted as a two-dimensional diffraction lattice which behaved as a light scattering medium for the light propagated in wave guiding modes within the device. Strongly scattered light emission through the front surface of the devices was observed. An increase in the device coupling-out factor for electroluminescent efficiency through the use of the scattering structure is demonstrated.

In [Vassiliev *et al.*, 1998] a new modification of external optical feedback (OFB) was used to narrow the line of a diodelaser (DL). A 'WGM' of a high-Q microsphere was excited by means of frustrated total internal reflection while the feedback for the optical locking of the laser was provided by the intracavity Rayleigh backscattering. The $\sim$600 MHz beatnote of the two laser diodes optically locked to a pair of orthogonally polarized modes of the same microresonator had the indicated spectral width of 20 kHz, as well as the stability of 2×10^{-6} over an averaging time of 10 s. The feasibility of miniature sub-kHz-line width laser is discussed.

In [Vernooy *et al.*, 1998] the measurements of the quality factor $Q \sim 8 \cdot 10^8$ are reported for the WGMs of quartz microspheres for the wavelengths 670, 780, and 850 nm; these results correspond to finesse $F \sim 2.2 \cdot 10^6$. The observed independence of Q from wavelength indicates that losses for the WGM are dominated by a mechanism other than bulk absorption in fused silica in the near infrared. Data obtained by atomic force microscopy combined with a simple model for surface scattering suggest that Q can be limited by residual surface inhomogeneities. Absorption by absorbed water can also explain why the material limit is not reached at longer wavelengths in the near infrared.

In [Viana *et al.*, 2002], the authors use a single microscope objective lens to optically trap, illuminate, and collect backscattered light of a dielectric microsphere. They then measure the temporal-intensity-autocorrelation functions (ACFs) and intensity profiles to obtain the trap stiffness and friction coefficient of the bead. This is an interesting study of an harmonically bound Brownian particle with nanometer resolution. The authors extend the work of Bar-Ziv *et al.* [*Phys. Rev. Lett.*, 78, 154 (1997)] to more general situations allowing for the use of our simpler geometry in other applications. As examples, the authors present measurements of the parallel Stokes friction coefficient on the trapped bead as a function of its distance from a surface and the entropic force of a single lambda-DNA molecule.

In [Arai *et al.*, 2003] the microspheres of refractive index $n_D > 2$ have been investigated. The organic–inorganic hybrid microspheres of refractive index $n_D = 1.70$–1.72 were prepared at room temperature by the vibrating orifice technique using titanium alkoxide and silane coupling reagents as starting materials. Subsequently heating the microspheres at 400–450°C resulted in increasing their refractive indices to $n_D = 2.10$–2.25 while maintaining good spherical shape. Rhodamine $6G$-doped microspheres of $n_D = 1.72$ were also prepared at room temperature and the lasing from them was performed by pumping by second-harmonic pulses of the Q-switched Nd-yttritium-aluminum-garnet laser.

In [Loscertales I.G *et al.*, 2000a] the authors report on a method to generate steady coaxial jets of immiscible liquids with diameters in the range of micrometer/nanometer sizes. This compound jet is generated by the action of electro-hydrodynamic (EHD) forces with a diameter that ranges from tens of nanometers to tens of micrometers. The eventual jet breakup results in an aerosol of monodispersed compound droplets with the outer liquid surrounding or encapsulating the inner one. Following this approach, the authors have produced monodispersed capsules with diameters varying between 10 and 0.15 μm, depending on the running parameters.

In [Artemyev *et al.*, 2001b] the optical microcavities that confine the propagation of light in all three dimensions (3D) are fascinating research tools to study 3D-confined photon states, low-threshold microlasers, or cavity quantum electrodynamics of quantum dots in 3D microcavities. A challenge is the combination of complete electronic confinement with photon confinement, e.g. by linking a single quantum dot to a single photonic dot. Authors report on the interplay of 3D-confined cavity modes of single microspheres (the photonic dot states) with photons emitted from the quantized electronic levels of single semiconductor nanocrystals (the quantum dot states). Authors show how cavity modes of high cavity finesse are switched by single, blinking quantum dots. A concept for a quantum-dot microlaser operating at room temperature in the visible spectral range is demonstrated. Authors observe an enhancement in the spontaneous emission rate; i.e. the Purcell effect is found for quantum dots inside a photonic dot.

Spherical microcavities of a few micrometers in diameter show sharp, spectrally well-separated cavity modes in the visible spectral range. These eigenmodes of a 3D microcavity are characterized by the angular quantum number l and the radial quantum number n for the transverse electric (TE_l^n) and transverse magnetic (TM_l^n) field modes. The optical quality of a microcavity is defined by $Q = \omega_{cav}/\Delta\omega_{cav}$, the ratio between resonance frequency of a cavity and bandwidth $\Delta\omega_{cav}$ of a cavity mode. High Q values mean narrow modes and efficient light trapping inside the microcavity. When a light-emitting dipole, for example a semiconductor nanocrystal (quantum dot (QD)), is inserted into a 3D microcavity and its eigenfrequency ω_{QD} is resonant with a high-Q cavity mode ω_{cav}, then the confined photonic and electronic states interact and a new coupled quantum-mechanical system evolves with a coupling strength defined by the Rabi splitting Ω_R [Gaponenko, 1998; Woggon, 1996]. Semiconductor quantum dots inside a 3D microcavity (photonic dot (PD)), a structure we will designate from now on as QDs@PD, provide a fascinating artificial system to study light — matter interaction in confined systems. The next examples presented by authors [Artemyev *et al.*, 2001b] (see Figs. 2.1 and 2.2) apply to the regime of weak coupling, i.e. $\Omega_R < \Omega_R$.

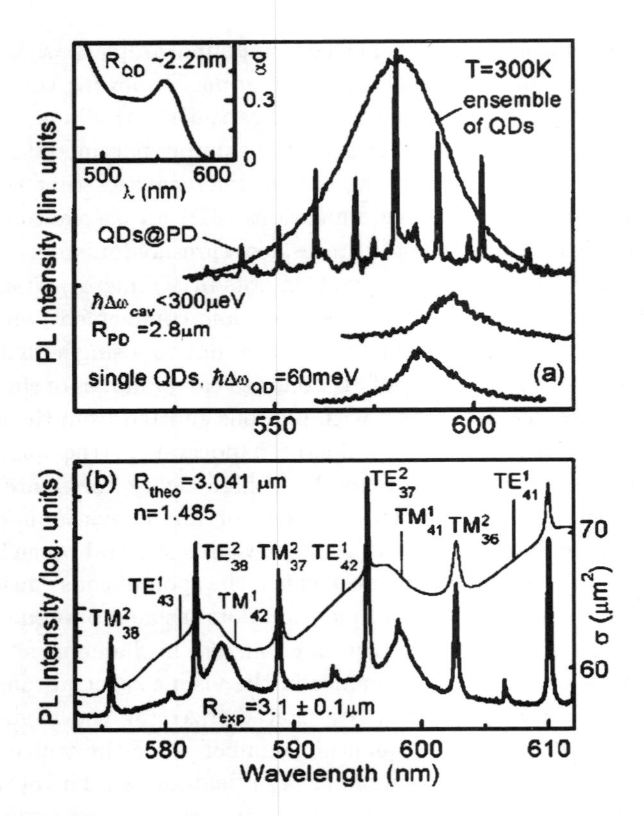

Fig. 2.1 (a) Room-temperature emission spectra of quantum dots before and after attaching them to a single photonic dot. Plotted are the spectra for an ensemble of QDs, single QDs, and an ensemble of QDs incorporated in a thin surface shell of a PD. Inset: Absorption spectrum of the QD ensemble. The QD emission is excited non-resonantly by a focused Ar-ion laser ($\lambda = 488\,\text{nm}$, $I_{pump} = 50\,\text{W/cm}^2$) and detected by the use of a microscope objective with a numerical aperture of 0.95, an imaging spectrometer and a CCD camera ($\sim$0.4 μm spatial resolution, $\sim$0.08 nm spectral resolution) for a single PD selected by a pinhole. (b) Emission spectrum of a $R_{PD} = 3.1\,\mu\text{m}$ PD taken with a factor of 4 higher spectral resolution for the spectral range around 595 nm (thick line). For the experimental cavity, Q values of $\sim$7500 have been determined. For comparison, the calculated scattering cross section characterizing the cavity eigenmodes is shown (thin line). The most pronounced PD eigenstates are labeled by their quantum numbers TE/TM, / and n. Within the detected spectral window the quantum numbers vary between $l = 32$ and $l = 42$ with $n = 1$, 2 for the sharp modes while modes with $n > 3$ form the weak background [Artemyev *et al.*, 2001b].

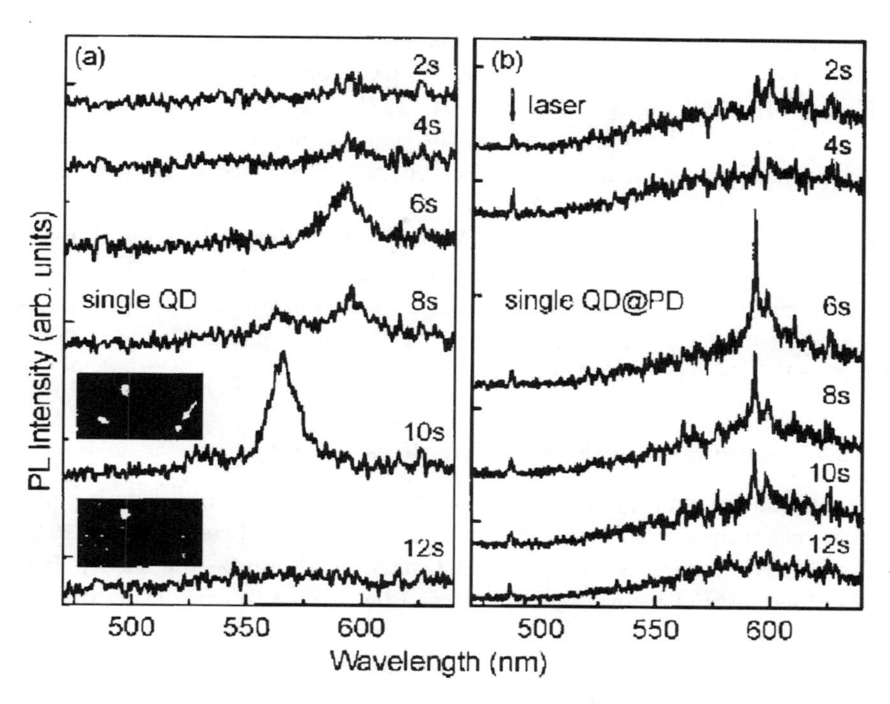

Fig. 2.2 (a) Room-temperature emission spectrum of a single *CdSe* QD (without cavity) taken at different observation times to illustrate the blinking effect in nanocrystals. (b) Same experiment for a single QD but now bound to a PD. The emission of the quantum dot switches two very sharp modes of high Q (mode width below spectral resolution). The signal from scattered laser light is shown too, to exclude any correlation with fluctuations in pump intensity [Artemyev *et al.*, 2001b].

2.2 Lorentz–Mie theory and its extensions

Investigations of the optical phenomena occurring in dielectric spheres has a long and vast history. A review of research in this field may be started from the rainbow studies described briefly in [Wang & van de Hulst, 1991]. The first ray optics theory of light refraction and reflection in a single spherical drop was elaborated by Descartes in 1639 [Descartes, 1965]. Two hundred years later, Airy improved this theory taking into diffraction [Airy, 1838; Wang & van de Hulst, 1991]. A strict analytical solution to the problem of scattering of electromagnetic wave by a homogeneous sphere of arbitrary radius and with arbitrary dielectric constant was independently obtained by four authors, Lorenz [Lorenz, 1890], Love [Love, 1899], Mie [Mie, 1908], and Debye [Debye, 1909] (see Refs. [van de Hulst, 1946; Kerker, 1969; Wang & van de Hulst, 1991]). This theory is known as the Lorenz–Mie theory. At about the same time, Lord Rayleigh studied propagation of sound over a curved gallery surface [Rayleigh, 1910; Oraevsky, 2002]. This work by Rayleigh deserves to be mentioned since the term whispering-gallery mode (WGM) is traditionally used with reference to electromagnetic surface oscillation in optical cavities. During the past decade, it was published hundreds of papers devoted to the properties and optical effects involving WGMs of a dielectric sphere [Datsyuk, 2001; Datsyuk & Izmailov, 2001; Oraevsky, 2002].

2.2.1 *Lorentz–Mie theory of elastic scattering*

The formalism employed in the solution to the problem of scattering of a plane electromagnetic wave by a homogeneous sphere [van de Hulst, 1946; Born & Wolf, 1980; Kerker, 1969] is termed the Lorentz–Mie theory. It is described in brief below. This theory can be applied to calculating parameters of scattered radiation. Using the Lorentz–Mie theory, other optical phenomena such as the morphology-dependent resonances (MDRs) in fluorescence spectra can be understood. Very recently, the theory has been generalized to handle other much more complex cavities such as concentric multilayered structures [Smith & Fuller, 2002].

In the framework of classical electrodynamics, vectors of the electric field $\vec{E}$ and the magnetic field $\vec{H}$ can be expressed through the Debye electric and magnetic potentials v^p. Hereinafter, the superscript p is used to indicate the polarization of electromagnetic field. Let $p = 0$ for a transverse electric (TE) field, and $p = 1$ for a transverse magnetic (TM) field. The Debye potentials satisfy the scalar wave equation

$$\Delta\, v^p + \frac{\epsilon\mu}{c^2}\frac{\partial^2}{\partial t^2}\, v^p = 0, \tag{2.1}$$

where Δ is the Laplace operator, $\dfrac{\partial}{\partial t}$ denotes the time derivative, c is the velocity of light in vacuum. It is assumed that the dielectric and magnetic primitivities of the sphere ($f = 2$) and the surrounding medium ($f = 1$) are constant:

$$\epsilon = \begin{cases} \epsilon_2, & r \le a, \\ \epsilon_1, & r > a \end{cases}, \qquad \mu = 1. \tag{2.2}$$

Here r is the distance from the center of the sphere. The relative index of light refraction is

$$n = \sqrt{\epsilon_2/\epsilon_1}. \tag{2.3}$$

The general solution of Eq. (2.1) in the spherical-coordinate system r, θ, ϕ at $r \leq a$ can be represented as follows

$$v^p = \sum_{l,\,m} a_{p\,l\,m} \frac{1}{r} \psi_l(z\xi)\, Y_l^m(\theta,\, \phi)\, e^{-\,i\,\omega t}, \tag{2.4}$$

where $z \equiv k_2\,a$, k_2 is the light wave number in the sphere, $\xi \equiv \dfrac{r}{a}$, $\psi_l(z) \equiv \sqrt{\dfrac{\pi\,z}{2}}\, J_{l+\frac{1}{2}}(z)$ is the Riccati–Bessel function, $J_{l+\frac{1}{2}}(z)$ is the Bessel function, and $Y_l^m(\theta,\, \phi)$ is the spherical harmonic. For $r > a$, the field consists of the field of a wave incident on the sphere:

$$v_0^p(\vec{r},\, t) = v_0^p\, e^{i\,(\vec{k}\vec{r} - \omega t)}, \tag{2.5}$$

and the scattered radiation field

$$v^p = \sum_{l,\,m} b_{p\,l\,m} \frac{1}{r} \zeta_l(x\xi)\, Y_l^m(\theta,\, \phi)\; e^{-\,i\,\omega t}, \tag{2.6}$$

where $x \equiv k_1\,a$, k_1 is the light wave number in the surrounding medium, $\zeta_l(x) \equiv \sqrt{\dfrac{\pi\,x}{2}}\, H^{(1)}_{l+\frac{1}{2}}(x)$ is the Riccati-Hankel function, $H^{(1)}_{l+\frac{1}{2}}(x)$ is the Hankel function of the first kind. The algorithm of computation of both the Riccati-Bessel and the Riccati-Hankel functions is described in detail in Ref. [Wang & van de Hulst, 1991]. If the time dependence of the electromagnetic field is taken in the form of $\exp(i\,\omega t)$, then the Hankel function of the first kind $H^{(1)}_{l+\frac{1}{2}}(x)$ should be replaced by the Hankel function of the second kind $H^{(2)}_{l+\frac{1}{2}}(x)$ [Shifrin & Zolotov, 1993; Roll *et al.*, 1999]. Coefficients $a_{p\,l\,m}$ and $b_{p\,l\,m}$ in Eqs. (2.4) and (2.6) are found from the continuity conditions for the tangential components of vectors $\vec{E}$ and $\vec{H}$ at the boundary of the surface. It is important that $a_{p\,l\,m}$ and $b_{p\,l\,m}$ being determined, the electromagnetic field can be calculated at any point in space.

The Lorentz–Mie theory is focused on the calculation of the following characteristics of the scattered light. The extinction cross section σ_{ext} is used as an integral characteristics of scattering. It is introduced as the ratio of the absorbed and scattered emission power to the electromagnetic-wave energy flux. If light absorption inside the sphere is absent, $\sigma_{\rm ext}$ coincides with the light scattering cross section. The ratio of $\sigma_{\rm ext}$ to the cross section of the sphere πa^2 is denoted by $Q_{\rm ext}$. According to the Lorenz-Mie theory, the normalized extinction cross section, often called extinction efficiency, is given by

$$Q_{\rm ext} = \frac{2}{x^2} \sum_{p=0}^{1} \sum_{l=1}^{\infty} (2l+1) \, {\rm Re}\{-\alpha_{p\,l}\} \,. \tag{2.7}$$

where

$$\alpha_{p\,l} = -\frac{n^{1-p}\,\psi_l(x)\,\psi_l'(z) - n^p\,\psi_l'(x)\,\psi_l(z)}{D_{p\,l}}\,, \tag{2.8}$$

$$D_{p\,l} = n^{1-p}\,\zeta_l(x)\,\psi'(z) - n^p\,\zeta_l'(x)\,\psi_l(z), \tag{2.9}$$

hereinafter the prime labels a derivative of a function with respect to its argument. Coefficients $\alpha_{p\,l}$ are called the Mie coefficient for the scattered field or Mie scattering function. They also allow to calculate the scattering efficiency $Q_{\rm sc}$ and radar backscattering efficiency $Q_{\rm back}$,

$$Q_{\rm sc} = \frac{2}{x^2} \sum_{p=0}^{1} \sum_{l=1}^{\infty} (2l+1)\; |\alpha_{p\,l}|^2 \,. \tag{2.10}$$

$$Q_{\rm back} = \frac{1}{x^2} \left| \sum_{p=0}^{1} \sum_{l=1}^{\infty} (2l+1)\; (-1)^{p+l}\, \alpha_{p\,l} \right|^2 \,. \tag{2.11}$$

The ratio of the scattering efficiency to the absorption efficiency is called single-scattering albedo [Smith & Fuller, 2002].

To illustrate the dependence of $Q_{\rm ext}$ on the size parameter x we built graph of Fig. 2.3. It shows the result of numeric calculation executed for a sphere with a refractive index of $n = 1.5$. In fact, the curve $Q_{\rm ext}(x)$ in Fig. 2.3 is a set of points $(x,\, Q_{\rm ext})$ spaced at $\Delta x = 0.005$. The sharp periodically spaced peaks seen in the region of $x = 15 - 20$ are MDRs appeared because of excitation of WGMs within the sphere. These MDRs are not found in analogous figures of the textbooks [van de Hulst, 1946, Fig. 32] and [Born & Wolf, 1980]. The MDRs were missed because of their small widths. Computations fulfilled by Hill and Benner [Hill & Benner, 1986] for micrometer-sized droplets even resulted in a conclusion that "many of the MDRs have computed linewidths far too narrow to be observed in elastic

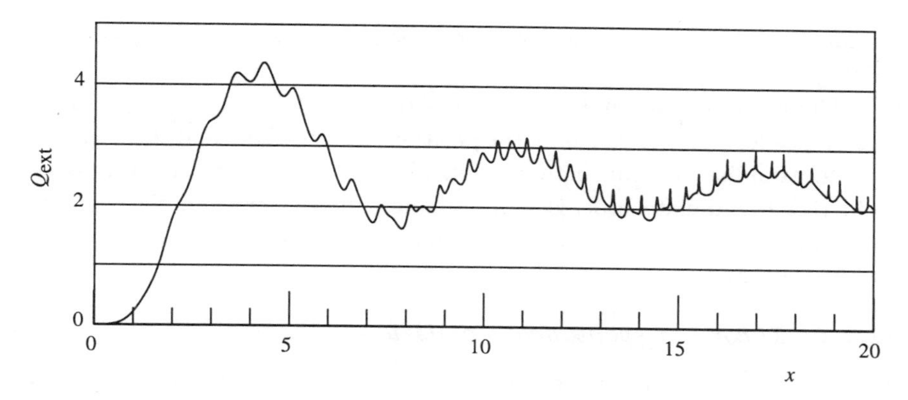

Fig. 2.3 Normalized extinction cross section for a dielectric sphere with refractive index $n = 1.5$ versus the size parameter x.

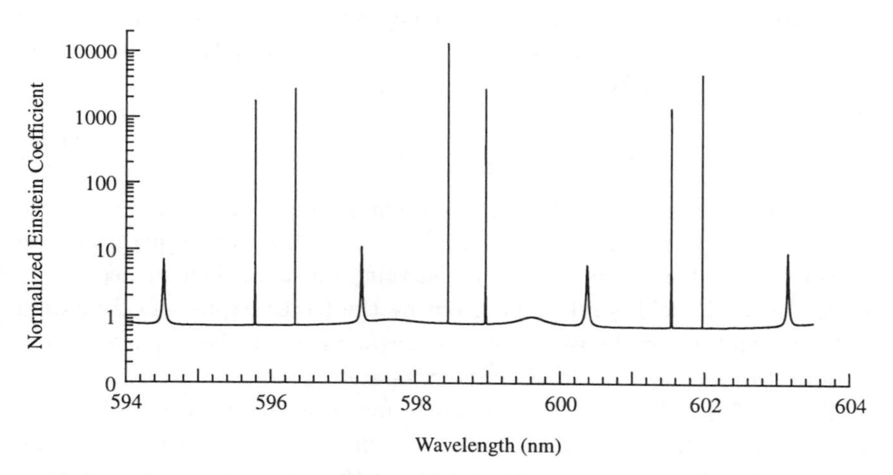

Fig. 2.4 Normalized Einstein coefficient for spontaneous emission calculated for a 15.306 μm diameter ethanol droplet.

scattering spectra." This statement was confirmed by many observations of elastic-scattering and fluorescence spectra. However, lines correspondent to excitation of these hidden WGMs were observed in spectra of lasing and stimulated Raman scattering (SRS). Different theoretical models predicted that transition rates for spontaneous emission increase in hundreds and thousands times in comparison with the free-space values even in droplets with moderate quality factors [Chew, 1987; Ching *et al.*, 1987a; Chew, 1988]. For example, Fig. 2.4 presents a calculated fluorescence spectrum for an ethanol droplet in air that was studied experimentally in Ref. [Chylek *et al.*, 1991].

The above examples and remarks show that it is necessary (i) to get analytical formulas to calculate the positions of the MDRs and their widths, (ii) to establish relation between optical cavities and emission processes, and (iii) to simulate processes in realistic spheres, for example, allowing for light absorption within a sphere and surface deviation from an ideal spherical shape.

2.2.2 *Theory of spontaneous emission*

Recently, a theory of spontaneous emission in absorbing and dispersive media has been developed [Xu *et al.*, 2000b; Dung *et al.*, 2000; Dung *et al.*, 2001; Knoll *et al.*, 2000]. This theory establishes a relation between the rate $\mathcal{A}$ of spontaneous emission and the Green tensor G_{ij} of the Maxwell wave equation [Wylie & Sipe, 1984; Chew, 1987; Xu *et al.*, 2000b; Dung *et al.*, 2000; Dung *et al.*, 2001; Knoll *et al.*, 2000],

$$\mathcal{A}(\omega) = \frac{2k_d^2 \mu_i \mu_j}{\hbar \epsilon_0} \Im G_{ij}(\vec{r}_d, \vec{r}_d, \omega). \tag{2.12}$$

here a two-energy-level dipole at position $\vec{r}_d$ has the transition dipole moment $\vec{\mu}$ at transition frequency ω, $k_d = \omega/c$, $\hbar$, ϵ_0 are respectively the Planck constant and the dielectric constant, the time dependence of the electromagnetic field is set to be given by the factor $\exp(-\imath\omega t)$. Equation (2.12) is valid under the weak coupling regime and can be applied when a medium has a complex refractive index $m = m_\mathrm{r} + \imath\, m_\mathrm{i}$.

Forms of the Green tensor are known for an ideal homogeneous sphere [Dung *et al.*, 2001] and a three-layered concentric spherical structure [Dung *et al.*, 2000]. The latter can be referred to as the simplest core-shell system. Thus, the spectral Einstein coefficient can be calculated for the open optical cavities of these kinds. For the sake of simplicity, all the emission rates will be averaged over emitting-dipole orientation in an open cavity.

A formula for the normalized averaged spontaneous emission rate inside a homogeneous sphere, according to [Dung *et al.*, 2001], to be

$$\frac{\mathcal{A}(\xi)}{\mathcal{A}_\mathrm{h}} = \frac{1}{3} + \frac{1}{2} \Re \Bigg\{ \frac{1}{(z\,\xi)^2} \sum_{l=1}^{\infty} \Big[\breve{\alpha}_{0\,l}\,(2l+1)\,\psi_l^2(z\,\xi) + \breve{\alpha}_{1\,l}\,\big((l+1)\,\psi_{l-1}^2(z\,\xi) + l\,\psi_{l+1}^2(z\,\xi)\big)\Big] \Bigg\}, \tag{2.13}$$

where $\mathcal{A}_{\rm h} = \mathcal{R}_{\rm h}$ is the Einstein coefficient for spontaneous emission in the homogeneous bulk volume of the resonator-core material, $\xi \equiv r/a$, r is a distance from the excited particle to the center of the sphere, $z \equiv k_2\, a$, $x \equiv k_1\, a$, $\breve{a}_{p\,l}$ is the coefficient for the outward-going spherical wave reflected by a concave boundary [Fuller, 1993; Smith & Fuller, 2002],

$$\breve{\alpha}_{p\,l} = -\,\frac{n^{1-p}\,\zeta_l(x)\,\zeta_l'(z) - n^p\,\zeta_l'(x)\,\zeta_l(z)}{D_{p\,l}}. \tag{2.14}$$

In order to reduce a number of the input parameters of the theoretical models, a value of $\mathcal{A}$ can also be averaged over ξ,

$$\mathcal{A} = 3\int_0^1 \mathcal{A}(\xi)\,\xi^2\,d\xi. \tag{2.15}$$

The right-hand side of the above equation contains the integral

$$\int_0^1 \psi_l^2(z\,\xi)\,d\xi = \frac{1}{2}\,\psi_l^2(z)\left(1 + \frac{1}{p_l^2} - \frac{2l+1}{z\,p_l}\right), \tag{2.16}$$

where $p_l \equiv \psi_l(z)/\psi_{l-1}(z)$ is a term incorporated into the algorithm of the computation of the Riccati–Bessel functions [Wang & van de Hulst, 1991].

For a three-layered spherical structure shown in Fig. 2.5, the normalized rate averaged over dipole orientation is

$$\begin{aligned}\frac{\mathcal{A}}{\mathcal{A}_{\rm h}} = \frac{1}{3} + \frac{1}{2}\,\Re\Bigg\{\frac{1}{z_5\,\xi}\sum_{l=1}^{\infty}\Big[&C_{0\,l}\,(2l+1)\,\psi_l^2(z_5\,\xi) \\ &+ C_{1\,l}\,\big((l+1)\,\psi_{l-1}^2(z_5\,\xi) + l\,\psi_{l+1}^2(z_5\,\xi)\big)\Big]\Bigg\},\end{aligned} \tag{2.17}$$

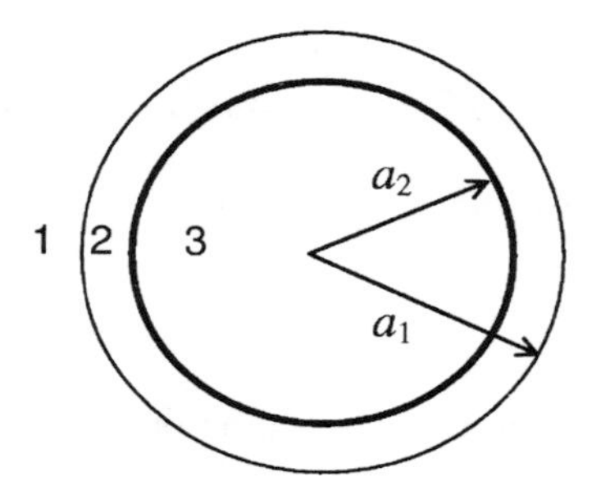

Fig. 2.5 Three-layered spherical structure. Index f = 1, 2, 3 labels a layer of the structure.

where $\xi \equiv r/a_2$, $z_{2f} \equiv k_f\, a_f$, $z_{2f+1} \equiv k_{f+1}\, a_f$.

$$C_{pl} = \frac{T^{pl}_{F2}}{T^{pl}_{P2}} \frac{T^{pl}_{P1}\, R^{pl}_{P2} + T^{pl}_{F1}\, R^{pl}_{P1}}{T^{pl}_{P1} + T^{pl}_{F1} R^{pl}_{P1} R^{pl}_{F2}}, \tag{2.18}$$

$$R^{pl}_{Pf} = \frac{k_{f+1-p}\, \zeta_l(z_{2f})\, \zeta_l'(z_{2f+1}) - k_{f+p}\, \zeta_l'(z_{2f})\, \zeta_l(z_{2f+1})}{k_{f+1-p}\, \psi_l(z_{2f})\, \zeta_l'(z_{2f+1}) - k_{f+p}\, \psi_l'(z_{2f})\, \zeta_l(z_{2f+1})}, \tag{2.19}$$

$$R^{pl}_{Ff} = \frac{k_{f+1-p}\, \psi_l(z_{2f})\, \psi_l'(z_{2f+1}) - k_{f+p}\, \psi_l'(z_{2f})\, \psi_l(z_{2f+1})}{k_{f+1-p}\, \zeta_l(z_{2f})\, \psi_l'(z_{2f+1}) - k_{f+p}\, \zeta_l'(z_{2f})\, \psi_l(z_{2f+1})}, \tag{2.20}$$

$$T^{pl}_{Pf} = \frac{k_f\, [\psi_l(z_{2f+1})\, \zeta_l'(z_{2f+1}) - \psi_l'(z_{2f+1})\, \zeta_l(z_{2f+1})]}{k_{f+1-p}\, \psi_l(z_{2f})\, \zeta_l'(z_{2f+1}) - k_{f+p}\, \psi_l'(z_{2f})\, \zeta_l(z_{2f+1})}, \tag{2.21}$$

$$T^{pl}_{Ff} = \frac{k_f\, [\psi_l(z_{2f+1})\, \zeta_l'(z_{2f+1}) - \psi_l'(z_{2f+1})\, \zeta_l(z_{2f+1})]}{k_{f+p}\, \psi_l(z_{2f+1})\, \zeta_l'(z_{2f}) - k_{f+1-p}\, \psi_l'(z_{2f+1})\, \zeta_l(z_{2f})}. \tag{2.22}$$

The rate of spontaneous emission from a sphere has also been found in the framework of classical electrodynamics [Chew, 1987; Chew, 1988; Lange & Schweiger, 1994; Datsyuk, 2002a]. The normalized rate averaged over orientation of a dipole in a sphere is given by the following formula

$$\frac{\mathcal{R}(\xi)}{\mathcal{R}_{\rm h}} = \frac{1}{2\,|n|} \left\{ \frac{1}{\xi^2} \sum_{l=1}^{\infty} \Big[|D_{0\,l}|^{-2}\, (2l+1)\, |\psi_l(z\,\xi)|^2 \right.$$
$$\left. + |D_{1\,l}|^{-2} \left((l+1)\, |\psi_{l-1}(z\,\xi)|^2 + l\, |\psi_{l+1}(z\,\xi)|^2 \right) \Big] \right\}. \tag{2.23}$$

When averaging $R(\xi)$ over ξ at a complex quantity z the following equation can be used

$$\int_0^1 |\psi_l(z\,\xi)|^2\, d\xi = \frac{1}{2}\, |\psi_l(z)|^2\, \frac{\Im(z\, p_{l+1})}{\Im(z)\, \Re(z)}. \tag{2.24}$$

The rate R allows to find the power of spontaneous emission radiated by an optical cavity,

$$\mathcal{P}_{\rm sp} = h\nu\, \mathcal{R}\, N, \tag{2.25}$$

where $h\nu$ is the photon energy, N volume density of the emitters. On the contrary, the value of $\mathcal{A}$ is a local characteristics of the spontaneous-emission process. The equality $\mathcal{R} = \mathcal{A}$ holds true only for emission in media with real refractive indexes.

2.2.3 *Mie scattering by concentrically stratified spheres*

In a recent paper by Smith and Fuller [Smith & Fuller, 2002], a simple iterative approach for modeling light scattering by N-layer spherical structure has been proposed. In the algorithm, the infinite medium was considered as the first layer ($f = 1$) and the innermost sphere was referred to as the last layer ($f = N$, $N = 3$ in Fig. 2.5).

2.3 Peculiarities of the modes of an open spherical cavity

The whispering gallery modes are modes of an open system. In such a system a dissipation of energy occurs because the electromagnetic field leaks from the optical cavity. In other words, open optical cavities are not-Hermitian systems [Ching *et al.*, 1987a; Ching *et al.*, 1987b; Lai *et al.*, 1990b]. In principle, all optical cavities with some degree of output coupling, say all laser cavities, belong to this category. If the radiative leakage of the energy is small, the common concept of closed, Hermitian, systems can be applied. It is not the case for the considered spherical microparticles. Hence, we must take into account the peculiarities of the WGMs, which are not orthogonal, do not form a full set of orthogonal functions, and have complex eigenfrequencies.

2.3.1 *Indexes and order of a whispering-gallery mode*

The whispering gallery modes are commonly imagined as closed-trajectory rays confined within the cavity by almost total internal reflections from the spherical surface. The most part of the energy of these modes is presumed to be under the surface $r = a$ with negligible contribution of the evanescent field that exists at $r > a$. The electromagnetic field in this region is described by divergent spherical waves. Namely, for the electric field of an eigen TE mode at $r > a$ one has (see Eq. (2.6))

$$\vec{e} \propto \frac{1}{r}\,\zeta_l(k_1\,r)\;e^{-\,i\,\omega t}\;\hat{\vec{L}}\,Y_l^m(\theta,\,\phi), \tag{2.26}$$

where $\hat{\vec{L}} = -\imath\,\hbar\,\vec{r}\times\vec{\nabla}$ is the orbital momentum operator. Similarly, for the magnetic field of an eigen TM mode at $r > a$ one has

$$\vec{b} \propto \frac{1}{r}\,\zeta_l(k_1\,r)\;e^{-\,i\,\omega t}\;\hat{\vec{L}}\,Y_l^m(\theta,\,\phi). \tag{2.27}$$

In this picture, each WGM is numbered by a set of the indices and the order: $s = p, l, m, q$. The first index (p) describes the polarization of the oscillation, that can be transverse electric (TE) or transverse magnetic (TM). The mode number l indicates the order of the spherical function $Y_l^m(\theta, \phi)$ that describes the angular field distribution. This function is the eigen function of the square of the orbital momentum operator,

$$\hat{\vec{L}}^2\, Y_l^m(\theta,\, \phi) = \hbar^2\, l\,(l+1)\, Y_l^m(\theta,\, \phi), \tag{2.28}$$

$$\hat{L}_z\, Y_l^m(\theta,\, \phi) = \hbar\, m\, Y_l^m(\theta,\, \phi). \tag{2.29}$$

Therefore, the mode number is sometimes called the angular momentum index or the orbital quantum number. The index m is called the azimuthal mode number or the azimuthal quantum number. It can take $2\,l + 1$ values from $-l$ to l. Since the considered spherical cavity has a spherical symmetry, many properties of the WGMs, such as frequencies or energies, are independent of m. Finally, the mode order q indicates the number of maxima in the radial dependence of the internal field distribution.

The mode indices $s = p, l, m$ are in fact quantum numbers which appear because the field satisfy certain boundary conditions. For example, $p = 0$ corresponds to the condition $\vec{r}\vec{E} = 0$ and $p = 1$ means that $\vec{r}\vec{H} = 0$. Meantime, the mode order q is not a quantum number since it is not related with any restricting condition for the field. When modeling processes occurring in a sphere, one can meet the problem of normalization of the electromagnetic mode energy. To solve this problem, one can introduce a quantizing well at $r = \Lambda$ with $\Lambda \to \infty$. In this case, a radial quantum number v appears and must be used instead of q. The concept of the infinitely-distant quantizing well is described in brief below.

2.3.2 *The problem of normalization of the whispering-gallery modes*

In order to turn open cavities into Hermitian systems, suitable conditions at the boundary of universe can be imposed. Thus, the authors of [Ching *et al.*, 1987a; Lai *et al.*, 1990b; Klimov *et al.*, 1999] studied the resonant properties of the electromagnetic field in a dielectric sphere with real n, placed inside a concentric conducting spherical surface of radius Λ at $\Lambda \to \infty$. On the infinitely distant surface, the tangential component of the electric field was set to zero. This method allowed to fulfill the correct quantization of the electromagnetic field, taking into account the energy dissipation from the open optical cavity. The same approach was used earlier to model resonant properties of media inhomogeneous in one dimension [Vlasenko *et al.*, 1973; Pekar, 1975].

In Ref. [Ching *et al.*, 1987a], vectors $\vec{E}(\vec{r}, t)$ and $\vec{H}(\vec{r}, t)$ were expanded over the real electric $\vec{e}(s, \vec{r})$ and magnetic $\vec{b}(s, \vec{r})$ eigenfunctions,

$$\vec{E} = \sum_s \omega_s^{-1} \frac{d\,a(s, t)}{dt}\, \vec{e}(s, \vec{r}), \tag{2.30}$$

$$\vec{H} = \sum_s a(s, t)\vec{b}(s, \vec{r}), \tag{2.31}$$

$$\vec{\nabla} \times \vec{e}(s, \vec{r}) = -\frac{\omega_s}{c}\, \vec{b}(s, \vec{r}), \tag{2.32}$$

where ω_s denoted frequencies. The eigenvector $\vec{e}(s, \vec{r})$ is the solution to the equation

$$\vec{\nabla} \times \left[\vec{\nabla} \times \vec{e}(s, \vec{r})\right] = \frac{\omega_s^2}{c^2}\, \epsilon(\vec{r})\, \vec{e}(s, \vec{r}), \tag{2.33}$$

where $\epsilon(\vec{r}) = n^2$ at $r \leq a$ and $\epsilon(\vec{r}) = 1$ at $r > a$. The decomposition coefficients $a(s, t)$ satisfy the equations

$$\frac{d^2\, a(s, t)}{dt^2} = -\omega_s^2\, a(s, t). \tag{2.34}$$

The normalization of $\vec{e}$ and $\vec{b}$ was chosen to hold the relations

$$I^{(1)}(s, s') = I^{(2)}(s, s') = \hbar\omega_s\, \delta_{s\, s'}, \tag{2.35}$$

with

$$I^{(1)}(s, s') = \frac{1}{4\pi} \iiint\limits_{r<\Lambda} \epsilon(\vec{r})\, \vec{e}(s, \vec{r})\, \vec{e}(s', \vec{r})\, dV, \tag{2.36}$$

$$I^{(2)}(s, s') = \frac{1}{4\pi} \iiint\limits_{r<\Lambda} \vec{b}(s, \vec{r})\, \vec{b}(s', \vec{r})\, dV. \tag{2.37}$$

Decompositions (2.30) and (2.31) yield the total energy of the electromagnetic field in the form of the following expression

$$\frac{1}{8\pi} \iiint (\epsilon\, \vec{E}\, \vec{E} + \vec{H}\, \vec{H})\, dV = \sum_s N_s \iiint u_s(\vec{r})\, dV.$$

In the right-hand side of the above equation, the integral of motion

$$N_s = \frac{1}{2}\left\{\frac{1}{\omega_s^2}\left[\frac{d\,a(s,\,t)}{dt}\right]^2 + a^2(s,\,t)\right\}$$

is the number of quanta in the mode s, and

$$u_s(\vec{r}) = \frac{1}{8\pi}\left[\epsilon(\vec{r})\,\vec{e}(s,\,\vec{r})\,\vec{e}(s,\,\vec{r}) + \vec{b}(s,\,\vec{r})\,\vec{b}(s,\,\vec{r})\right]$$

is the energy density of the electromagnetic field in which there is one quantum per mode s. According to conditions (2.35), the function $u_s(\vec{r})$ is normilized to $\hbar\omega_s$, so that $\frac{1}{\hbar\omega_s}\,u_s(\vec{r})$ is the probability to find one photon in the mode s at point $\vec{r}$ in space. Thus, the local density of states of the electromagnetic field in the unit frequency interval about ω is given by

$$\rho(\omega,\,\vec{r}) = \frac{1}{\hbar\omega}\sum_s \delta(\omega-\omega_s)\,u_s(\vec{r}). \tag{2.38}$$

In the right-hand side of formula (2.38), summation is over the mode indices p, l, m, and the radial quantum υ of the system consisting of the cavity and the conducting sphere. The distance between the neighbouring modes with different υ is equal to about $\pi\,c/\Lambda$ and tends to zero in the limit of $\Lambda\to\infty$. It means that the summation over υ can be replaced by integration over ω_s:

$$\sum_\upsilon \to \frac{\Lambda}{\pi\,c}\int d\omega_s. \tag{2.39}$$

After integration of expression (2.38) over the frequency ω_s and the angles θ, ϕ and summation over m, Ching *et al.* [Ching *et al.*, 1987a] undertook a detailed analysis of the following ratios

$$g(x,\,n;\,\xi) \equiv \frac{\rho(\omega,\,r)}{\rho_{\rm v}(\omega)}, \quad G(x,\,n) \equiv \frac{\rho(\omega)}{\rho_{\rm v}(\omega)\,V}$$

where

$$\rho(\omega) \equiv \int_0^a \rho(\omega,\,r)\,r^2\,dr, \tag{2.40}$$

$$V \equiv \frac{4\pi}{3}\,a^3, \;\rho_{\rm v}(\omega) = \frac{\omega^2}{\pi^2\,c^3}, \;\xi \equiv \frac{r}{a}.$$

The quantity g characterizes the alteration of the local density of states at a distance r from the center of the cavity as compared with an infinite vacuum. The quantity G is the ratio of the spectral density of states of the electromagnetic field in the whole cavity to that of the infinite-vacuum volume V. The found sharp spikes in the frequency dependence allowed to explain the correspondent sharp MDRs in fluorescence spectra of micrometer-sized dielectric spheres (see, for example, [Campillo *et al.*, 1991; Lin *et al.*, 1992; Campillo *et al.*, 1992]). Indeed, a formula for $G(x, n)$ can be written in the following form [Datsyuk & Izmailov, 2001]:

$$G = \frac{3n}{x^2} \sum_{p=0}^{1} \sum_{l=1}^{\infty} (2l+1) \left| \frac{n^{1-2p} j_l(z)}{D_{pl}(x, n)} \right|^2 \times \left[1 - \frac{l(l+1)}{z^2} + \left(\frac{j_l'(z)}{j_l(z)} \right)^2 \right]. \tag{2.41}$$

The MDRs in the dependence of G on x are caused by the factor $|D_{pl}(x, n)|^{-2}$, that also enters to the equations for Q_{ext} (2.7), Q_{sc} (2.10), Q_{back} (2.11), $\mathcal{A}$ (2.13), and $\mathcal{R}$ (2.23).

2.4 Quality factor of a whispering-gallery mode

Many recently observed optical effects can be explained in terms of variation of the quality factor. This important parameter characterizes energy dissipation from the resonator and is defined by the formula

$$\frac{1}{Q} \equiv \frac{1}{\omega \langle W(t) \rangle} \left\langle \frac{d}{dt} W(t) \right\rangle = \frac{2\omega''}{\omega'}. \tag{2.42}$$

Here $W(t)$ is the energy of electromagnetic mode, $\omega = \omega' + i\omega''$ is the angular frequency, and brackets $< \ldots >$ denote averaging over the oscillation period $T = 2\pi/\omega'$.

2.4.1 *Radiative quality factor of an ideal dielectric sphere*

The quality factor can be calculated readily for resonators of simple form. Thus, for a resonator in a form of ideal sphere, the Lorenz-Mie theory of elastic light scattering is traditionally used. This theory use Mie expansion of the extinction cross section σ_{ext} that requires numeric calculations of many Riccati-Bessel and Riccati-Hankel functions of high orders. Instead of the Lorenz-Mie theory, one can use the formulas of the Einstein coefficients for spontaneous emission $\mathcal{A}$. The resultant value of Q shall be the same since the sharp resonant frequency dependence of both $\sigma_{\rm ext}$ and $\mathcal{A}$ is determined by the denominator of the Mie coefficient.

Understanding of the origin of the peaks in $\sigma_{\rm ext}(\omega)$ and $\mathcal{A}(\omega)$ allows to get formulas for Q in a simple analytical form. Namely, the MDRs appear in spectra of elastic scattering and spontaneous emission (see Eqs. (2.7), (2.10), (2.11), (2.13) and (2.23)) due to poles of the function $|D_{p\,l}|^{-2}$ in the complex frequency plain. (Note that in the case of real parameters x and z the following equation takes place $\Re\{-\alpha_{p,\,l}\} = |\alpha_{p,\,l}|^2$.) In order to calculate factor $|D_{p\,l}|^{-2}$ asymptotic formulas for the Riccati-Bessel and Riccati-Hankel can be used. Namely, the WKB approximation yields the following formulas for ψ_l:

$$\psi_l(z) = \sqrt{\pi\,\beta}\,\mathrm{Ai}(t_q + y), \tag{2.43}$$

$$\psi_l'(z) = -\sqrt{\pi/\beta}\,\mathrm{Ai}'(t_q + y), \tag{2.44}$$

where $z = l + \dfrac{1}{2} - (t_q + y)\,\beta$, Ai is the Airy function, t_q is the qth root of the equation

$$\mathrm{Ai}(t_q) = 0, \quad q = 1, 2, 3, \ldots, \quad t_q < 0, \tag{2.45}$$

the number q is called the mode order or radial mode index, and $\beta \equiv \left[\frac{1}{2}\left(l + \frac{1}{2}\right)\right]^{1/3}$, $|t_q + y|\,\beta \ll l$. For the Riccati-Hankel function, the Debye expansion gives the following approximations:

$$\zeta_l(x) = -\imath/\sqrt{\sinh\eta}\,e^T\left(1 + \frac{\imath}{2}\,e^{-2\,T}\right), \tag{2.46}$$

$$\zeta_l'(x) = \imath\,\sqrt{\sinh\eta}\,e^T\left(1 - \frac{\imath}{2}\,e^{-2\,T}\right), \tag{2.47}$$

where $x = \left(l + \frac{1}{2}\right)/\cosh\eta$, $T \simeq \left(l + \frac{1}{2}\right)(\eta - \tanh\eta)$, $\left(\frac{3}{2}\,T\right)^{2/3} \gg 1$.

On substitution of the complex $n \equiv n_{\rm r} + \imath n_{\rm i}$, functions (2.43), (2.44) with argument $z = l + 1/2 - t\,\beta + \imath\,n_{\rm r}\,x$, and functions (2.46), (2.47) with argument $x = (l + 1/2)/\cosh\eta$ into Eq. (2.9) one gets

$$|D_{pl}|^{-2} = 2\,x_s\left|x\,\frac{d}{dx}\,j_l(n_{\rm r}x)\right|^{-2} F(\omega), \tag{2.48}$$

where $x_s = \dfrac{\omega_s\,a}{c}\,n_1$ is the resonant size parameter, s is a set of the mode indexes p, l, q, and ω_s is the resonant frequency of the mode [Vainstein, 1969;

Datsyuk *et al.*, 1990]:

$$\omega_s = \frac{c}{n_1\,a}\left[l + \frac{1}{2} - (t_q + \Delta t)\beta\right], \tag{2.49}$$

where

$$\Delta t = n_{\rm r}^{1-2p}\,\frac{1 + i\exp(-2T)}{\beta\sqrt{n_{\rm r}^2 - 1}},$$

$$\cosh\eta = n_{\rm r}\left[1 - \frac{1}{l + \dfrac{1}{2}}\left(t_q\beta + \frac{n_{\rm r}^{1-2p}}{\sqrt{n_{\rm r}^2 - 1}}\right)\right]^{-1}.$$

In Eq. (2.48), $F_s(\omega)$ is a form factor of the MDR

$$F_s(\omega) \equiv \frac{1}{Q_{s0}}\left[4\left(\frac{\omega - \omega_s}{\omega}\right)^2 + \left(\frac{1}{Q_s}\right)^2\right]^{-1} \tag{2.50}$$

Here Q_{s0} is the radiative quality factor of an ideal dielectric sphere with $n_{\rm i} = 0$ [Vainstein, 1969; Datsyuk *et al.*, 1990; Datsyuk, 1992]

$$Q_{s0} = \frac{1}{2}\left(l + \frac{1}{2}\right)n^{1-2p}\,(n^2 - 1)^{1/2}\,e^{2T}, \tag{2.51}$$

$$\frac{1}{Q_s} \equiv \frac{1}{Q_{s0}} + \frac{1}{Q_{\rm a}}, \tag{2.52}$$

$$Q_{\rm a} \equiv \frac{n_{\rm r}}{2\,n_{\rm i}}. \tag{2.53}$$

A good accuracy of the approximate equation (2.51) at $n_{\rm i} = 0$ has been confirmed by making comparisons with numerical values obtained using the Lorenz-Mie scattering theory [Buck & Kimble, 2003].

Formulas for calculation of the frequencies of the WGM were also proposed in [Chylek, 1990; Lam *et al.*, 1992; Schiller, 1993; Geints *et al.*, 1999]. In particular, formulas for the frequencies, more precise than Eq. (2.49), were published in [Lam *et al.*, 1992; Schiller, 1993] and cited by Oraevsky [Oraevsky, 2002]. An excellent agreement of the observed spectra with those predicted by the Lorenz-Mie theory was noted in [Schiller & Byer, 1991].

Equation (2.51) can be applied to evaluate the quality factor of a WGM with l of the order of hundred. For spheres with radii smaller than several λ/n_2, approximations (2.43)–(2.47) become invalid and WGMs are not exist. The radiative Q rises approximately exponentially as the radius a

increases. For example, for a 15 μm radius fused-silica sphere and a wavelength of about $\lambda = 852$ nm, a value of Q_{s0} of about 2×10^{21} is obtained [Buck & Kimble, 2003]. Therefore, the quality factor Q_s of "large" spheres is dominated by other energy loss mechanisms, for example, light absorption.

2.4.2 *Effect of light absorption on the quality factor*

A decomposition of the classical electric field over the modes of high-Q cavity [Vainstein, 1969] allows to take into account various processes of quality factor degradation. One of the most important process is light absorption by the cavity material. Neglecting an influence of $n_{\rm i}$ on the mode structure and an inter-mode energy exchange, the cavity quality factor is found in form (2.52) with more precise factor $Q_{\rm a}$:

$$Q_{\rm a} = \frac{\iiint \epsilon_{\rm r}(\vec{r}) \left|\vec{E}_s(\vec{r})\right|^2 dV}{\iiint \epsilon_{\rm i}(\vec{r}) \left|\vec{E}_s(\vec{r})\right|^2 dV}, \tag{2.54}$$

where $\epsilon_{\rm i}$, $\epsilon_{\rm r}$ are the imaginary and real part of the permittivity, respectively, $\vec{E}_s(\vec{r})$ is the electric field of the resonant mode. Integrations in the right-hand side of Eq. (2.54) are over the resonator volume. The above formula is in fact Eq. (1.1) of a QED study [Lai *et al.*, 1990b] given without derivation.

Formula (2.54) allows to take into consideration a not-uniform light-absorbent inclusions in the cavity or surrounding medium. Note that even in the case of a homogeneous distribution of the absorption inside a spherical cavity, Eq. (2.54) yields a value of $Q_{\rm a}$ that differs from the factor (2.53) well-known in the literature, [Datsyuk, 2002a]. This distinction appears due to the evanescent part of a WGM that gives a small but noticeable contribution to the numerator of the right-hand side of Eq. (2.54). The evanescent field can also provide a dependence of the MDR intensity on the light absorption (or amplification) beyond the microcavity surface. This effect can be used in spherical microsensors or microlasers.

2.4.3 *Light scattering on inhomogeneities of the refractive index*

The question "How is the complex eigen frequency is changed when the cavity is perturbed by a small change of a dielectric constant?" was addressed in a paper by Lai *et al.* [Lai *et al.*, 1990b]. In this paper, the work by

Zeldovich [Zeldovich, 1961] for scalar field in one dimension was generalized and a time-independent perturbation theory for leaking electromagnetic modes in open systems was elaborated. The formalism was applied to study the influence of shape perturbations of a dielectric microdroplet on characteristics of WGMs. The shifts of the MDR frequencies were analytically calculated.

The surface perturbations were theoretically found to remove the azimuthal degeneracy of WGMs in a multiplet. The predicted effect was confirmed by observations of emission spectra from traveling deformed droplets. The measurements of stimulated Raman scattering [Chen *et al.*, 1991; Chen *et al.*, 1992], fluorescence [Chen *et al.*, 1993b], and lasing [Chen *et al.*, 1993a] were fulfilled with high temporal and frequency resolution. It is simple to calculate a shift $\Delta\omega$ of the mode frequency relative to the initial value of ω for a small shape distortion, when the droplet surface can be approximated by an ellipsoid. For such a distortion, the frequency shift of each WGM is given by the following formula

$$\frac{\Delta\omega_m}{\omega} = \frac{\epsilon}{6}\left[\frac{3\,m^2}{l\,(l+1)} - 1\right], \tag{2.55}$$

where $\epsilon = (r_{\text{pol}} - r_{\text{eq}})/a$, r_{pol}, r_{eq} are the polar and equatorial radii, respectively, the small parameter ϵ is positive for a prolate ellipsoid, and negative for an oblate one. The frequency shift is independent of the polarization (TE or TM), mode order q, and droplet radius a.

The study of [Lai *et al.*, 1990b] as well as the consequent one [Lai *et al.*, 1990a] did not predict a broadening of an individual component of a splitted multiplet. This disagree with a theoretical study of [Tsipenyuk, 1965]. It is noted that a small probe body with a real dielectric constant inserted into an open cavity results in not only a frequency shift but also a degradation of the quality factor. The theory of [Lai *et al.*, 1990a,b] is consistent with experimental findings if a radiation process involves all members of the multiplet in a coherent fashion. However, the Q-factor of a deformed sphere should not decrease according to [Lai *et al.*, 1990a,b] upon an excitation of a single mode, for example, with $m = l$, regardless of other modes of the multiplet.

A simpler model [Datsyuk, 1992] used the theory of diffraction of electromagnetic waves. Using this theory, a degradation of the quality factor because of light scattering on inhomogeneities of the resonator surface can be calculated in the following manner.

Let

$$\vec{E}(t,\,\vec{r}) = \Re\left[\vec{E}(\omega,\,\vec{r})\,\exp(-\imath\,\omega t)\right], \tag{2.56}$$

$$\vec{H}(t,\,\vec{r}) = \Re\left[\vec{H}(\omega,\,\vec{r})\,\exp(-\imath\,\omega t)\right], \tag{2.57}$$

be known electric and magnetic fields of a mode in a medium with the dielectric and magnetic permittivities (2.2). Below, a value of $\epsilon_1 = 1$ is used for simplicity. Our aim is to find the electric $\vec{E}'$ and magnetic $\vec{H}'$ fields in a medium with ϵ' and μ', which differ weakly from ϵ and μ. According to the light diffraction theory, the vectors $\vec{E}'$ and $\vec{H}'$ can be found from the approximate equations

$$\mathrm{rot}\,\vec{H}' + i\,k\,\vec{E}' = \frac{4\pi}{c}\,\vec{J}, \quad \mathrm{div}\,\vec{E}' = 4\pi\,\rho, \tag{2.58}$$

$$\mathrm{rot}\,\vec{E}' - i\,k\,\vec{H}' = \frac{4\pi}{c}\,\vec{J}_{\mathrm{m}}, \quad \mathrm{div}\,\vec{H}' = 4\pi\,\rho_{\mathrm{m}}, \tag{2.59}$$

$$\mathrm{div}\vec{J} + \frac{\partial\rho}{\partial t} = 0, \quad \mathrm{div}\vec{J}_{\mathrm{m}} + \frac{\partial\rho_{\mathrm{m}}}{\partial t} = 0 \tag{2.60}$$

where $\vec{J} \equiv -\,i\,\frac{\omega}{4\pi}\,(\epsilon' - 1)\,\vec{E},\ \vec{J}_{\mathrm{m}} \equiv -\,i\,\frac{\omega}{4\pi}\,(\mu' - 1)\,\vec{H}$. Equations (2.59) are obtained from the exact Maxwell equations, if one sets

$$(\epsilon' - 1)\,\vec{E}' \simeq (\epsilon' - 1)\,\vec{E}, \quad (\mu' - 1)\,\vec{H}' \simeq (\mu' - 1)\,\vec{H}.$$

The solutions to Eqs. (2.59) were studied by many authors [Stratton & Chu, 1939]. According to the diffraction theory

$$\vec{E}'(\vec{r}\,') = -\,i\,\frac{4\pi}{\omega}\iiint\limits_{V}\left[k^2\,\vec{J} + (\vec{J}\vec{\nabla})\,\vec{\nabla} + i\,k\,\vec{J}_{\mathrm{m}}\times\vec{\nabla}\right]\,\Psi(\vec{r},\,\vec{r}\,'), \tag{2.61}$$

where

$$\Psi(\vec{r},\,\vec{r}\,') = -\,\frac{1}{4\pi\,|\vec{r} - \vec{r}\,'|}\,\exp(i\,k\,|\vec{r} - \vec{r}\,'|). \tag{2.62}$$

The vectors $\vec{J}$ and $\vec{J}'$ in Eq. (2.61) are functions of $\vec{r}$. A formula for $\vec{E}'(\vec{r}\,')$ is obtained from Eq. (2.61) with the following substitution $\vec{E}' \rightarrow \vec{H}'$, $\vec{J} \rightarrow \vec{J}_{\mathrm{m}}$, and $\vec{J}_{\mathrm{m}} \rightarrow \vec{J}$. At $r \gg a$ vectors $\vec{E}'(\vec{r}\,')$, $\vec{H}'(\vec{r}\,')$ describes a divergent spherical wave.

Let us define the form of ϵ' and μ' in the equations for $\vec{J}$ and $\vec{J}'$ more explicitly. Assume that the difference of ϵ' and μ' from ϵ and μ is caused by a deviation of the resonator surface from an ideal spherical shape,

$$r = a + b\,y(\theta,\,\phi), \tag{2.63}$$

where $r,\,\theta,\,\phi$ are coordinates of points on the resonator surface ,

$$\int\!\!\int y(\theta,\,\phi)\,d\Omega = 0, \quad \int\!\!\int y^2(\theta,\,\phi)\,d\Omega = 1, \tag{2.64}$$

$d\Omega = \sin\theta\,d\theta\,d\phi$. In this case, the electric field is given by the following equation

$$\vec{E}'(\vec{r}\,') = \vec{E}(\vec{r}\,') + \int\!\!\int\!\!\int [\epsilon - \epsilon' + (\epsilon - 1)\,(\mu - \mu')]\;k^2\,\vec{E}(\vec{r})\,\Psi(\vec{r},\,\vec{r}\,')\,dr. \tag{2.65}$$

The obtained equation allows to calculate the Q factor of the resonator. It is defined as the ratio of the electromagnetic energy stored in the optical cavity, multiplied by 2π, to the energy lost by the resonator during an oscillation period T: $Q = 2\pi\,\mathcal{E}/(P\,T)$, where

$$P = \frac{c}{4\pi}\int\!\!\int \overline{\left[\vec{E}' \times \vec{H}'\right]}\,\vec{e_r}\,r^2\,d\Omega, \quad r \gg a, \tag{2.66}$$

the overbar denotes averaging over oscillation period. When calculating P for a TE mode, the following approximation was adopted [Datsyuk *et al.*, 1993]:

$$P = \frac{c}{8\pi}\int\!\!\int \left|\vec{E}'\right|^2 r^2\,d\Omega \le P_0 + \frac{c}{8\pi}\int\!\!\int \left|\vec{E}' - \vec{E}\right|^2 r^2\,d\Omega. \tag{2.67}$$

Here P_0 defines the energy loss at $b = 0$,

$$\left|\vec{E}' - \vec{E}\right|^2 \le \left[(n_{\mathrm{r}}^2 - 1)\,\frac{a\,k\,b}{4\pi\,r}\right]^2 \int\!\!\int \left|\vec{E}(\omega,\,\vec{a})\right|^2 d\Omega, \tag{2.68}$$

where $\vec{a}$ is a vector with the spherical coordinates a, θ, and ϕ. Similar approximations were made when calculating P for a TM mode. Finally, the Q factor was found in the following form [Datsyuk *et al.*, 1993]:

$$1/Q \simeq 1/Q_{s0} + 1/Q_{\rm sc}, \tag{2.69}$$

where Q_{s0} is given by Eq. (2.51) and

$$Q_{\rm sc} = 2\pi \left(l + \frac{1}{2}\right) \left(\frac{\epsilon}{\mu}\right)^{p-1/2} (\epsilon\mu - 1)^{-1} (kb)^{-2}. \tag{2.70}$$

Distinct formulas for $Q_{\rm sc}$ were proposed in [Vernooy *et al.*, 1998; Oraevsky, 2002].

2.4.4 *Effect of a spherical submicrometer-size inclusion*

An effect of a small probe body on resonant properties of an open cavity was studied theoretically in [Tsipenyuk, 1965]. This study applied a decomposition of the classical electromagnetic field over the quasinormal modes of an arbitrary open cavity. The presence of a small dielectric sphere with radius b was considered using a perturbation theory. A shift of the resonant frequencies, also known for a closed optical resonator, was found. This shift was proportional to b^3:

$$\frac{\Delta\omega_s}{\omega_s} = -\frac{b^3}{2N_s} \left(\frac{\epsilon_{\rm b} - 1}{\epsilon_{\rm b} + 2} \vec{E}_s^2 - \frac{\mu_{\rm b} - 1}{\mu_{\rm b} + 2} \vec{H}_s^2\right)^2, \tag{2.71}$$

where $\epsilon_{\rm b}$ and $\mu_{\rm b}$ are the relative permittivity and magnetic permeability of the probe, correspondingly. Its radius was presumed to be small:

$$k\, b \ll 1, \quad k\sqrt{\epsilon_{\rm b}\mu_{\rm b}}\, b \ll 1. \tag{2.72}$$

Vectors $\vec{E}_s$ and $\vec{H}_s$ determines the resonant electric and magnetic field as follows:

$$\vec{E}(t, \vec{r}) = \Re\left\{\vec{E}_s(t, \vec{r})\, e^{-\imath\omega_s t}\right\}, \tag{2.73}$$

$$\vec{H}(t, \vec{r}) = \Re\left\{\vec{H}_s(t, \vec{r})\, e^{-\imath\omega_s t}\right\}, \tag{2.74}$$

ω_s is the complex frequency,

$$\omega_s = \omega_s' - \imath\,\omega_s''. \tag{2.75}$$

The quality factor

$$Q_{s0} = \frac{\omega_s'}{\omega_s''} \tag{2.76}$$

was presumed to be known. Factor N_s in Eq. (2.71) is the norm of the mode defined as [Vainstein, 1969]:

$$N_s = \frac{1}{4\pi} \lim_{R\to\infty\,\exp(\imath\,\gamma)} \iiint_{r\leq R} \epsilon\,\vec{E}_s^2\,dV$$
$$= -\frac{1}{4\pi} \lim_{R\to\infty\,\exp(\imath\,\gamma)} \iiint_{r\leq R} \mu\,\vec{H}_s^2\,dV, \tag{2.77}$$

where integration is over the volume of a ball with a complex radius R, the angle γ should provide the convergence of the integrals. An approximate calculation of N_s can be fulfilled with integration over a finite volume of an open cavity [Vainstein, 1965]. The problem of the normalization of the modes of an open cavity was also discussed by Lai *et al.* [Lai *et al.*, 1990b]. If an effect of splitting of the degenerate modes can be neglected, then the norm N_s has to be calculated as [Lai *et al.*, 1990b]:

$$N_s = \frac{1}{4\pi} \iiint_{r\leq R} \epsilon\,\vec{E}_s^2\,dV + \frac{\imath}{8\pi\,\omega_s} \iint_{r=R} \vec{E}_s^2\,dS, \tag{2.78}$$

Here R is an arbitrary large distance, integration in the second integral in the right-hand side of the above equation is over surface of a sphere with radius R. The both formulae (2.77) and (2.78) allow for that an open cavity is not a Hermitian system.

It was established in [Tsipenyuk, 1965] that a small dielectric body in an open cavity also causes a degradation of the Q-factor described with the following equation:

$$\frac{1}{Q_s} = \frac{1}{Q_{s0}} + \frac{1}{Q_\mathrm{b}}, \tag{2.79}$$

where

$$\frac{1}{Q_\mathrm{b}} = \frac{2\,k^3\,b^6}{3\,N_s} \left[\left(\frac{\epsilon_\mathrm{b}-1}{\epsilon_\mathrm{b}+2}\right)^2 \left|\vec{E}_s\right|^2 + \left(\frac{\mu_\mathrm{b}-1}{\mu_\mathrm{b}+2}\right)^2 \left|\vec{H}_s\right|^2 \right]. \tag{2.80}$$

This effect of the additional radiative energy loss can not be found in a closed cavity.

2.4.5 *Comparison of different WGM-scattering models*

Using the above equation (2.10), one can evaluate the energy losses in a cavity with multiple perturbations of the dielectric constant. Let us follow [Oraevsky, 2002] and consider cavity surface having inhomogeneities with linear sizes much smaller than the wavelength. Since each inclusion scatters light independently of others, a value of $1/Q_{sc}$ can be found by summarizing contributions from all subwavelength-sized balls. When using in Eq. (2.10) values $\epsilon_{\rm b} = \epsilon$ and $\mu_{\rm b} = 1$ we get

$$\frac{1}{Q_{\rm sc}} = 12\,\pi^2 \left(\frac{\epsilon - 1}{\epsilon + 2}\right)^2 \lambda^3 \frac{N_{\rm b}\, V_{\rm b}^2\, \langle \vec{E}_s^2 \rangle}{\iiint \epsilon\, \vec{E}_s^2\, dV}, \tag{2.81}$$

where $N_{\rm b} \simeq 4\pi\, a^2/(\pi\, B^2)$ is a number of inhomogeneities on the surface,

$$V_{\rm b} = \frac{4\pi}{3}\, d\, B^2, \tag{2.82}$$

B, d are an averaged longitudinal dimension and an average height of the inhomogeneities, respectively, $\langle \vec{E}_s^2 \rangle$ is an average value of $\vec{E}_s^2(\vec{r}_{\rm b})$ on the cavity surface. In the above equation the following estimation can be applied

$$\iiint \epsilon\, \vec{E}_s^2\, dV \simeq 4\pi\, a^2\, h\, \epsilon\, \langle \vec{E}_s^2 \rangle. \tag{2.83}$$

After substitution to Eq. (2.81) the above estimate for N and Eq. (2.83) with a rough estimate $h_s \simeq \sqrt{2a\,\lambda}$ [Vernooy *et al.*, 1998], one gets

$$Q_{\rm sc} = \frac{3\,\epsilon\,(\epsilon + 2)^2}{(4\,\pi)^3\,(\epsilon - 1)^2}\, \frac{\lambda^{7/2}\,(2\,a)^{1/2}}{d^2\, B^2}. \tag{2.84}$$

The obtained equation differs from a conventional formula [Oraevsky, 2002; An & Moon, 2003; Buck & Kimble, 2003] for $Q_{\rm sc}$ of [Gorodetsky *et al.*, 1996; Vernooy *et al.*, 1998] by a factor of $(\epsilon - 1)^{1/2}$. The latter has the following form

$$Q_{\rm sc} = \frac{3\,\epsilon\,(\epsilon + 2)^2}{(4\,\pi)^3\,(\epsilon - 1)^{5/2}}\, \frac{\lambda^{7/2}\,(2\,a)^{1/2}}{d^2\, B^2}. \tag{2.85}$$

Equation (2.85) was given in [Gorodetsky *et al.*, 1996; Vernooy *et al.*, 1998] without derivation and was grounded in a very recent review [Oraevsky, 2002]. However, a different formula was obtained [Oraevsky, 2002],

$$Q_{\rm sc} = \frac{\sqrt{\epsilon}}{12\,\pi^2} \left(\frac{\epsilon + 2}{\epsilon - 1}\right)^2 (4\,\pi)^3\,(\epsilon - 1)^{5/2}\, \frac{\lambda^3\, h_s}{\Omega\, d}, \tag{2.86}$$

where Ω is the average volume of the inhomogeneity, h_s is the effective width of the mode s.

In [Gorodetsky *et al.*, 2000], Gorodetsky *et al.* obtained the factor $Q_{\rm sc}$ for a TE mode in the following form:

$$Q_{\rm sc} = \frac{K}{1+K}\,\frac{3\,\lambda^3\,a}{8\,m_{\rm r}\,\pi^2\,d^2\,B^2}, \tag{2.87}$$

where K is the ratio of the complete scattered power over the power scattered into the angles satisfying definite angular cutoff conditions.

From the above derivation and conditions (2.72), it follows that formulas (2.85) and (2.86) can be applied only if the following inequalities are satisfied:

$$k\,d \ll 1, \quad k\,B \ll 1. \tag{2.88}$$

Hence, these formulas are inappropriate in the case of capillary vibrations of droplets for which a value of B is as large as $a \gg \lambda$.

The distinct theoretical formulas (2.70), (2.85), (2.86), and (2.87) for $Q_{\rm sc}$ can be compared with each other and with experimental data making mention of [Vernooy *et al.*, 1998]. The latter paper reported a very weak dependence of the Q factor on λ measured in the range of 679 to 850 nm (see also [Oraevsky, 2002]). Taking into account that $l + 1/2 \simeq k\,a\,m_{\rm r}$, the following proportionality is found from Eq. (2.70)

$$Q_{\rm sc} \propto \frac{a\,\lambda}{b^2}. \tag{2.89}$$

Assuming that factor $Q_{\rm sc}$ is caused by the multiplet splitting examined in Ref. [Lai *et al.*, 1990a,b], one would get from Eq. (2.55) that

$$Q_{\rm sc} \propto \frac{a}{b}. \tag{2.90}$$

Both Eq. (2.70) and estimate (2.90) are in a reasonably good agreement with the experimental findings of [Vernooy *et al.*, 1998]. On the contrary, the formulas (2.85), (2.86), and (2.87) do not describe the wavelength dependence of the Q factor.

2.4.6 *Q factor of a loaded cavity*

The effect of coupling of a microcavity with other components of a optoelectronic device can be allowed for in terms of degradation of the Q factor. A rigorous electromagnetic wave theory can be built only for simple composite systems, such as bi-spheres [Barton *et al.*, 1991; Fuller, 1991; Ilchenko *et al.*, 1994] or sphere-plate complex [Brevik *et al.*, 2003]. Even for these simplest systems the theories are quite complicated. An alternative wave of calculation of the partial Q factor can apply a perturbation theory [Oraevsky, 2002]. In the framework of this theory, the electromagnetic fields of separated sphere and other dielectric body are first determined. Then, the overlaps of these fields are calculated that allows us to find amplitudes of the modes. Hence, the problem of extraction of WGM energy by an outer probe is hence similar to the problem of pumping of a WGM by an external source. A review of [Oraevsky, 2002] describes and analyzes various mechanisms of excitation of WGMs, starting from excitation by a plane wave and giving a detail consideration of a planar-waveguide wave excitation, excitation via the evanescent field of a prism, and excitation by waves of fibers, including tapered and half-block ones. In the same paper, a formula for the partial Q factor of coupled sphere and planar waveguide, with reference to [Gorodetsky & Ilchenko, 1999; Little *et al.*, 1999], is given.

The loaded Q values measured in experiments of [An & Moon, 2003] were as high as the expected bare Q values. The numeric estimates fulfilled in this experimental paper showed that the most important energy-loss processes in the microsphere laser were the radiative losses of an ideal sphere and the absorption loss in the sphere medium (fused silica). The surface scattering loss of the microsphere and the absorption loss in the surrounding medium that was an ethanol solution were appeared to be negligible. A similar assessment was made in a theoretical study devoted to strong coupling of an external atom and microsphere [Buck & Kimble, 2003]. The latter theoretical paper studied a balance between the radiative losses for an ideal dielectric sphere and other material loss mechanisms: namely, losses from cavity surface inhomogeneities, absorption losses due to water on the surface of the sphere and bulk absorption in the fused silica.

CHAPTER 3

Electromagnetic Eigen Oscillations and Fields in a Dielectric Microsphere with Multilayer Spherical Stack

3.1 Introduction

In recent years, one can see a trend in the use of microcavities and microspheres for photonic applications as well as new approaches to solve various problems, both in optics and photonics. Using a microcavity and a microsphere allows for a new view of various effects and interactions in the structured and layered media. In [Yamasaki *et al.*, 2000] strongly scattered light emission was observed from organic light-emitting device with an ordered monolayer of silica microspheres. It was shown in [Zhang *et al.*, 2000] that photonic band gaps can be realized using metal or metal-coated spheres as building blocks. Authors [Hayata & Koshiba, 1992] presented a theoretical formulation of second harmonic generation from an optically trapped microsphere. Cao *et al.* [2000], reported on a large enhancement of second harmonic generation in polymer films by microcavities. By employing the microcavity formed by a distributed Bragg reflector, they observed up to 50 times increase of second-harmonic light intensity in polymer thin films. In [Yi & Stafsudd, 1999], the plane-wave electromagnetic scattering in active dielectric sphere was investigated and the existence of anomalous resonances that occur at discrete wavelengths was found.

Another interesting direction of investigation is to use the microcavities and microspheres for the change of the spontaneous emission of radiation by an atom as a result of it being placed into a cavity [Klimov *et al.*, 1996 and references therein]. This effect is due to the change of the mode density in the cavity in comparison with that in an open space. The inhibition of spontaneous emission of radiation in a periodic dielectric structure with an electromagnetic band has been discussed in [Yablonovich, 1987]. The problem of the influence of a dielectric sphere on the emission of radiation by an atom is much more involved. The classical problem is to find the field distribution created by an arbitrary source in the presence of a dielectric sphere. Chew [1987] and Kien *et al.* found a compact representation of the field in the presence of a dielectric sphere and used it to find the general expression for the change of the spontaneous lifetime of an atom in a dielectric sphere (or a cavity).

Klimov *et al.* [1996] have demonstrated that with high enough refractive index of the microsphere material, the size of the microsphere can be selected such that the spontaneous decay rate decreases (or increases) several times. In [Klimov & Letokhov, 1999] the explicit solution for 3-layered spherical system was found and the electrodynamic properties were studied. It was shown that the developed theory described the experimental data of the spontaneous emission observation sufficiently.

Nowadays the basic regime of the operation of microspheres is a whispering gallery mode (WGM) for a microsphere with a radius of the order of 100 μm and less. The extremely great values of a Q factor in a narrow frequency range are observed [Vassiliev *et al.*, 1998; Ilchenko *et al.*, 1998]. But by fabricating the dielectric spheres of the submicron sizes [Ueha, 1998; Laboratories & Inc., 2001], a problem arises, prompting researchers to investigate an optical oscillations in microspheres beyond the WGM regime for harmonics with a small number in a low mode regime. This direction has not sufficiently investigated yet. An application of the well-known idea on quarter-wave layer coating provides us with a possibility to increase the Q factor of such a system of up to 10^5.

It should be noted that the known analytical approaches as well as the results for an interpretation of physical picture in such systems and their analysis are promising for applications. At the same time they are rather sophisticated in exploring much more complicated structured systems similar to the multilayered microspheres in details.

Other important circumstances that should be taken into account are the frequency resonances of the electromagnetic oscillations in the spherical resonators. Such resonances were first discovered by Debye for the metallized spherical resonators [see Stratton, 1941]. For such sphere the Q factor of oscillations is determined only by losses in a metallic substrate and it can reach up to 10^6. In general case of a dielectric sphere there is the electromagnetic wave radiation (leakage) from the inner source placed in the center part of the sphere, into an external medium. This radiation leads to the loss of energy of the wave, and in this case the Q factor is rather small even for a case of a small material losses. Recently such resonances have been observed for systems in which the gain in the active dielectric compensates exactly the radiative losses of a mode [Yi & Stafsudd, 1999]. Optical pumping was used to provide for the gain, and the threshold of each resonant mode was found to be a function of the dye concentration and the radius of the dielectric sphere.

Another way of increasing the Q factor is the coating of the spherical resonator with the quarter-wave layers to create stop bands in the specific frequency range [Brady *et al.*, 1993; Burlak *et al.*, 2000]. This allows us to increase the reflectivity in such structures in the specified range of wavelengths. However the spectrum of the eigenfrequencies of such system depends strongly on the properties of the stack. In other words, the stack can strongly perturb the spectrum of eigenfrequencies of a spherical resonator. In this case, the problem becomes more complicated as long as one should solve the eigenfrequency problem for the composite system: the resonator coated by the stack.

The layered systems with a spherical geometry are quite complex and its electromagnetic oscillations have been investigated rather completely only for asymptotically great number of lossless layers [Brady *et al.*, 1993]. In [Sullivan, 1994; Sullivan & Hall, 1994] the original approach to study the multilayered spherical systems was developed on the basis of the technique of Green functions. Sullivan & Hall [1994] derived the coupled-amplitude equations, and they have analyzed the case of the Bragg regime of the layered structure in details. In [Li *et al.*, 1994] the dyadic Green function was constructed and it was applied up to the four-layer media. In [Burlak *et al.*, 2000] another approach was proposed based on generalizations of the well-known plane case matrix procedure of the calculations of reflectance [Born & Wolf, 1980; Solimeno *et al.*, 1986] for the spherical geometry case. This approach provides the possibilities to investigate numerically the properties of spherical multilayered systems with an arbitrary number of lossless or lossy layers of the arbitrary thicknesses.

The classical problem in finding the field set by an arbitrarily distributed source in the presence of a dielectric sphere of arbitrary dielectrics is rather complicated. Except for the limited cases, intensive computations are required for the solutions. The novel aspect of this problem is to calculate the eigenfrequencies and the electromagnetic fields in the case of the coated sphere. In such structures the radial distribution of the eigen fields is rather different for eigenfrequencies which are within and beyond the zone of high reflectivity (stop bands). Therefore the properties of a compound system: resonator coated by a stack, depend on total parameters both of a resonator and a stack. The knowledge of the eigenfrequencies with $Q \gg 1$ is very important, because namely they determine the frequency position of resonances of the electromagnetic field of the internal radiation source [Vainstein, 1988].

This chapter is organized as follows. In Sec. 2 we discuss the geometry and basic equations. In Sec. 3, the eigenfrequencies equation for optical modes in the spherical resonator coated by the multilayered dielectric stack is derived. The spectrum of the eigenfrequencies and their Q factors for a number of the first eigenmodes are calculated and discussed. We analyze the stability of the spectrum from the point of view of the frequency dependence of reflectivity (or transmittance) of the stack with respect to the number of layers in the stack. In Sec. 4, we calculate the radial distribution of the electromagnetic amplitudes and the density of energy of the electromagnetic field for some eigenfrequencies. We show strong localizations of the field energy in the kernel of multilayered resonator for the eigenfrequencies laying within the zone of high reflectivity. In Sec. 5 we discuss the obtained results and compare the various geometries of the spherical coated resonator.

3.2 Geometry and Basic Equations

Consider a dielectric sphere of a radius r_1 and a concentric system of spherical layers contacting with the sphere. The layers are localized at the distances r_k from the center (see Fig. 1.7) where $r_k - r_{k-1} = d_k$ is the thickness of a kth layer; $k = 2, \ldots, N-1$. The expressions for the components of fields H_φ, E_θ and E_r in the case of TM waves are written in (1.55)–(1.58). Both R, T and Z can be calculated by applying the general formula (1.172). Note that the above expressions are derived without any assumptions about the ratio of λ and r and therefore they include the case of a spherical Bragg structure as a particular case. Now we apply the general formulas (1.172)–(1.183) to explore the frequency properties of a spherical stack. The calculations are applied to determine both the nature of the waves leaving the stack and the waves reflected back into the spherical resonator. In an analogy with the plane case, we now consider a stack with $1/K_0 = \Lambda_0/2\pi$, where Λ_0 is the reference wave length of the structure and it defines the periodicity in the stack. Below we consider the case $d_k n_k = \Lambda_0/4$ that corresponds to a quarter-wave case, in detail. The following parameters have been used in calculations: the geometry of a system is $ABCBCBC \ldots BD$, where letters A, B, C, D indicate the materials in the system, $\Lambda_0 = 1.75\,\text{mkm}$ ($K_0 = 2\pi f_0/c$, $f_0 = 171.5\,\text{THz}$). We have used the parameters of materials for [Hodgson & Weber, 1997] A(spherical substrate): glass, $n_A = 1.5$, B: SiO_2, $n_B = 1.45$, C: Sub2, $n_C = 2.05$ [Kumar *et al.*, 1999], D (outer medium): air, $n_D = 1$. Figure 3.1 shows the modules and phases of the complex coefficient reflectivity R and the coefficient

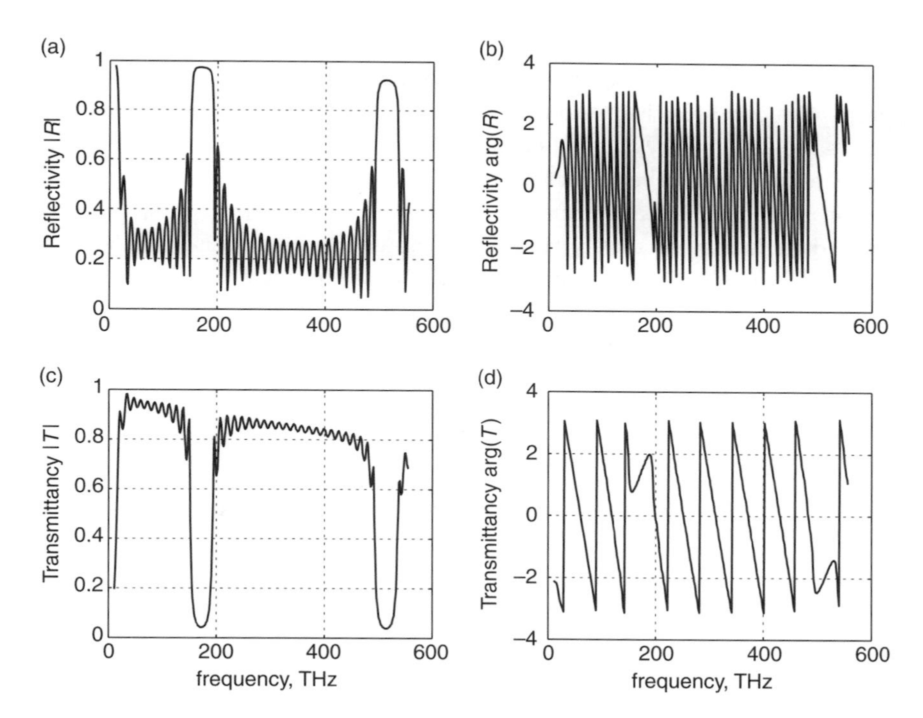

Fig. 3.1 The modules and phases of reflectivity R and transmittance T of the spherical quarter-wave stack versus the initial frequency f of electromagnetic radiation. The number of layers in the stack is 24 (see details in the text).

of transparency T as a function of initial frequency f for the case of 24 quarter-wave layers stack. Note that $|R|^2 + |T|^2 + K = 1$, where K is a loss coefficient, which characterizes the material losses in a medium. For the lossless layers $K = 0$, we have $|R|^2 + |T|^2 = 1$.

Furthermore we will concentrate on calculations of the spectrum of eigenfrequencies and also on calculations of the corresponding radial electromagnetic field distributions.

3.3 Eigenfrequencies of the Spherical Resonator Coated by the Stack

At first we derive the eigenfrequency equation for microsphere coated by the dielectric stack (Fig. 1.7). For that one should separate the problem into external and internal parts. The external part serves to calculate the properties (R and Z) of the multilayered stack and it has been described in Chapter 1, see also [Burlak *et al.*, 2000a]. Here we discuss the internal problem to derive the eigenfrequency equation. We take into account the properties of a spherical resonator, namely the impedance of the spherical resonator $Z_{res} = E_\theta / H_\varphi$. We consider the resonator as a dielectric sphere with radius r_1 and the refraction index n_{res} (see Fig. 1.7). In the internal region of the spherical resonator (at $r \to 0$), the fields H_φ, E_θ must be limited. It is easy to satisfy this if one should consider in Eqs. (1.57) and (1.58) $a = b$. Then the singular parts of Hankel function are canceled. For example, in the case of the lowest spherical mode ($m = 1$), from Eqs. (1.57) and (1.58) one obtains [Ramo *et al.*, 1994]

$$H_\varphi = a\frac{in_{res}}{y}\left[\frac{\sin y}{y} - \cos y\right], \tag{3.1}$$

$$E_\theta = a\frac{1}{y^2}\left[\frac{y^2-1}{y}\sin y + \cos y\right], \tag{3.2}$$

$$Z_{res} = in_{res}y\frac{\sin y - y\cos y}{(y^2-1)\sin y + y\cos y}, \tag{3.3}$$

where $y = k_0 r n_{res}$, $k_0 = \omega/c$, $0 \le r \le r_1$. In the general case Z_{res} has a much more complicated form. For real k_0, Eqs. (3.1) and (3.2) define the standing wave in a resonator. To derive the eigenfrequencies equation, it is necessary to utilize the boundary conditions of the continuity of fields in the interface of the stack and spherical resonator at $r = r_1$. However, it is easier to use the equality of the impedances at this boundary in the form

$$Z_{res}(\omega) = Z_{stack}(\omega), \tag{3.4}$$

where $Z_{res}(\omega)$ and $Z_{stack}(\omega)$ are the impedances of the resonator and stack (see Eq. (1.184)), respectively. Equation (3.4) is the eigenfrequencies equation for the microsphere coated by the dielectric stack.

For a sake of brevity, we have written Eq. (3.4) in a rather compact form, but for a system with the given number of layers, it can be easily unwrapped by a substitution of expressions (1.174)–(1.177) and (1.184) into Eq. (3.4). However the appropriate expression appears too long already for a system with 3–4 layers. As a result, it seems impossible to write down the derived expression for the eigenfrequencies in some explicit observable form. However it is not necessary to do it, because such system can be easily subjected to the numerical analysis as the sequence of the algorithmized steps (see Part III).

It is not difficult to calculate the spectrum of eigenfrequencies for the case of dielectric sphere (cavity) with perfect conductive boundaries. This geometry was analyzed by Debye [Stratton, 1941, p. 561], who found that empty spheres have the lowest value of $2\pi r_1/\lambda_0 = 2.75$ where $\lambda_0 = 2\pi/k_0$. But in a general case for spheres with deposited dielectric stack such calculations become much more difficult as it requires the frequency properties of the stack on each step of the calculations. Therefore, the solution of Eq. (3.4) requires rather intensive computations. As a result, we obtain the set of complex roots for ω which determines the eigenfrequencies spectrum of the spherical resonator coated by the stack. The real part of ω defines the eigenfrequencies of oscillations and the imaginary part of ω determines a decrement of the attenuation. The attenuation is due to material losses and also due to the leakage of waves from the resonator into the surrounding medium. We define the Q factor of oscillations by the standard expression [Ramo *et al.*, 1994]

$$Q = \frac{\mathrm{Re}(\omega)}{2\,\mathrm{Im}(\omega)}. \tag{3.5}$$

We study the dependencies of the eigenfrequencies and the corresponding Q factors of the multilayered system versus the number of layers in the spherical stack. Results are presented in Figs. 3.2–3.4. In the following calculations we pay attention to the case of $m = 1$ of the spherical harmonic. Figure 3.1 shows the modules and phases of the reflectivity R and the transmittance

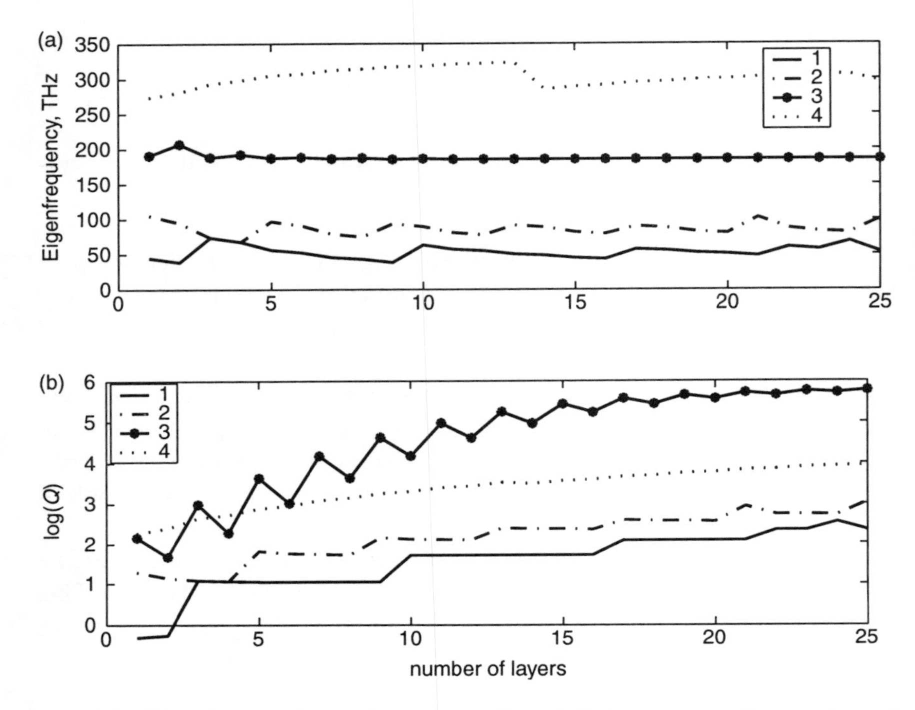

Fig. 3.2 Eigenfrequencies and corresponding Q factors versus the number of layers in the quarter-wave stack. Only one eigenfrequency (3rd line) is within the high reflectivity zone.

T of the spherical quarter-wave stack versus the initial frequency f of the electromagnetic radiation. The number of layers in the stack is 24.

In Fig. 3.2, the eigenfrequencies spectrum and the quality factor Q of the system versus the number of lossy layers for the microsphere with radius $r_1 = 1\,\mu\text{m}$ are presented. The plots show a few of the lowest roots of Eq. (3.4). One can see that the shift of frequencies is quite sensitive to the number of layers in the stack. Only one eigenfrequency (184 THz) is stable while the number of layers changes. From Fig. 3.1 we can see that the corresponding frequency is situated within the first stop band. The Q factor exponentially grows with an increasing number of layers in the stack. It corresponds to [Brady *et al.*, 1993], where exponential growth of the Q factor was found for the case of $N \gg 1$. Moreover, we have found that the Q factor of such frequency grows exponentially while the number of layers in the stack is of an order of magnitude $N = 10$. Since $N = 15$, the Q factor has a tendency towards saturation due to the material losses of the layers. One

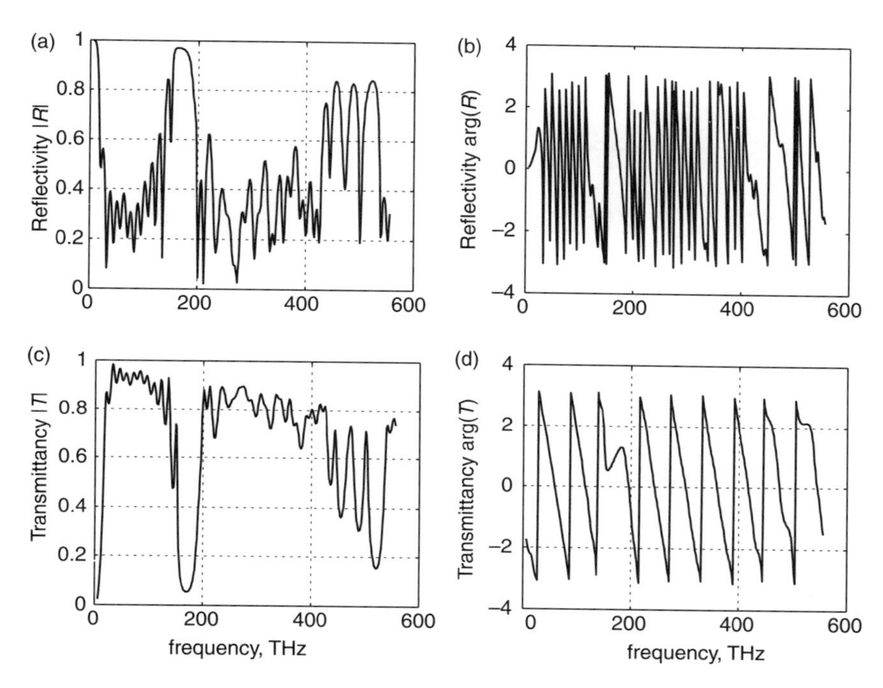

Fig. 3.3 The modules and phases of the reflectivity R and the transmittance T of the spherical quarter-wave stack versus the initial frequency f of electromagnetic radiation in the case of the 20% random deviation of the thicknesses of layers. The number of layers in the stack is 24.

can see that the other eigenfrequencies have quite low Q factors, because such frequencies are beyond the zone of high reflectivity, where $|R| \sim 1$ (see Fig. 3.1). Some oscillations of Q with their global growth are due to an alternation of even and odd numbers of layers in a stack [Hodgson & Weber, 1997]. The time multiplier $\exp(i\,\omega t)$ is used here. In the case of the complex frequency $\omega = \omega_r + i\,\omega_i$ (where the imagine part $\omega_i \geq 0$) the fields in the layered resonator decrease with time as $\exp(-\omega_i t)$. Therefore the retrieval of the eigenfrequencies with a small imaginary part $|\omega_i| \ll |\omega_r|$ or high $Q = \omega_r/2\omega_i \gg 1$ is very important for applications. The eigenfrequencies and Q factors of some lowest numbers l were calculated from Eq. (3.4) and they are written in the Table 3.1.

Although we pay much attention to the purely quarter-wave case, we have also explored what happens if the thicknesses of layers in the stack possess random deviations from the ordered $\lambda/4$ length. In this case the far order periodicity in the system is broken and every particular layer has

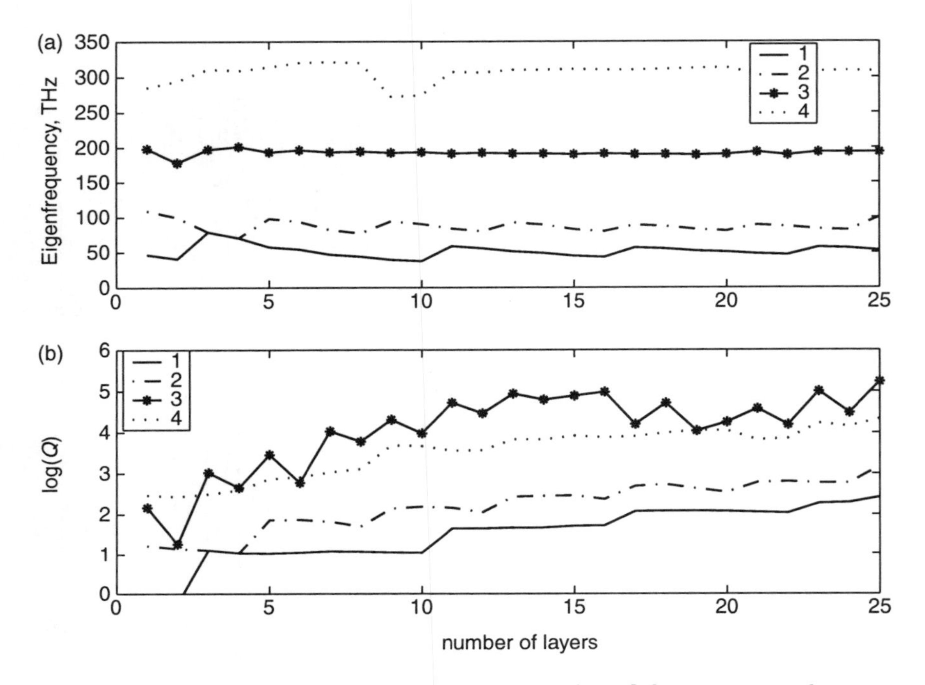

Fig. 3.4 Values of eigenfrequencies and corresponding Q factors versus the number of layers in the quarter-wave stack in the case of 20% random deviation of the thicknesses of layers. Only one eigenfrequency (3rd line) is within the high reflectivity zone.

Table 3.1 The eigenfrequencies and Q factors of some lowest numbers calculated from Eq. (3.4).

l	f_n, THz	Q	r_1/λ	l	f_n, THz	Q	r_1/λ
1	68.1	12.7	0.23	5	184	302	0.61
2	80.6	15.2	0.27	6	196	68	0.66
3	92.9	17.9	0.31	7	281	48	0.94
4	105	21	0.35	8	294	43.3	0.99

influence on the total properties of the stack. Figures 3.3 and 3.4 show the same results as Figs. 3.1 and 3.2, except that the thickness of layers have the 20% random deviation. One can see that in this case the high order frequency stop bands are destroyed completely. Only the eigenfrequency within the first stop band keeps its value with the change of the number of layers in the stack (see line No. 3 in Fig. 3.4). We can explain this in the following way: Forming the stop band is the collective effect which is generated by the influence of a great number of individual layers. Therefore, the small random deviations may be statistically recompensated in such zones. The high order stop bands are more sensitive to the random perturbations. Besides, the position of the interference maximums located between the zones of high reflectivity can vary with the change of the number of layers in the stack. Therefore, the values of eigenfrequencies change in such a range, correspondingly. The locations of the stop bands are more stable, and therefore, the eigenfrequencies located in this range of frequency are stable too.

3.4 Radial Distribution of Fields

In this section we discuss the radial distribution of the amplitudes of electromagnetic field in the system: a sphere coated by the dielectric stack. We apply the above developed theory to find out the radial distribution of the amplitudes and the local density of the energy of the field.

The idea of calculation is as follows: since the $\overrightarrow{U}_k$ on the external boundary of the kth layer has been calculated, one can readily obtain the value of $\overrightarrow{U}_k(y)$ in any points within this kth layer, following [Burlak *et al.*, 2000], in the form

$$\begin{bmatrix} H_\varphi(y) \\ E_\theta(y) \end{bmatrix}_k = \overrightarrow{U}_k(y) = \widehat{D}(y) \cdot \widehat{D}_k^{-1} \cdot \overrightarrow{U} = \widehat{D}(y) \cdot \widehat{D}_k^{-1} \begin{bmatrix} H_\varphi(y_k) \\ E_\theta(y_k) \end{bmatrix}, \tag{3.6}$$

where $y_{k-1} \leq y \leq y_k$, $y = k_0 n_k r$ and the matrix $\widehat{D}(y)$ is given by

$$\widehat{D}(y) = \begin{bmatrix} inP_m^{(2)}(y)e^{iy} & inP_m^{(1)}(y)e^{-iy} \\ G_m^{(2)}(y)e^{iy} & G_m^{(1)}(y)e^{-iy} \end{bmatrix}, \quad \widehat{D}_k = \widehat{D}(y_k).$$

We follow the next steps to calculate the radial fields distributions for the layered sphere:

(i) Calculate the eigenfrequency f_n by means of solving the eigenfrequency equation (3.4). We note that this step is not trivial, because this equation is a complex one and in general it has a set of different complex roots. Some of them are not stable with respect to the change of properties of the stack (see Figs. 3.2, 3.4).

(ii) Calculate the radial field distributions using the expression (3.6).

With that, the properties of the continuity of fields in the boundaries of the spherical layers and the limitation of fields in the center of sphere obey automatically. We have calculated the spectrum of eigenfrequencies and the corresponding radial distributions of fields within the coated spherical resonator. The general behavior of the eigen fields distribution outside the spherical open resonators is described in [Vainstein, 1969, 1988], and this has been confirmed under our calculations.

In Figs. 3.5–3.8, the normalized field distributions for some eigenfrequencies from Table 3.1 are shown, and r/r_1; r_1 is the radius of the dielectric sphere. In Table 3.1, the real parts of the eigenfrequencies are listed in

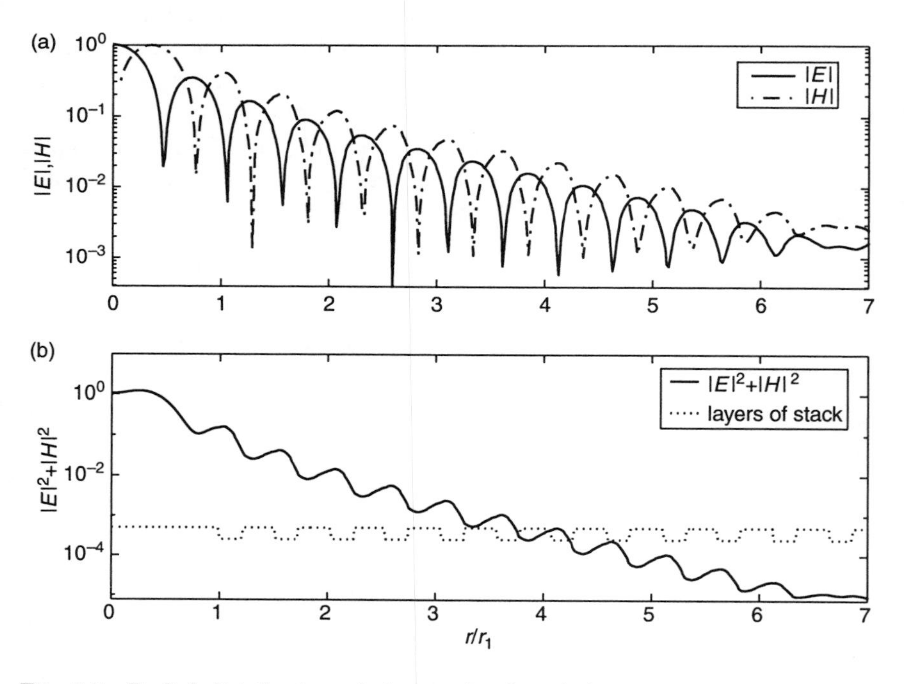

Fig. 3.5 Radial distribution of the amplitudes of electromagnetic field for the eigenfrequency $f = 184\,\text{THz}$ ($l = 5$ in Table 3.1) in the zone of the high reflectivity $|R| = 1$. (a) Amplitudes of electric and magnetic fields; (b) The average energy of field (solid line) and the structure of the stack (dash line). The number of layers in the stack is 24.

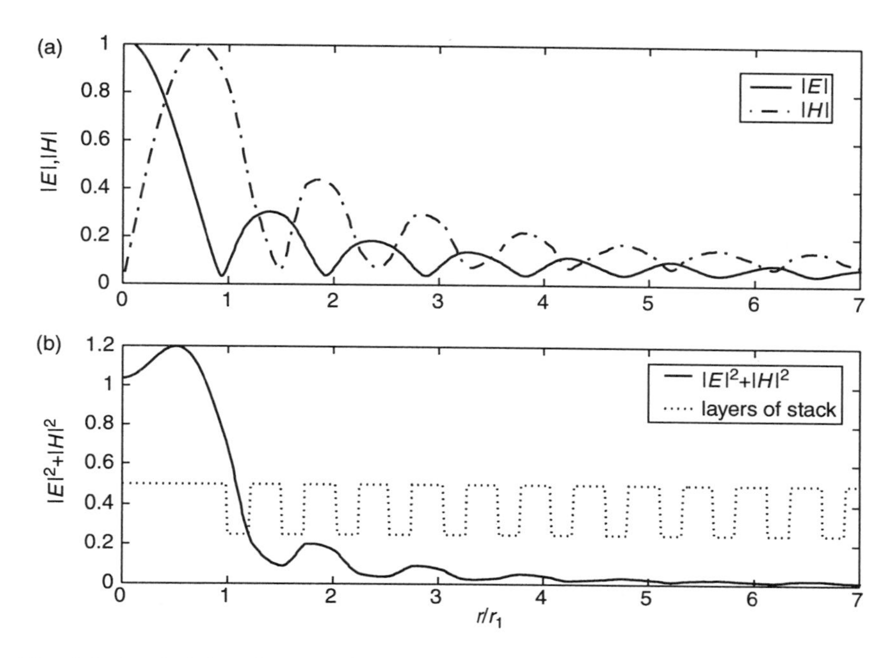

Fig. 3.6 Radial distribution of the amplitudes of electromagnetic field for eigenfrequency $f = 92.9\,\text{THz}$ ($l = 3$ in Table 3.1) beyond the zone of high reflectivity. (a) Amplitudes of electric and magnetic fields; (b) The average energy of field (solid line) and the structure of the stack (dash line). The number of layers in the stack is 24.

ascending order. Such classification is rather convenient because it is associated with a number of nodes of electric eigen field in the kernel of the dielectric resonator (see below the figures for fields). Figure 3.5 shows the radial dependence of modules of electric E_θ and magnetic H_φ fields for the eigenfrequency $f_5 = 184\,\text{THz}$. From Fig. 3.1 one can see that the value f_5 belongs to the stop band range. We found that in this case the fields exponentially decrease with the radii r growth within the coated resonator. To present more details in Fig. 3.5a, we have used the logarithmic scale for fields. Figure 3.5b shows the radial distribution of the field energy, with the dash line representing the structure of the thickness of layers in the stack. Figure 3.6 presents the analogous dependencies for eigenfrequencies $f_3 = 92.9\,\text{THz}$, which is beyond the stop bands. In this case, we have used the linear scale. For this frequency and other out-stop-band frequencies, the fields penetrate into the stack to a rather long distance before it is radiated to the external medium.

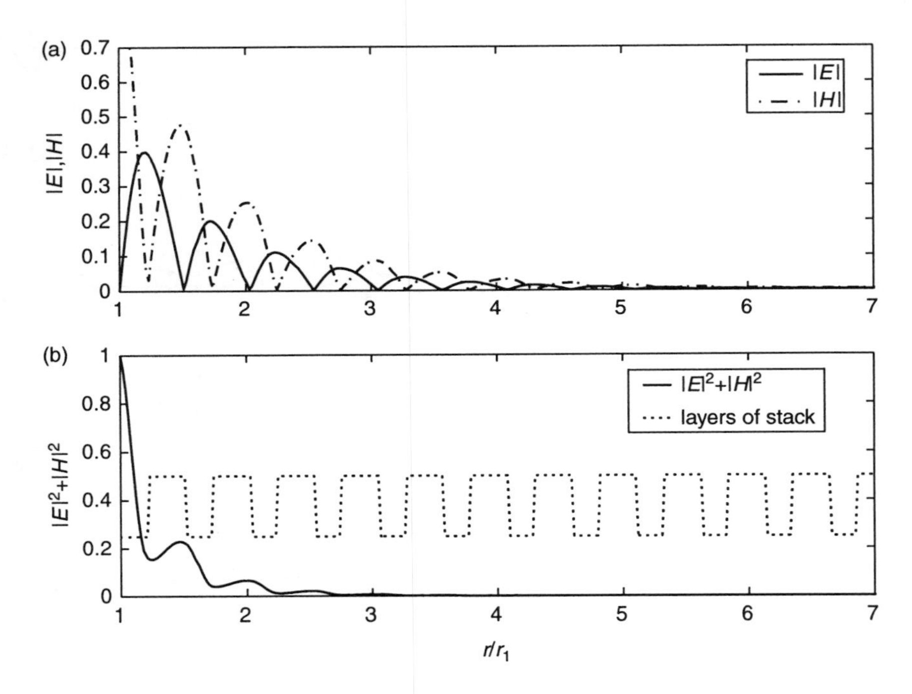

Fig. 3.7 Radial distribution of the amplitudes of electromagnetic field for eigenfrequency $f = 173.5\,\text{THz}$ in the case of the metallized spherical substrate. (a) Amplitudes of electric and magnetic fields; (b) The average energy of field (solid line) and structure of the stack (dash line).

In the center of dielectric sphere $r = 0$ the fields must have limited values. From (1.57)–(1.58) for $r \to 0$ and $a = b$ one can easily obtain $E_\theta = \frac{4}{3}a$ and $H_\varphi = 0$ for $m = 1$.

To compare these results with other representative geometries, we have investigated the electromagnetic eigen oscillations of the metallic sphere coated by the dielectric stack. In this case $E_\theta = 0$ on the surface of the sphere $r = r_1$. Figure 3.7 demonstrates both frequency and field dependencies for the stack deposited on the metallized sphere for $f = 173.5\,\text{THz}$.

In Fig. 3.8 above mentioned information is summarized for eigenfrequencies of the coated dielectric sphere from Table 3.1. One can see that only in the case of $f_5 = 184\,\text{THz}$ the electric field is concentrated in the center of the spherical resonator and it almost does not penetrate into the stack.

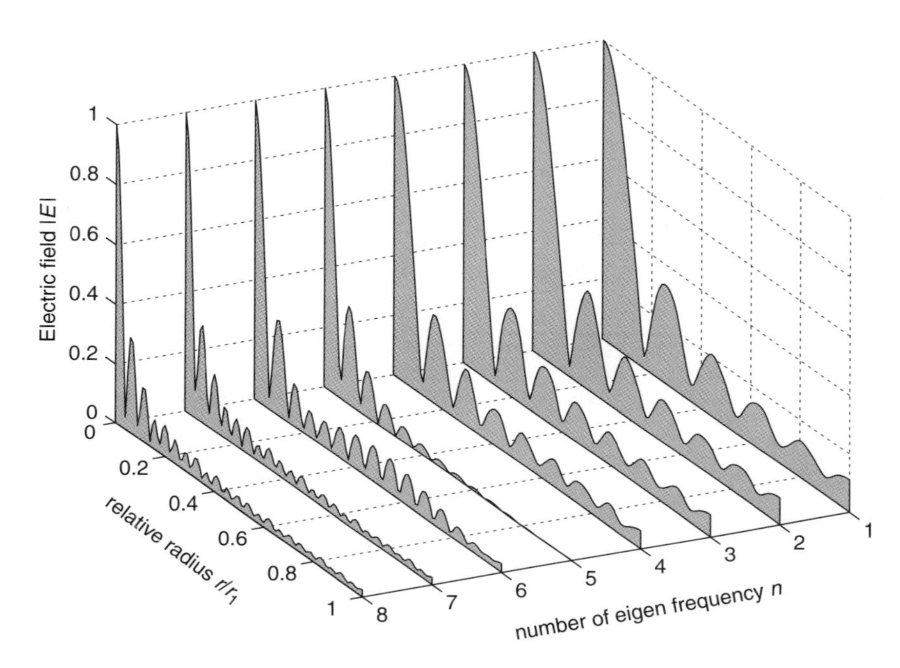

Fig. 3.8 Radial distributions of the normalized amplitudes of electric field $|E|$ for the different eigenfrequencies. Corresponding values of eigenfrequencies with the number l and the Q factors are written in the Table 3.1.

3.5 Discussions

The system of periodic dielectric layers with equal optical lengths (stack) allows formation of a zone of high reflectivity with sharp slopes (band edges). This fact is rather attractive in various optical applications [Fogel *et al.*, 1998]. Quarter-wave stack is traditionally used for the design of reflecting layered systems and in the one-dimensional case nowadays it is well known as the distributed Bragg reflector (DBR) [Convertino *et al.*, 1999]. A simple DBR consists of multiple layers of alternating high- and low-index materials arranged periodically. From the physical point of view there are no fundamental differences between the quarter-wave and halfwave stacks, and therefore, the use of them in the optical applications is frequently dictated by technological reasons only.

In Fig. 3.9, the frequency dependence of the modules of reflection coefficient $|R|$, transmittance $|T|$, and the coefficient of losses $K = 1 - |T|^2 - |R|^2$

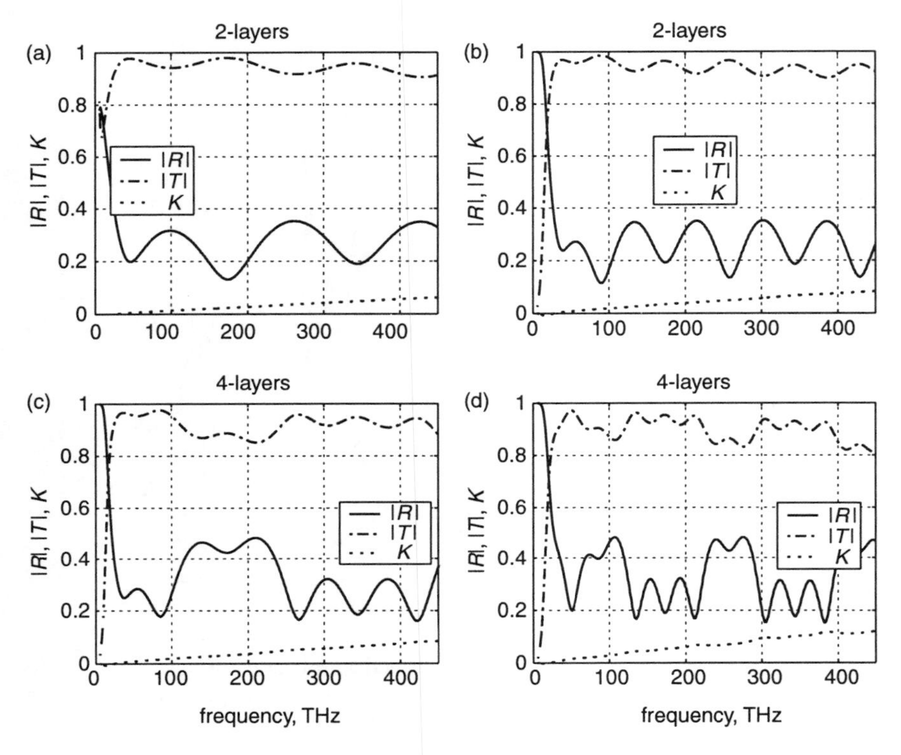

Fig. 3.9 The modules of reflectivity R, transmittance T, and the coefficient of losses $K = 1 - |R|^2 - |T|^2$ of the spherical stack versus the initial frequency f of electromagnetic radiation. (a) and (c) Case of $\lambda/4$ stack; (b) and (d) Case $\lambda/2$ stack. Number of layers in the stack are (a) and (b) $N = 2$; (c) and (d) $N = 4$ (see details in the text).

for the $\lambda/4$ and $\lambda/2$ cases for various $N = 2, 4$-layers in a stack are presented. One can see that in those small-number stacks the band edges of reflectivity are rather smooth even for $N = 4$, and therefore the selective properties of such the stack are rather poor.

In Fig. 3.10, the dependence is the same as in Fig. 3.9 but it is specially for the $N = 24$ alternative layers in stack. One can see that in the $\lambda/4$ case (Fig. 3.10a) zones of high reflectivity are almost twice wider, 3.9 in comparison with the $\lambda/2$ case (Fig. 3.10b). The same result holds also for the distance between the neighboring band gaps, respectively. The coefficients

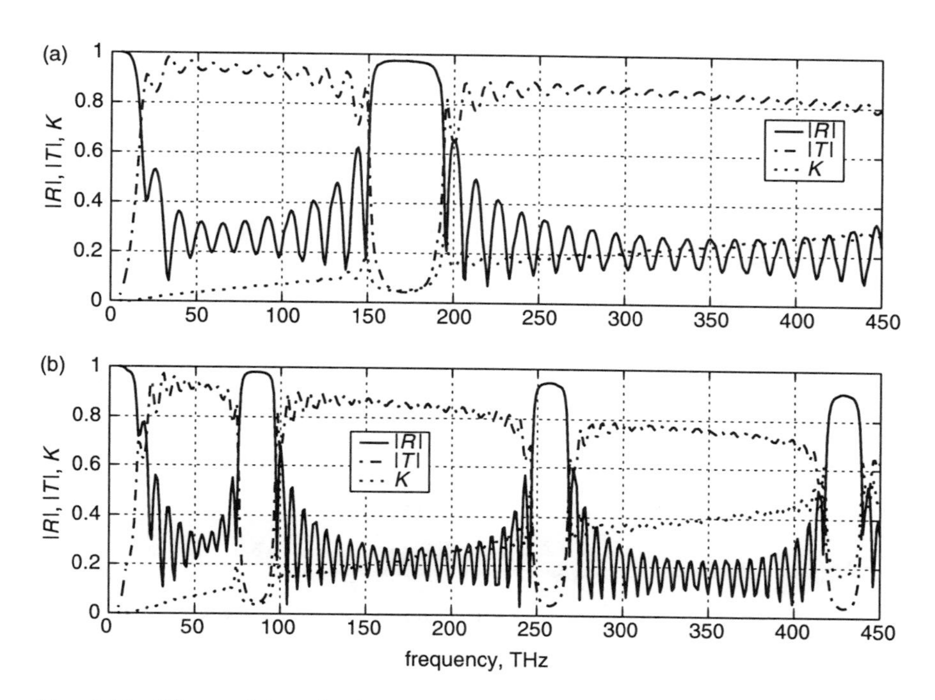

Fig. 3.10 The modules of reflectivity R, transmittance T, and the coefficient of losses $K = 1 - |R|^2 - |T|^2$ of the spherical stack versus the initial frequency f of electromagnetic radiation. (a) Case of $\lambda/4$ stack; (b) Case of $\lambda/2$ stack. Number of layers in the stack is 24 (see details in the text).

of losses K in Fig. 3.10 are greater in comparison with the case of the $N = 2, 4$ stacks (Fig. 3.9), because of the bigger number of layers in the stack. The band edge in the $N = 24$ spherical stack is rather sharp. More details of forming the stop bands in a spherical stack under increasing number of layers can be found in Sec. 1.4.

The deviation of the perfect periodicity of layers in a stack results in the destruction of the higher zones of reflectivity and also in the narrowing of the main zone. It is shown in the example of the 20% randomly perturbed quarter-wave stack, see Fig. 3.4.

With increasing number of layers in the stack, the Q factor of oscillations grows very sharply (exponentially) for the eigenfrequencies which are inside the zone of high reflectivity. We note that only the number and the

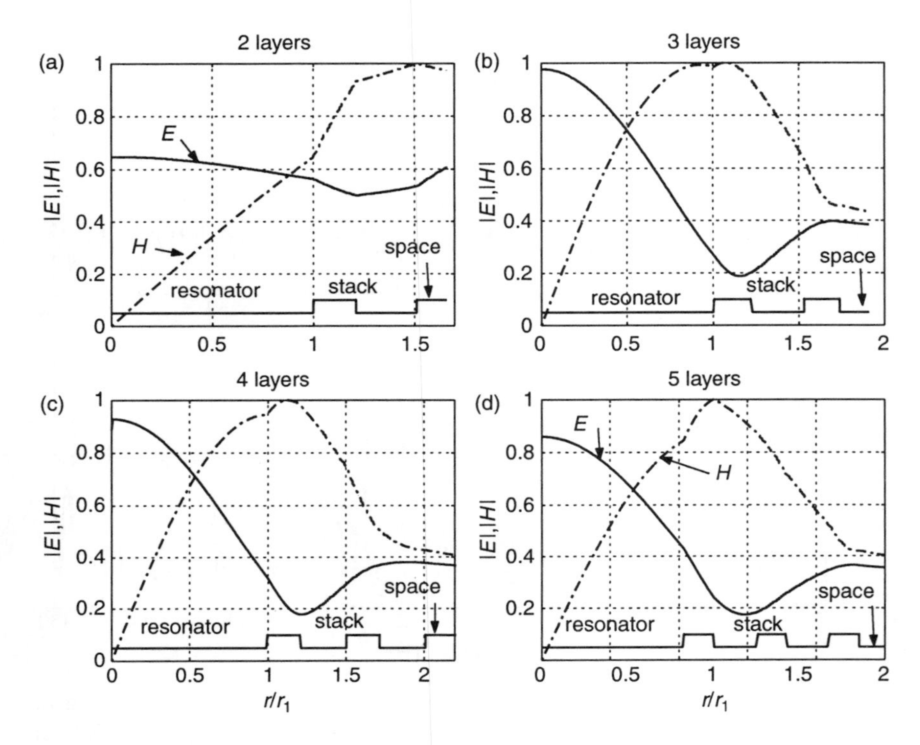

Fig. 3.11 Radial distributions of the normalized amplitudes of electric $|E|$ and magnetic $|H|$ eigen fields for the different numbers of layers in the stack versus the normalized distance for the lowest eigenfrequencies (see details in the text).

thickness of layers determine both location and the steepness of such zones, which therefore are the internal characteristics of the spherical stack. This correlates with a shape of the spatial distribution of the eigen fields. From Figs. 3.11 and 3.8, it is seen that the Q factors of the small-layers system are much lower in comparison with the multi-layered stack. But this difference is large only for those oscillations which are located in the areas of stop bands of the stack.

Figure 3.11 presents the radial electric and magnetic eigen fields distributions for the lowest eigenfrequencies for the stack with a few layers. Each subFig. (a,b,c,d) in Fig. 3.11 shows the fields in the linked zones: in the dielectric core of the spherical resonator, in the volume of the stack, and in the zone of the nearby outer space. The used parameters of materials are the same as those pointed above. The eigenfrequencies and the Q factors have been found as: (a) $38.63 + i\,24.77$, THz, $Q = 0.78$; (b) $73.74 + i\,12.65$ THz, $Q = 2.91$; (c) $67.52 + i\,11.87$ THz, $Q = 2.84$; (d) $65.2 + i\,9.92$ THz, $Q = 2.83$. One can see that the stack sufficiently modifies both eigenfrequencies and the eigen fields of the spherical layered system. We note that such a problem has a large number of parameters, and its more detailed study requires more complex computations.

3.6 Conclusion

In this chapter we have analyzed the eigenfrequencies and the eigen field distribution of the submicron size sphere coated by the multilayered dielectric stack in the optical frequency range. The eigenfrequency equation for optical modes has been derived. The quarter-wave length layers have been studied in details. The spectrum of the eigenfrequencies and their Q factors for a number of the first modes are calculated and discussed. We have analyzed the stability of the spectrum from the point of view of the frequency dependencies both of reflectivity and transmittance of the stack for a different number of layers. We have found that only eigenfrequencies which are in the range of a high reflection (stop band) can retain their stable values. Taking into account the random deviations of thickness of the layers does not change this conclusion. The Q factor exponentially grows with the increase of the number of layers in the stack, but only for eigenfrequencies which are in the zone of the high reflection of the stack. Such behavior is due to a simple physical reason: the considerable reduction of the amplitudes of the spherical waves in that frequency band, where the collective strong wave reflection from all alternative spherical quarter-wave layers takes place. For a great number of layers, the exponential growth of Q tends to saturation due to material loss in layers. The eigenfrequencies which are out of a stop band have low Q values and they are not of a practical interest.

The radial distribution of the electromagnetic amplitudes and the local density of the average energy of the electromagnetic field for some eigenfrequencies are calculated. It is found that the strong localization of energy of the field in the kernel of multilayered resonator for the eigenfrequencies within the stop bands. The amplitudes of oscillations decrease exponentially with a removal from the center of the sphere up to external boundary.

Therefore the influence of nonlinearity is the most essential in the central part of the resonator. We note that the possibility of lowering threshold conditions for nonlinear effects for stimulated Raman scattering in microspheres in WGM regime was pointed out in [Braunstein *et al.*, 1996] and has been observed in [Spillane *et al.*, 2002; Vahala, 2003 and references therein].

CHAPTER 4

Transmittance and Resonance Tunneling of the Optical Fields in the Microspherical Metal-Dielectric Structures

4.1 Introduction

In this chapter we consider the frequency transmittance and the radial distribution of the optical waves in the spherical metallo-dielectric structure deposited on a dielectric microsphere. We are mainly interesting in the range of frequency where the dielectric permittivity $\varepsilon(\omega)$ of a metal is negative.

It is well known [Kittel, 1976] that electromagnetic waves in a range $\varepsilon(\omega) < 0$ cannot propagate in a thick metal film. However for films of thickness less than a wavelength the situation is changed, depending on the incident angle at which the electromagnetic wave can leak or be reflected by such a film. This phenomenon is well known for the case of plane film at a non-normal initial angle [Berreman, 1963]. Scalora *et al.* [1998] predicted the transmission of light over a tunable range of frequencies of metallo-dielectric, one-dimensional, photonic band-gap structures. The transparent metallic structure is composed of a stack of alternating layers of metal and a dielectric material, such that the complex index of refraction alternates between a high and a low value. Scalora *et al.* [1998] note that in such structures the transmission at visible wavelengths can be controlled more effectively, with a metallo-dielectric, periodic structures that has more than two metal layers.

In general the properties of the structure of the thin metal-dielectric layers in a frequency area where the dielectric permittivity of a metal is negative have some unique features. The metal layers where the electric field penetrates at a skin-depth play a role of the barriers sharing the dielectric layers with strong oscillating fields. With the increase of thickness of the metal layers the height of the equivalent barrier is also increased. It means that the electromagnetic oscillations in every dielectric layer become weakly coupled with oscillations in the other dielectric layers of the stack. In such structures a peculiar competition between layers with $\varepsilon(\omega) < 0$ and layers with $\varepsilon(\omega) > 0$ occurs. The layers with $\varepsilon(\omega) < 0$ act as barrier layers and they cause the frequency-band extension of the area of strong reflection, and the layers with $\varepsilon(\omega) > 0$ are responsible for formation of an interference pattern.

We study such a structure from a point of view achieving the tunneling of optical radiation through the spherical metallo-dielectric layers. In general such tunneling appears in a formation of the frequency spectrum of the resonances with almost complete transmittance [Yeh, 1988]. However, this phenomenon has been poorly studied for microstructures similar to spherical multilayers.

It is well known that a dielectric sphere has a complex spectrum of electromagnetic low quality eigenoscillations because of the leakage in the outer space [Stratton, 1941]. In Chapter 3 we have shown a compound structure: the dielectric sphere coated by an alternative stack possesses more the manifold properties. The Q factor (quality) of such oscillations strongly depends on the properties of the stack. Q becomes large in the areas of high reflectivity and beyond these zones Q remains small [Brady *et al.*, 1993; Sullivan, 1994; Sullivan & Hall, 1994; Burlak *et al.*, 2000]. The combination of such factors causes the large variety of optical properties of microspheres with a multilayered stack. In particular, such a system can serve as a spherical symmetric photonic band gap (PBG) structure which possesses the strong selective transmittance properties. For applications of PBG structures for a case of the microspheres see [Zhang *et al.*, 2001; Miyazaki *et al.*, 2000 and references therein].

This chapter is organized as follows. In Sec. 2 we discuss basic equations for a spherical resonator coated by a multilayered metal-dielectric stack. Section 3 presents the results of the numerical study of the spectrum of the eigenfrequencies for a range $\varepsilon(\omega) < 0$, their Q factors and the radial distribution of the electromagnetic amplitudes for such eigenfrequencies.

4.2 Geometry and Basic Equations

Consider a dielectric microsphere and a periodic system of multilayers deposited on it (see Fig. 1.7). We use $\varepsilon(\omega)$ as a complex dielectric permittivity of a layer. Keeping in mind that we use the complex exponential multiplier in the form $\exp(i\,\omega t)$, we assume here a case of a stack with metallo-dielectric alternating layers. In this chapter we consider that the frequency dependence of ε of the dielectric layers can be neglected, but in the metal layers the dispersive properties are essential and the complex dielectric permittivity for the Drude model [Kittel, 1976] takes the form (see Sec. 1.1)

$$\varepsilon(\omega) = 1 - \frac{\omega_p^2}{\omega(\omega - i\,\nu)}, \tag{4.1}$$

where ω_p is a plasma frequency and ν is a frequency of collision. For metals such as aluminum, copper, gold, and silver, the density of free electrons is on the order of $10^{23}\,\mathrm{cm}^{-3}$. This means that $\omega_p \sim 2 \cdot 10^{16}\,\mathrm{s}^{-1}$ [Yeh, 1988, p. 44] and that for visible and infrared radiation with $\omega < \omega_p$ and $\omega \gg \nu$, the value of $\varepsilon(\omega) < 0$ lies according to Eq. (4.1). The value $\varepsilon(\omega)$, in general, is complex when ν is finite.

In this chapter we apply the theory for spherical stack with metal layers having material dispersion (4.1).

On Fig. 4.1(a), the frequency dependence of refractive indexes of layers (n_1-metal, n_2-MgF_2) in the area where the real part of the dielectric permittivity of the metal is negative (and the index refraction of a metal is almost imaginary) are shown. (In the figures we use the frequency $f = \omega/2\pi$ instead of ω.) We apply the approach developed in Sec. 1.4 for such a system. However now we use Eq. (1.79), and we take into account the fact that $\varepsilon = \varepsilon(\omega)$ has a frequency dispersion. In Fig. 4.1(b) the frequency dependence of the transmittance coefficient of the spherical stack is shown. One can see in Fig. 4.1(b) the series of the equidistant resonances with a high transmittance.

The zone of resonances has an oscillatory interference structure. In Fig. 4.1(b) the narrow resonance of transparency is seen at $f = f_p = \omega_p/2\pi$ in area, where $\varepsilon(f) = 0$. Such a resonance was also mentioned in a case of

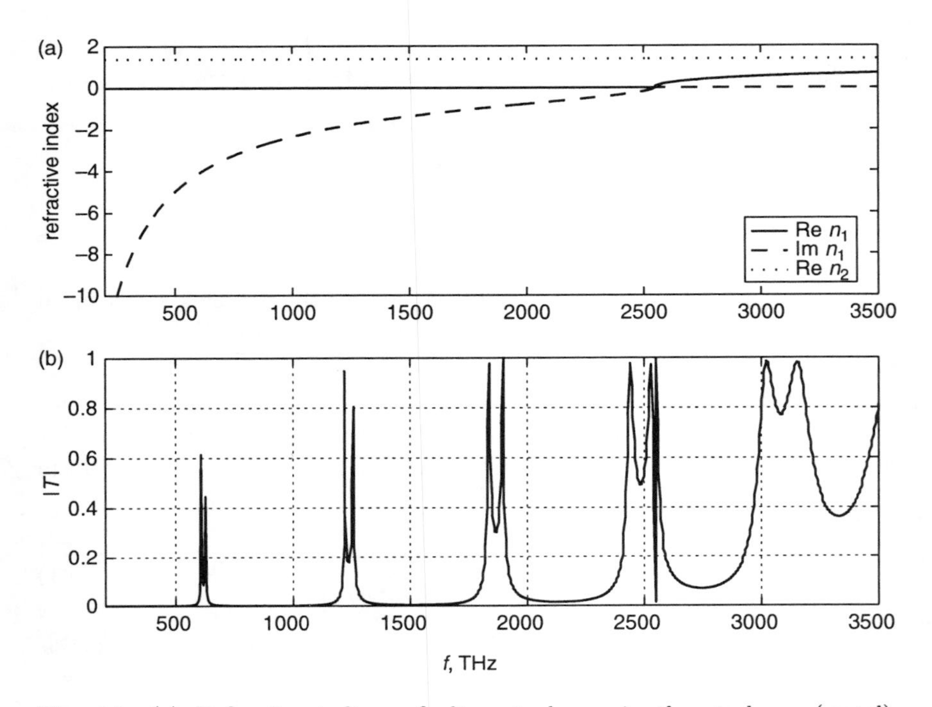

Fig. 4.1 (a) Refractive indices of alternate layers in the stack n_1 (metal), n_2 (MgF_2); (b) modulo of the transmittance coefficient of 5-layer metallo-dielectric stack. Thickness of the metal layers and MgF_2 layers is $0.05\,\mu$m and $0.14\,\mu$m respectively.

a metallo-dielectric plane stack [Scalora *et al.*, 1998]. This peak is not due to any periodicity in the system and it corresponds to a well-known peak of a transmission of a thin film at an abnormal initial angle [Berreman, 1963], see also [Kittel, 1976, p. 307]. In this region the longitudinal wave of polarization or polariton is excited. For more information about surface polaritons see [Ruppin, 2000; Bozhevolnyi *et al.*, 2001 and references therein].

Our calculations show that the frequency shift between the centers of resonance approximately equals to $\sigma \simeq \mu c/2n_2d_2$, where n_2 and d_2 are the refractive index and the thickness of the dielectric layers in the stack accordingly, a multiplier $\mu \simeq 0.77$. Really, substituting $n_2 = 1.37$ and $d_2 = 0.14\,\mu$m one can obtain $\sigma \simeq 600\,$THz, what approximately corresponds to the first peak of $|T|$ in Fig. 4.1(b). Our calculation shows that this value is practically independent of the thickness of the metal layers (see

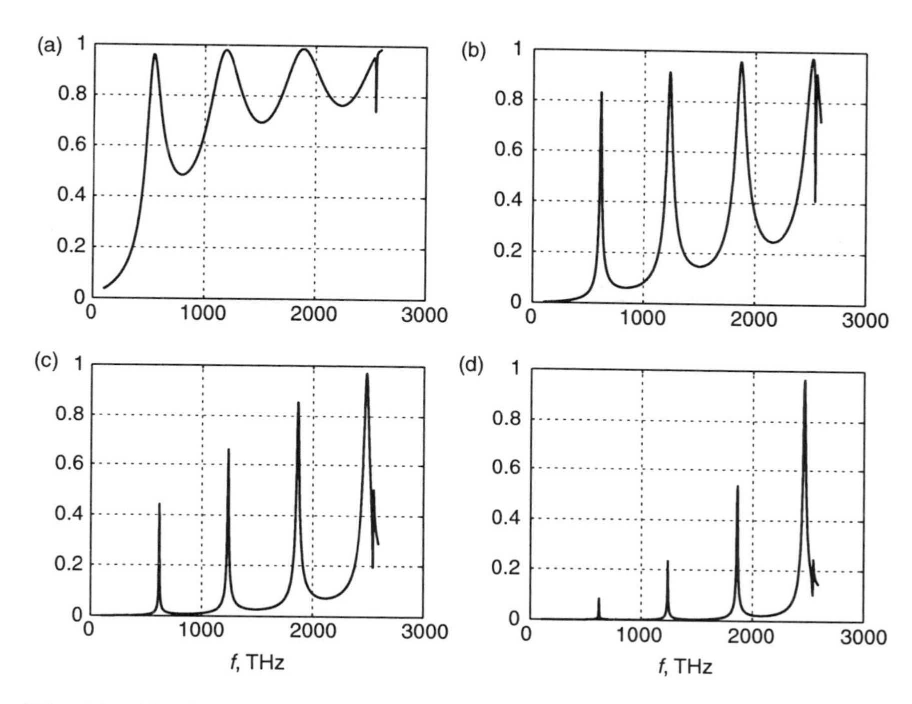

Fig. 4.2 The frequency transmittance of the metalo-dielectric stack at different values of thickness of the metal layers: (a) 0.01 μm; (b) 0.05 μm; (c) 0.07 μm; (d) 0.1 μm. Designed 5 layers in the stack.

Fig. 4.2). The thickness of the metal layer defines only the width of the peak of transmittance. Further we derive the expression for σ from a simple estimation.

The eigenfrequencies equation for a case of spherical stack was derived in [Burlak *et al.*, 2001b; Kalitievski *et al.*, 2001]. Generalizing to a stack with dispersive layers of positive or negative dielectric permittivity was made in [Burlak *et al.*, 2001a]. Here we follow such an approach for a case of a metallo-dielectric stack. We suppose that for the absence of singularity in the center of internal substrate $r \to 0$ one can see that such condition is satisfied if the amplitudes of forward and backward waves in the internal substrate are the same: $a_0 = b_0 \neq 0$ [Burlak *et al.*, 2001b]. (In this case the superposition of Hankel functions is described by the spherical Bessel function [Korn & Korn, 1961].) Then in (1.181) the reflection coefficient becomes $R = 1$. Substituting $R = 1$ in the first equation in (1.185), we obtain the eigenfrequencies equation in the closed form

$$\Delta(\omega, n(\omega)) = Q_{11}(\omega) - Q_{21}(\omega) = 0, \tag{4.2}$$

where Q_{ij} is written in (1.180).

This approach is in accordance with the approach developed in Sec. 3.3 [see also Burlak *et al.*, 2001b]. However it allows us to write down the eigenfrequencies equation in closed form. Such a form of eigenfrequency equation is slightly modified with respect to the approach described in Chapter 3. Now we apply the above-developed theory to calculate the reflectance, transmittance, eigenfrequencies and eigen fields for a spherical metallo-dielectric stack.

4.3 Results and Discussions

The following parameters have been used in calculations: the geometry of a system is $SLNLNLN\ldots NV$ where letters S, L, N, V indicate the materials in the system with the radius of the internal substrate at $r_1 = 1\,\mu\text{m}$. We have used the parameters of materials [Hodgson & Weber, 1997] S(spherical substrate): glass, $n_S = 1.5$, L: metal layer, $\omega_P = 1.6 \cdot 10^{16}\,\text{s}^{-1}$ [Yeh, 1988, p. 44] ($f_p = 2.55 \cdot 10^3\,\text{THz}$). For calculations we used $\nu = 1.6 \cdot 10^{13}\,\text{s}^{-1}$ and thickness of the metal layers was varied in the range of 0.01 to 0.2 μm, N: MgF_2, the thickness 0.14 μm, $n_N = 1.37$ [Kumar *et al.*, 1999], V (outer medium): *air*, $n_V = 1$. The number of spherical harmonic is $m = 1$. In Fig. 4.1(a) the refractive index for both layers: metal and MgF_2 are shown.

Further we use frequency $f = \omega/2\pi$ instead of ω. For frequencies $f < f_p$ in metal the permittivity $\varepsilon(f) < 0$, and $n = n(f) = \varepsilon(f)^{0.5} = n_r + i\,n_i$ is nearly purely imaginary $n = i\,n_i$ (see Fig. 4.1(a)). This changes the behavior of the complex spherical Hankel functions in (1.165) completely. It also completely changes the frequency dependence in both reflective and transmittance coefficients. Figure 4.1(b) shows T for the frequency range where $\varepsilon(f) < 0$. As we have previously mentioned, in this frequency range there is a peculiar competition of layers in the stack. The layers with $\varepsilon(f) < 0$ cause the frequency-band extension of the area of strong reflection, and the layers with $\varepsilon(f) > 0$ are responsible for the formation of an interference pattern. Such competition results in the formation of peaks of high transparency at what one can treat as the effective tunneling of optical waves in the area of strong reflectivity (see Fig. 4.1(b) below of 2600 THz). Beyond such resonances the reflection appears to a near total of $R \approx 1$ and $T = 0$.

Figure 4.2 shows how the width of the resonant passbands varies with a change of thickness of the metal layers in the stack while the thickness of the dielectric MgF_2 is a constant. In this case we consider the 3-layered stack with a dielectric layer in the center. One can see that when the thickness of the metal layers is $d_1 = 0.1\,\mu m$ the first resonance practically disappears. This means that the transmittance of the stack is insignificant when the barrier metal layers are thick enough. The positions of such resonance do not depend on the thickness of the metal layers.

Figure 4.3(a) shows the details of the transmittance of a 5-layered metallo-dielectric stack. Two peaks appear at $f = 1188\,\text{THz}$ and at

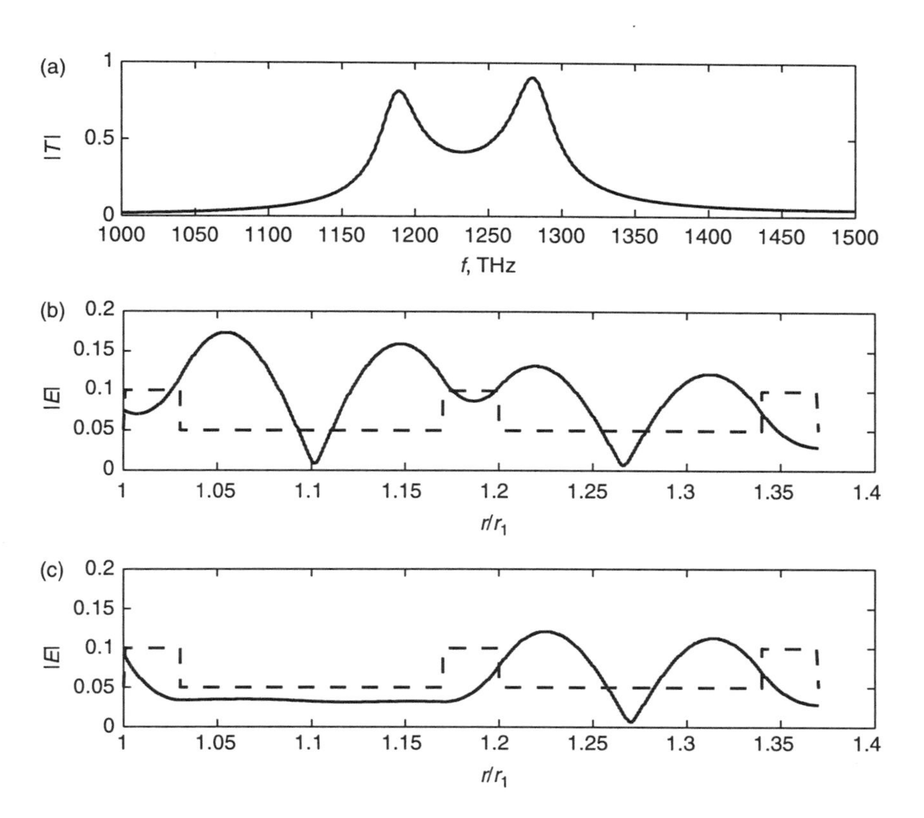

Fig. 4.3 (a) The frequency transmittance of 5-layer metallo-dielectric stack; (b) and (c) are the radial distribution of the optical electric field in the stack for frequency $f = 1188\,\text{THz}$ (b), and $f = 1231\,\text{THz}$ (c). The thickness of the metal layers is $0.03\,\mu m$. The dashed line shows the position of layers in the stack and r_1 is the radius of the internal spherical substrate.

$f = 1280\,\text{THz}$. Figure 4.3(b,c) shows the radial distribution of the electric field at frequencies $f = 1188\,\text{THz}$ and $f = 1231\,\text{THz}$. The field distributions at those frequencies are different. In Fig. 4.3(c) the field is located at the peripherical layers, and in Fig. 4.3(b) the field is distributed approximately similarly in both dielectric layers. Such a field structure can be treated again as a dynamical tunneling of the optical fields from the internal substrate to the external layers. Such a picture can arise as radiation from an atom (dipole) located in the center of coated microsphere. The amplitude of the radiation increases when the frequency of the atomic radiation is close to one of the eigenfrequencies of given compound the system.

Figure 4.4 shows the radial distribution of the optical electric field in a 3-layered stack for various thickness of the metal layers at frequency

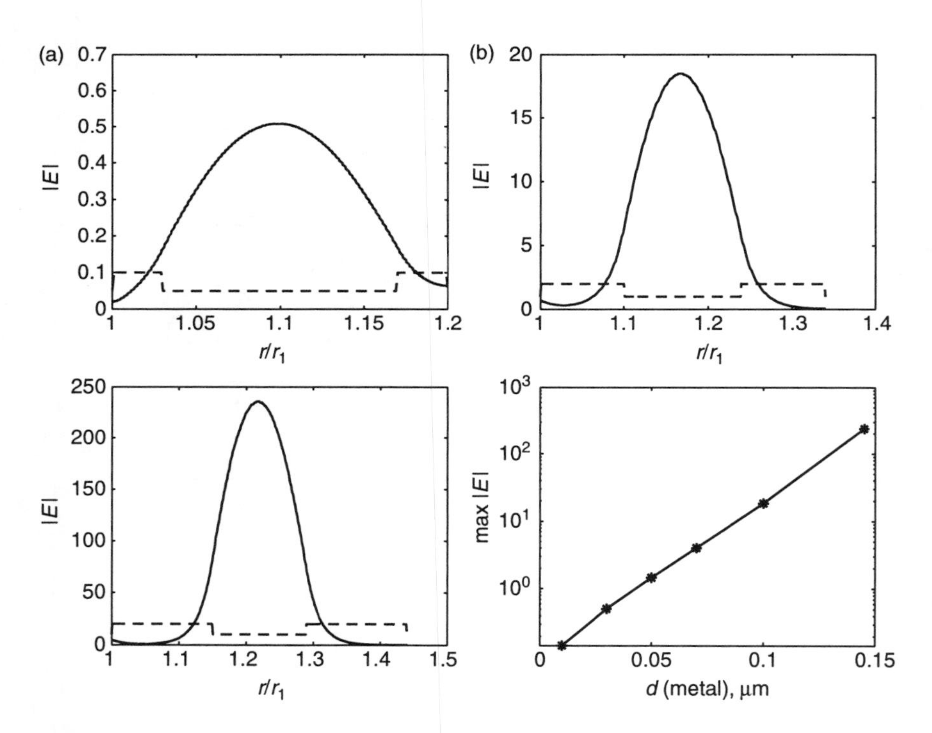

Fig. 4.4 Radial distribution of the optical electric field at frequency $f = 618\,\text{THz}$ in a 3-layered stack at various thickness d of the metal layers: (a) $d = 0.03\,\mu\text{m}$; (b) $d = 0.1\,\mu\text{m}$; (c) $d = 0.15\,\mu\text{m}$; (d) dependence of the maximum field in the stack versus the thickness of the metal layer. The dashed line shows the position of layers in the stack and r_1 is a radius of the internal spherical substrate.

$f = 618\,\text{THz}$ that corresponds to the first maximum of the transmission (see Fig. 4.1(b)). From Fig. 4.4 one can see that the amplitudes of field grow very rapidly (exponentially) while the thickness of the metal layers grows slowly. In Fig. 4.4(d) the maximums of the fields versus the thicknesses of the metal layers in the stack are shown.

We study the spectrum of eigenfrequencies and the details of the spatial distribution of the eigen field. Such problem is more difficult, as it first requires solving a complex eigenfrequency problem (the eigenfrequencies equation) and only then one can calculate the correct radial field distribution for the calculated eigenfrequency. We now discuss the eigenfrequencies, and also the eigenfield radial distribution for an interval of $f < f_p$.

From a numerical solution of Eq. (4.2) $\Delta(f, n(f)) = 0$ we find a number of eigenfrequencies in such a frequency range. In Table 4.1 the eigenfrequencies $\text{Re}(f)$ and their Q factors for 6-layered metal-dielectric stacks on the spherical substrate are written down. One can see that the typical values Q are in a range $(1 \text{ to } 3) \cdot 10^3$ and only for some frequencies Q decreases on the order of magnitude up to 100 to 150. From Fig. 4.1(b) one can see that the appropriate eigenfrequencies $f_6 = 592\,\text{THz}$ and $f_{11} = 1268\,\text{THz}$ belong to the resonant zone of the high transmittance of the stack.

In Fig. 4.5 a 5-layered stack is shown: three metal layers of thickness $0.03\,\mu\text{m}$ are shared by MgF_2 layers with thickness $0.14\,\mu\text{m}$. In this a case the frequency band of a transmission has 2 maximums (see Fig. 4.3(a)).

In Fig. 4.5(a) the radial distributions of the eigenfields in the compound internal substrate and spherical stack are shown. Figure 4.5(b) provides more details of such a distribution in the stack.

Table 4.1 Eigenfrequencies and Q factors for the metallo-dielectric stack deposited on the microsphere (see details in the text).

No.	$\text{Re}(f)$, THz	Q-factor, 10^3	No.	$\text{Re}(f)$, THz	Q-factor, 10^3
1	85	1.22	7	687	2.01
2	191	2.74	8	784	1.89
3	291	1.96	9	882	1.78
4	389	1.68	10	980	1.67
5	447	1.43	11	1268	0.159
6	592	0.13	12	1378	2.15

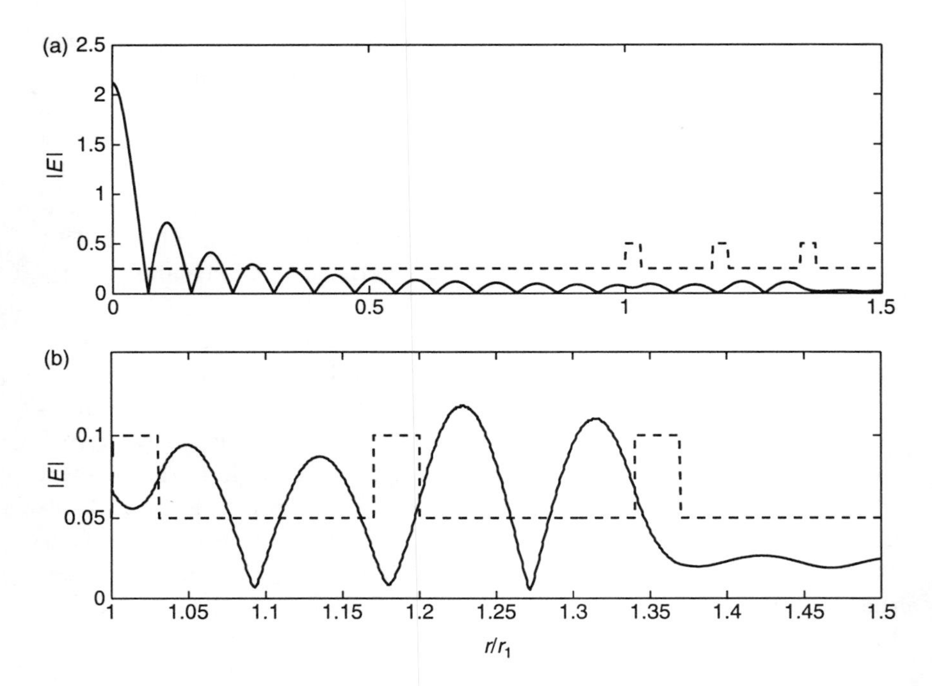

Fig. 4.5 (a) Radial distribution of the eigen electric optical field in the compound spherical structure: internal substrate + stack; (b) the details of the field distribution in the stack. Stack with 5 layers. The dashed line shows the position of layers in the stack and r_1 is the radius of the internal spherical substrate.

Further we explore the distribution of eigen field at frequency $f = f_6 = 592\,\text{THz}$, that corresponds to $\lambda = 0.51\,\mu\text{m}$ and $Q = 130$. In this a case there is an excitation of a rather strong field in the dielectric layers of stack. While the thickness of the metal layer is about $0.1\,\mu\text{m}$ the field penetrates poorly into metal and the electromagnetic field is almost completely located in the dielectric layers (see Fig. 4.4).

For a simple estimation of the position of resonances of transmission we consider, that on boundaries of dielectric layers field $E_\theta \to 0$ because of contact with metal. Then neglecting the spherical geometry we have a standard spectral problem, which gives the series of eigenfrequencies according to a half-wave layers $\lambda/2 = ld_2n_2/2c, l = 1, 2, \ldots$. From here one can readily obtain the above mentioned estimate of the frequency shift between resonances $\mu c/2n_2d_2$, where we recall that n_2 and d_2 are the refractive index and the thickness of the dielectric layers in the stack respectively, and $\mu \simeq 0.77$. Such relation provides a condition for the resonant tunneling in the stack. We believe that the multiplier $\mu = 0.77$ takes into account the finite depth of penetration of the field into the metal and it has the order of the magnitude: one minus the ratio of skin-depth to the thickness of the metal layer. The finite depth of penetration of the optical field in the metal layers causes the interferential interaction of the oscillations between shared dielectric layers (potential walls). Therefore the interference picture inside the resonances arises as it shown at Fig. 4.6(a) for 25-layer stack. But such oscillations already do not interfere when the thickness of the metal layers increases. This results in the narrowing of the band of a resonances as it was observed on our numerical experiments (see Fig. 4.2).

While a large number of layers in the stack the interference pattern inside the resonances and the radial field distribution become more complex

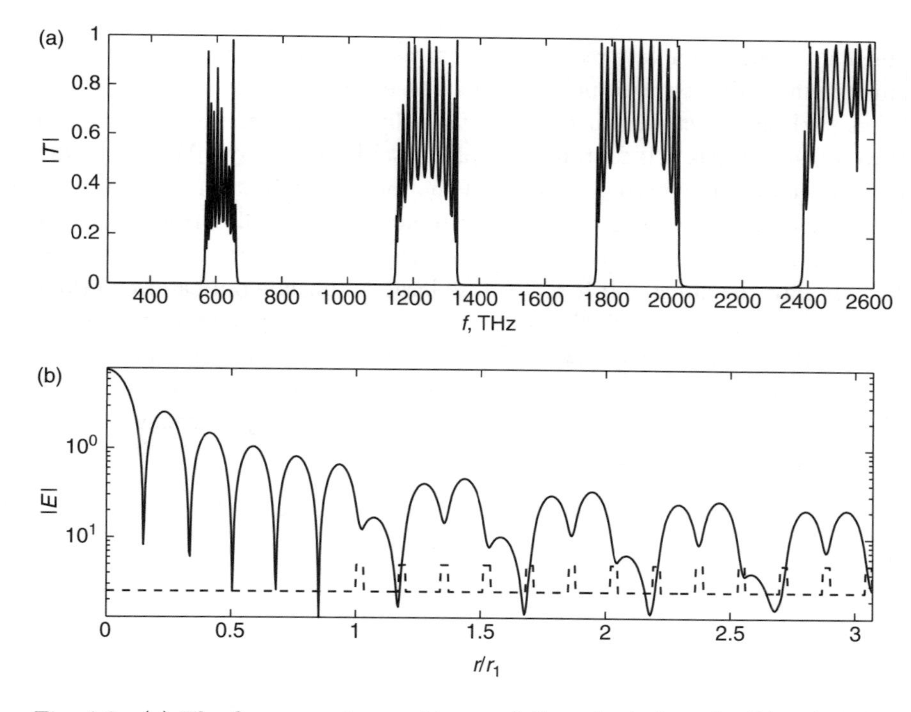

Fig. 4.6 (a) The frequency transmittance of the spherical stack; (b) radial distribution of the eigen optical electric field for eigenfrequency $f = 586\,\mathrm{THz}$. Stack with 25 layers. The dashed line shows the position of layers in the stack and r_1 is the radius of the internal spherical substrate.

(see Fig. 4.6). However it does not change the main features of transmittance, field distribution, and the tunneling of the optical waves through the microspherical structures.

4.4 Conclusion

In this chapter we have studied the optical oscillations in the periodic metallo-dielectric spherical stack deposited on a dielectric microsphere when the dielectric permittivity of the metal layers in the stack is essentially dispersive and is negative. We have found a spectrum of equidistant narrow peaks of resonances of the optical transmittance in such a frequency range. Both thickness and dielectric permittivity of dielectric layers in the stack define the position of the peaks. The width of the resonances depends on the thickness of the metal layers. Therefore the desired width of the resonance line can be achieved by varying the thickness of the metal layers. In lossless media a nearly complete resonant transmission occurs, but the value of the maximum of the resonant peaks will decrease in a lossy material. We believe that such an effect with a resonant tunneling of optical waves through a layered structures will contain multiple metal barrier layers. We calculated the spectrum of eigenfrequencies and the radial distribution of eigen optical fields for a structure with various numbers of metallo-dielectric layers and we found that in the zone of resonance the electric field penetrates deeply into the stack and it may cause large amplitude optical oscillations in the dielectric layers. Such an effect provides the opportunity to use the microspheres coated by metallo-dielectric multilayers for an effective surface harmonic generation in the optical range [Hayata & Koshiba, 1992; Centini *et al.*, 1999] and for a high-resolution frequency conversion in the bands of such resonances.

CHAPTER 5

Confinement of Electromagnetic Oscillations in a Dielectric Microsphere Coated by the Frequency Dispersive Multilayers

5.1 Introduction

In this chapter we consider the confinement of electromagnetic oscillations inside the multilayered spherical stack deposited on a dielectric microsphere. We assume that the layers in the stack have the essential frequency dispersion, typical for ionic crystals. It is known that such materials have two characteristic frequencies: ω_T is a frequency of transverse optical phonons and ω_L is a longitudinal optical frequency [Kittel, 1976]. The dielectric permittivity $\varepsilon(\omega)$ is negative, between its pole at $\omega = \omega_T$ and its zero at $\omega = \omega_L$, therefore a frequency gap between ω_T and ω_L arises.

Is well-known [Kittel, 1976] that electromagnetic waves with frequencies within the gap $\omega_T < \omega < \omega_L$ cannot propagate in a thick crystal. But for films of thickness less than a wavelength the situation is changed.

Depending on the incident angle the electromagnetic wave can leak or be reflected by such a film. This phenomenon is well known for a case of plane film at a non-nornal initial angle [Berreman, 1963]. In this chapter we study this phenomena for spherical microstructures. In such structures again a peculiar competition between layers with $\varepsilon(\omega) < 0$ and layers with $\varepsilon(\omega) > 0$ occurs. The layers with $\varepsilon(\omega) < 0$ cause frequency-band extension in the area of strong reflection, and the layers with $\varepsilon(\omega) > 0$ are responsible for the formation of an interference pattern.

In a case of $\omega \approx \omega_T$ the dielectric permittivity of a layer has a large negative value. In such conditions the field can only weakly penetrate into the stack. Accordingly even in a case of low order spherical harmonics the eigen oscillations of a field must have a large Q factor in this range, which usually takes place only for a dielectric microspheres with many multilayers [Brady *et al.*, 1993; Sullivan, 1994; Sullivan & Hall, 1994; Burlak *et al.*, 2000] or in a whistling gallery mode regime [Vassiliev *et al.*, 1998].

5.2 Basic Equations

Consider a dielectric microsphere and a periodic system of multilayers deposited on it (see Fig. 1.7). In every layer the set of Maxwell's equation satisfied with $\varepsilon(\omega)$ is a complex dielectric permittivity of a layer. We use complex exponential multiplier in the form of $\exp(i\,\omega t)$. We assume here a simple case of a stack with two types of alternative layers. In the first layer the dispersive properties are essential. We assume such a material is a type of ionic crystal [Kittel, 1976]:

$$\varepsilon(\omega) = \varepsilon_\infty \frac{\omega_L^2 - \omega^2}{\omega_T^2 - \omega^2}, \tag{5.1}$$

where ε_∞ is the optical dielectric constant.

As already was shown, Maxwell equations in the spherical coordinates (ρ, θ, φ) reduces to the Helmholz equation for Debye potential $\Pi(\rho, \theta, \varphi)$. We should join the solution for fields H_φ, E_θ on boundaries of neighboring layers using standard conditions of continuity of fields.

From (1.181) one easily obtains R and T in form of

$$R = \frac{Q_{21}(\omega)}{Q_{11}(\omega)}, \qquad T = \frac{1}{\sigma Q_{11}(\omega)}. \tag{5.2}$$

In (1.185) two relations relate three variables: R, T and frequency ω. Defining ω one can calculate the frequency dependence of $R(\omega), T(\omega)$ for the stack. In Fig. 5.1 the frequency dependencies of reflectivity and transmittance are shown: (a) for a dispersive LiH/SiO_2 layers stack, and

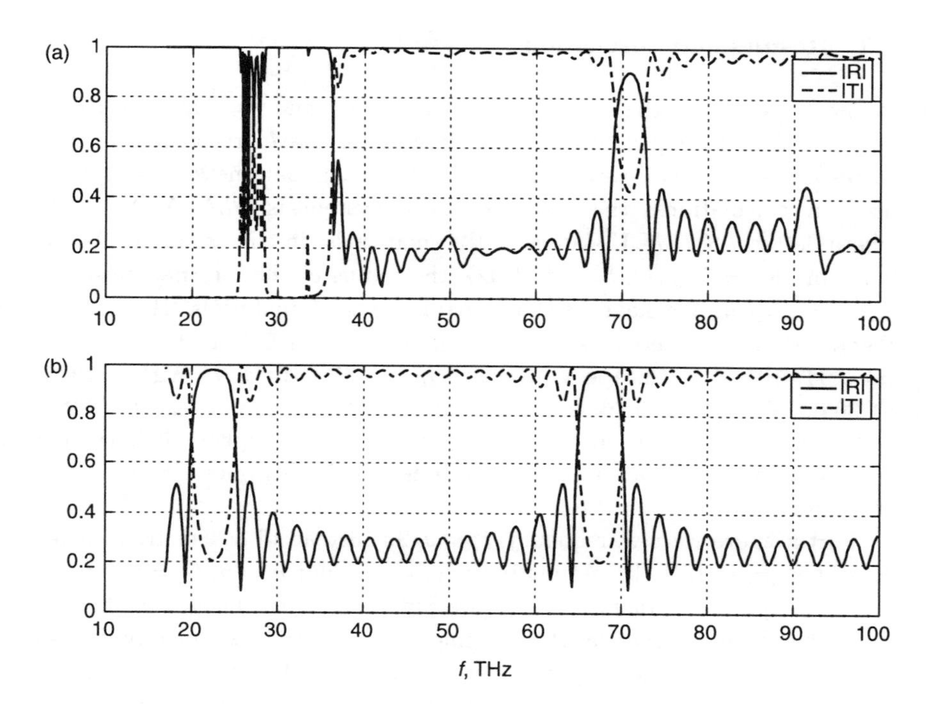

Fig. 5.1 Module of a reflection coefficient R and a transmittance coefficient T of spherical stack versus frequency (THz). (a) Dispersive case; (b) Dispersiveless case. Designed 15 LiH/SiO_2 half-wave layers.

(b) for material with the same parameters, but for the dispersionless case ($\omega_T = \omega_L$). (In the figures we use the frequency $f = \omega/2\pi$ instead of ω). One can see that in a high-frequency region, when $n(\omega) = n_\infty$, the plots are almost similar. However in the range of low frequencies when $\omega_T < \omega < \omega_L$, the behavior both R and T in Fig. 5.1(a) differs substantially from the dispersionless case of Fig. 5.1(b).

The eigenfrequency equation one can derive by similar way as in Chapter 4. The general form of such equation is given by

$$\Delta(\omega, n(\omega)) = Q_{11}(\omega) - Q_{21}(\omega) = 0, \tag{5.3}$$

where $n = \varepsilon(\omega)^{1/2}$ and $\varepsilon(\omega)$ is written in (5.1). We apply the above theory to calculate the reflectance, transmittance, eigenfrequencies and eigen fields for a spherical dispersive stack with a dielectric permittivity (5.1).

5.3 Results and Discussions

Below, we consider the case $d_k n_{\infty k} = \Lambda_0/2$ (d_k and $n_{\infty k}$ are the thickness and a high-frequency refractive index of a k-layer respectively) that corresponds to a half-wave stack. The following parameters have been used in calculations: the geometry of a system is $SLNLNLN\dots NV$ where letters S, L, N, V indicate the materials in the system, $\Lambda_0 = 6.67\,\mu\text{m}$ ($K_0 = 2\pi f_0/c, f_0 = 45\,\text{THz}$), the radius of the internal substrate is $r_1 = 15\,\mu\text{m}$. We have used the parameters of materials [Hodgson & Weber, 1997] S(spherical substrate): glass, $n_S = 1.5, L:\text{LiH}, n_\infty = 3.6$, $\omega_T = 11 \cdot 10^{13}\,\text{s}^{-1}$, thickness $1.75\,\mu\text{m}, \omega_L = 21 \cdot 10^{13}\,\text{s}^{-1}$ [Kittel, 1976, p. 309], $N:\text{SiO}_2$, thickness $2.28\,\mu\text{m}, n_N = 1.45$ [Kumar *et al.*, 1999], V (outer medium): air, $n_V = 1$. The number of the spherical harmonic is $m = 1$. In Fig. 5.2(a) the refractive index for both layers LiH and SiO_2 are shown.

Furthermore we use frequency $f = \omega/2\pi$ instead of ω. For frequencies $f_T < f < f_L$ in LiH the permittivity $\varepsilon(f) < 0$, and $n = n(f) = \varepsilon(f)^{0.5} = n_r + i\,n_i$ is pure imaginary $n = i\,n_i$ (see Fig. 3(a)). This changes the behavior of the complex spherical Hankel functions in (1.165) completely. It also completely changes the frequency dependence in both reflective and transmittance coefficients. Figure 5.2(b) shows R and T for the low frequency part of Fig. 5.1(a) with details. As we have already noted, in this frequency range there is the competition for layers in the stack. The layers with $\varepsilon(f) < 0$ cause the frequency-band extension of area of strong reflection, and the layers with $\varepsilon(f) > 0$ are responsible for the formation of an interference pattern. Such competition results in the formation of a zone with oscillation in the area of strong reflectivity (see Fig. 5.2(b) in the vicinity of 27 THz). Our calculations determinate that the number of minima in such a band is $2N_0/5$, where N_0 is a total number of layers in the stack. In the vicinity of $f \geq f_T$ the value n_i becomes large as $n_i \gg 1$. In this case the LiH layers are practically equivalent to metal layers. The reflection in this range appears nearly total $R \approx 1$.

In the vicinity of $f \leq f_L$ we have $n_i \ll 1$. This causes a decrease of reflection in a narrow bandwidth at $f \approx f_L$ and an increase of transmittance respectively. Similar behavior of R was already observed in thin films [Berreman, 1963, and Fig. 16] in [Kittel, 1976, p. 307]. In this area the longitudinal wave of polarization or polariton is excited. About surface polaritons see [Ruppin, 2000 and references therein].

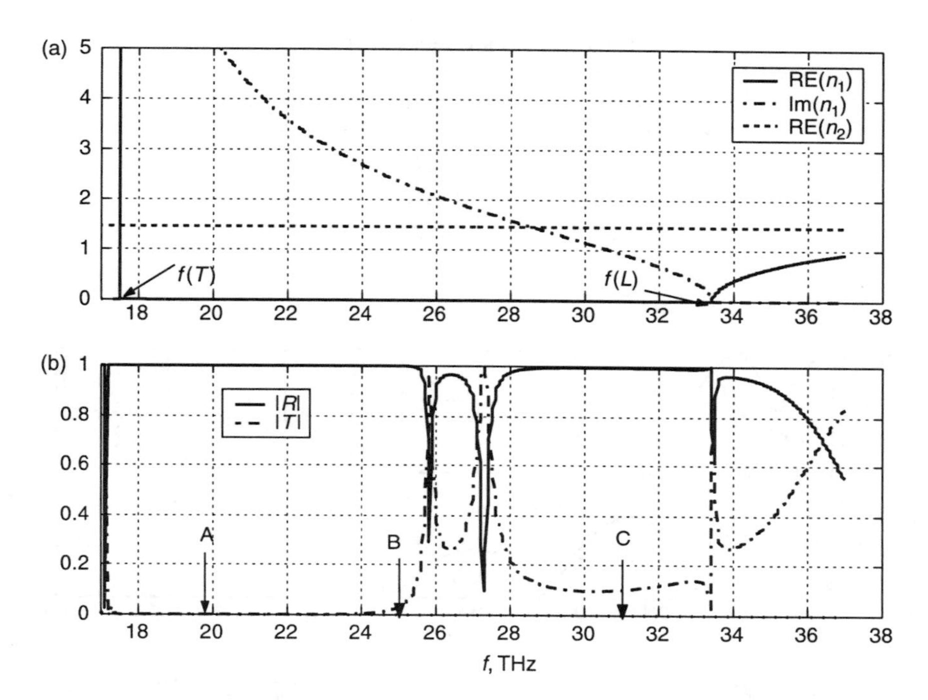

Fig. 5.2 (a) Refractive indices of alternative layers $n1$ (LiH), $n2$ (SiO_2) in stack; (b) Module of a reflection coefficient R and a transmittance coefficient T of stack versus frequency (THz). Designed 5 layers. Eigenfrequencies are marked as A, B, C. See text and Table 5.1 for details.

Table 5.1 Dispersive stack.

l	Re(*f*), THz	Im(*f*), THz	Q-factor
A	19.2	$1.3 \cdot 10^{-14}$	$7.38 \cdot 10^{14}$
B	25.1	$6.8 \cdot 10^{-4}$	$1.80 \cdot 10^{4}$
C	31.2	$4.0 \cdot 10^{-3}$	$3.27 \cdot 10^{3}$

Furthermore we discuss the eigenfrequencies and also the radial distribution of fields for eigenfrequencies inside interval $f_T < f < f_L$. The numerical solution of (5.3) $\Delta(f, n(f)) = 0$ brings three eigenfrequencies in this range (see Tables 5.1 and 5.2).

Table 5.2 Dispersiveless stack.

l	Re(f), THz	Im(f), THz	Q-factor
A	20.54	0.278	36.9
B	25.42	0.318	40
C	32.68	0.729	22.4

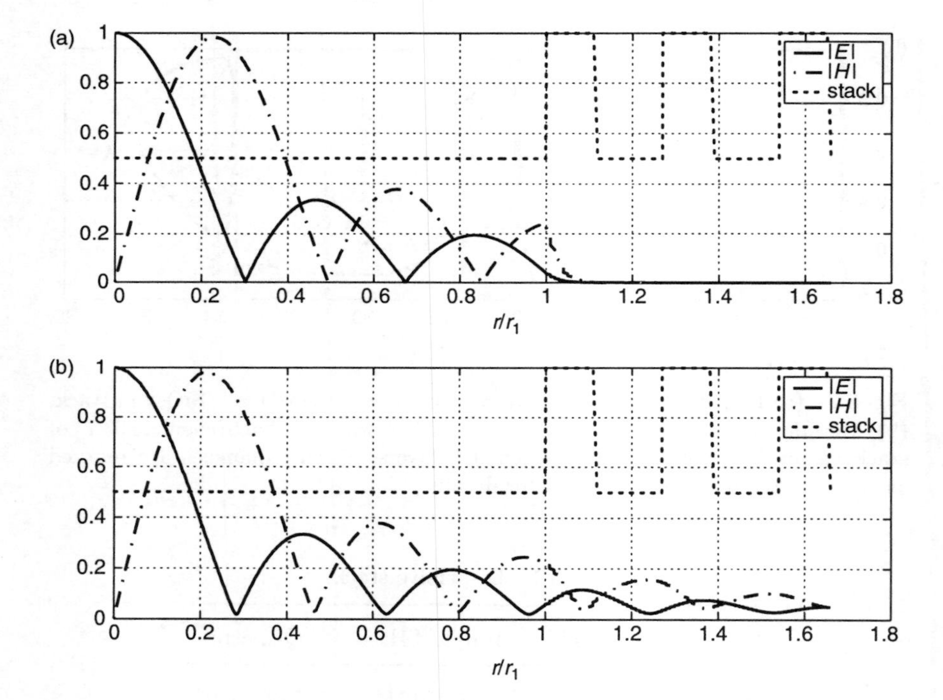

Fig. 5.3 Radial distribution of the amplitudes of electric and magnetic fields for eigenfrequency 19.2 THz. (a) Dispersive case; (b) Dispersiveless case. r_1 is the radius of the internal substrate.

In Table 5.1 the eigenfrequencies of a spherical multilayer resonator situated in a range between f_T and f_L for the dispersive stack are given. Table 5.2 lists the eigenfrequencies for the same stack, but for the dispersiveless case. These eigenfrequencies are enumerated as A, B, C and they are shown in Fig. 5.2(b). The identical initial guesses were used for the numerical calculation of the complex roots of Eq. (5.3) in both cases. Calculation accuracy was $5 \cdot 10^{-16}$.

Comparing both tables one can see that the material dispersion results in a minor shift in the values of real parts of frequencies. However values of Q factors differ about several orders of a magnitude and more. This difference is greatest for $f_A = 19.2\,\text{THz}$, nearest to f_T. Here Q factors differ on 12 orders of a magnitude. Such differences considerably decrease for f_C, nearly for f_L with $\varepsilon(f_L) \to 0$. The Q factors are very large though the spherical stack has few layers. We note that in other case [Brady *et al.*, 1993] one required much more layers to obtain such large Q factor.

We calculated the radial distribution of the amplitudes of electromagnetic field in such a structure. In Figs. 5.3 and 5.4 the normalized field distributions for eigenfrequencies from Tables 5.1 and 5.2 are shown. Figure 5.3 shows the radial dependence of electric E_θ and the magnetic H_φ fields for the eigenfrequency $f_A = 19.2\,\text{THz}$. One can see from Fig. 5.2(b) that this

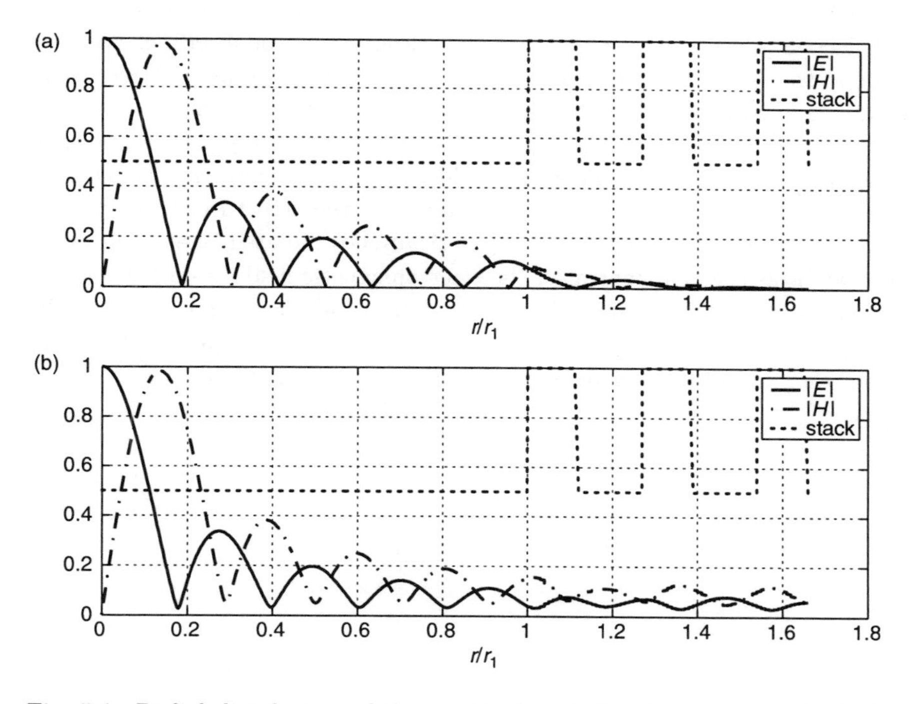

Fig. 5.4 Radial distribution of the amplitudes of electric and magnetic fields for eigenfrequency 31.2 THz. (a) dispersive case; (b) dispersiveless case. r_1 is the radius of the internal substrate.

value f_A belongs to the range where $n_i = \text{Im}(n(f)) \gg 1$. We found (see Fig. 5.3(a)) that in this case the fields exponentially decrease in the first layer and almost do not penetrate into the next layer in the stack.

To highlight the role of material dispersion for this effect in Fig. 5.3(b) we show the same as in Fig. 5.3(a), but only for a dispersiveless stack. In this a case the fields penetrate into the stack much more deeply. Figure 5.4 is similar to Fig. 5.3, but only that it is for eigenfrequency $f_C = 31.2\,\text{THz}$. For this frequency $n_i = \text{Im}(n(f))$ is much less and therefore the fields penetrate into the stack more deeply before being radiated to the external medium.

5.4 Conclusion

In this chapter we studied the electromagnetic oscillations in a multilayered spherical stack deposited on a dielectric microsphere when the dielectric permittivity of some layers in the stack have a substantially material dispersion. Special attention was given to the frequency range where the dielectric permittivity of such layers is close to zero or negative. We have found that such microspheres with dispersion in layers support electromagnetic oscillations with very high Q factor already for a small number of layers. We have calculated eigenfrequencies in this frequency range, and also the radial distribution of eigenfields. Our results show that these fields are localized in a vicinity of the substrate of the structure, weakly penetrate into the stack and, thus, are practically not radiated. We refer to such behavior as the confinement of the electromagnetic field inside a dispersive multilayered microsphere. We note that in a lossy case the values of the Q factors become less, but it does not change the general conclusion of the confinement of electromagnetic field in such a structure.

CHAPTER 6

Oscillations in Microspheres with an Active Kernel

Recently, the fabrication and optical properties of different kinds of microcavities containing semiconductor nanoclusters or quantum dots (QDs) was achieved. When semiconductor QDs are embedded in the spherical microcavity, the QD luminescence can couple with the eigen electromagnetic modes of microcavity and a lower threshold of stimulated emission (or lasing modes) of QDs may be realized. In recent works the coupling between the optical emission of embedded $CdSe_xS_{1-x}$, QDs and spherical cavity modes was investigated. The strong whispering gallery mode (WGM) resonances with high Q factors is registered in the photoluminescence spectra. Pendleton and Hill [1997] have studied the light scattering from an optically active sphere into a circular aperture.

In this chapter we consider the optical radiation of compound system: active microsphere with the spherical stack deposited on it. Our main assumption is that the internal spherical substrate has gain in band of frequencies of the exciton resonance.

First we discuss the basic formulas and we derive the eigenfrequencies equation for active microsphere coated by the multilayered dielectric stack. We discuss a numerical study of the spectrum of the eigenfrequencies for a range nearly of a band gain. Also we discuss the Q factors and the radial distribution of the electromagnetic amplitudes for such eigenfrequencies. By means of the matrix method we have derived the eigenfrequencies equation and analyzed some representative cases. We discuss the dependence of the Q factor of the eigenfrequencies being in the resonance zone versus the number of layers in the stack. We discuss the conditions when Q may become negative when the number of layers in the stack is large enough. This may occur due to the compensation of the radiate losses by the gain of the optical field in the active substrate.

6.1 Basic Equations

We consider a dielectric microsphere and a periodic system of multilayers deposited on it (see Fig. 1.7) and we assume here a microsphere with active core coated by a stack of dielectric alternating layers. In substrate and layers one has to use the Maxwell's equation with a complex dielectric permittivity $\varepsilon = \varepsilon(\omega)$. We use the complex exponential multiplier in form of $\exp(i\,\omega t)$. We assume that a glass substrate is implemented as the semiconductor with a system of quantum dots, similar to the experiment by Artemyev & Woggon [2000] and Jia *et al.* [2001].

We use the simplest phenomenological model of a QD considering exciton as a Lorentzian contribution to the dielectric constant ε_h of the QD material [Maksimenko *et al.*, 2000] and restricting ourselves to the single-resonance approximation. Dielectric permittivity of such a compound substrate looks like

$$\varepsilon(\omega) = \varepsilon_h - \frac{G}{\omega^2 - \omega_0^2 - i\,2\omega\nu}, \tag{6.1}$$

where $G = 2g_0\omega_0$, ω_0 is the exciton resonant frequency and $\nu = 1/\tau$, τ is the exciton lifetime. Near to a resonance $\omega \simeq \omega_0$ the dielectric permittivity $\varepsilon(\omega)$ in (6.1) becomes a well-known form [Meystre & III., 1998]

$$\varepsilon(\omega) = \varepsilon_h - \frac{g_0}{\omega - \omega_0 - i\,\nu}. \tag{6.2}$$

In Eq. (6.1) the parameter g_0 is related to the oscillator strength per dot. In Part II we will derive such formula for $\varepsilon(\omega)$. The dielectric permittivity of a host material ε_h is assumed to be frequency independent. The case of $g_0 > 0$ corresponds to QD with inverted population of levels, thus characterizing optical gain. Outside of the band of the resonance at $\omega \neq \omega_0$ the dielectric permittivity $\varepsilon(\omega)$ is practically constant and equal to ε_h. As follows from (1.55) the optical fields for such a structure read as

$$E_\theta = \frac{\omega}{k_0^2 r \varepsilon(\omega)} \frac{\partial^2 \Pi(\rho,\theta,\varphi)}{\partial r \partial \theta}, \quad H_\varphi = -\frac{i}{r\mu_0} \frac{\partial \Pi(\rho,\theta,\varphi)}{\partial \theta}, \tag{6.3}$$

$$E_r = \frac{i\,\omega}{k_0^2 \varepsilon r^2} m(m+1)\Pi(r,\theta,\varphi) + \frac{i\,\omega\mu_0}{k_0^2\varepsilon} j_r(r), \tag{6.4}$$

where again $\varepsilon = \varepsilon(\omega)$. As it was noted in Sec. 1.1.6 that in the case of $m = 0$ fields H_φ, E_θ are zero and current $j_r(r)$ contributes in a spherical symmetric part in E_r only. Thus $j_r(r)$ can be responsible for the processes of the inverting of population in the active core of microsphere. But $j_r(r)$ does not contribute to the resonance multipole radiation. Therefore the $m = 0$ case is excluded from our analysis.

6.2 Results and Discussions

The following parameters have been used in calculations: the geometry of a system is $SLNLNLN\ldots NV$ where letters S, L, N, V indicate the materials in the $\lambda/4$ stack. The radius of the active core (internal substrate) is $r_1 = 16\,\mu\text{m}$. We have used the parameters of materials [Hodgson & Weber, 1997] S (internal substrate): glass, $n_S = 1.5$, L: SiO_2, the thickness $0.3\,\mu\text{m}$, $n_L = 1.46$. For calculations we used $\tau = 1.6 \cdot 10^{-12}\,\text{sec}$, N: Si, the thickness $0.122\,\mu\text{m}$, $n_N = 3.58$ [Kumar *et al.*, 1999], V (outer medium): air, $n_V = 1$. A small damping term $\sim 10^{-3}$ has been added to ε of layers. The number of the spherical harmonic is $m = 1$, or 2. We follow a simple estimate of Maksimenko *et al.* [2000] $g_0 \sim 10^{14}\,\text{sec}^{-1}$ and less.

The first question is how sufficient is the influences of both dispersion and gain in the internal substrate on the reflection coefficient in such a system.

In Fig. 6.1 the frequency dependence of the refraction indexes of the internal substrate and the layers of the stack for cases of amplification ($g_0 > 0$) (a) and absorption ($g_0 < 0$) (c) are shown accordingly. The total dielectric permittivity of such substrate looks like (6.1). In Fig. 6.1(a and c) we use $|g_0| = 7.74 \cdot 10^{11} \cdot \text{sec}^{-1}$. In Fig. 6.1(b and d) the appropriate coefficients of reflection $|R|$ and transmittance $|T|$, and also coefficient of losses $A = 1 - |R|^2 - |T|^2$ for $|g_0| = 3.7 \cdot 10^9 \cdot \text{sec}^{-1}$ are shown. One can see the arising of the narrow peak (in a gain case) and of the hollow (in case of absorption) in the vicinity of the exciton resonance on the reflection's curve. The losses coefficient A becomes negative in the gain band that corresponds to the regime of generation in microsphere.

In Fig. 6.2 the details of the frequency dependence of reflection coefficient are shown for several values of the gain factor g_0. One can see that as the band of a resonance $|R|$ grows sharply for $g_0 = 1.48 \cdot 10^{10} \cdot \text{sec}^{-1}$ the $|R|$

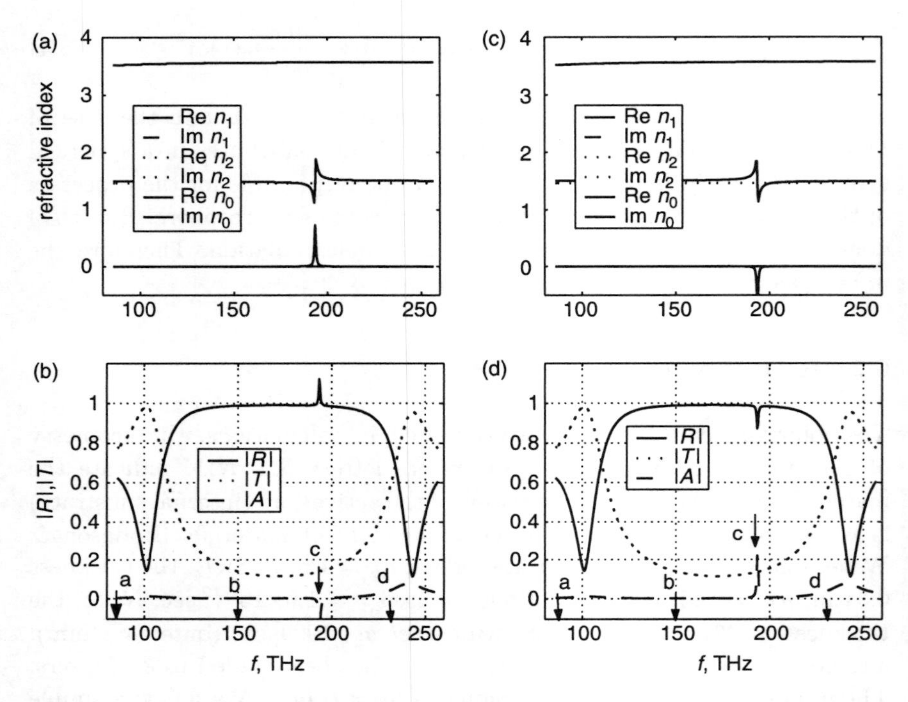

Fig. 6.1 (a) Real and imaginary parts of the refraction indexes of an internal substrate (n_0) and the layers of a stack (n_1 and n_2). Gain factor is, sec^{-1}: $g0 = 7.74 \cdot 10^{11}$. The narrow peak corresponds to a resonance of the inverted environment at the inside of the internal substrate (glass + semiconductor) on a frequency of 193.5 THz ($\lambda = 1.55\,\mu$m); (b) coefficients of reflection $|R|$, transmittance $|T|$ and losses $1 - |R|^2 - |T|^2$ for a $\lambda/4$ spherical stack, gain factor is, sec^{-1}: $g0 = 3.7 \cdot 10^9$. In (b) and (d) are shown eigenfrequencies, THz: a-82; b-150; c-193.5; d-231.

exceeds the unit at 80%. At the same time the transmittance $|T|$ varies a little bit only in the zone of the resonance (see Fig. 6.1(b and d)). It is due to the that the gain environment is beyond the stack and it cannot change the transmittance properties of the stack.

To elucidate the influence of structure of the coated microsphere on a regime of generation we have calculated a few representative eigenfrequencies, as the inside of the gain zone, and also beyond it. Such complex eigenfrequencies were calculated by solving the eigenfrequencies equation, which has a typical form (6.1), but for $\varepsilon(\omega)$ as given by Eq. (6.1). We study the influence of number of layers in the stack both on the real and

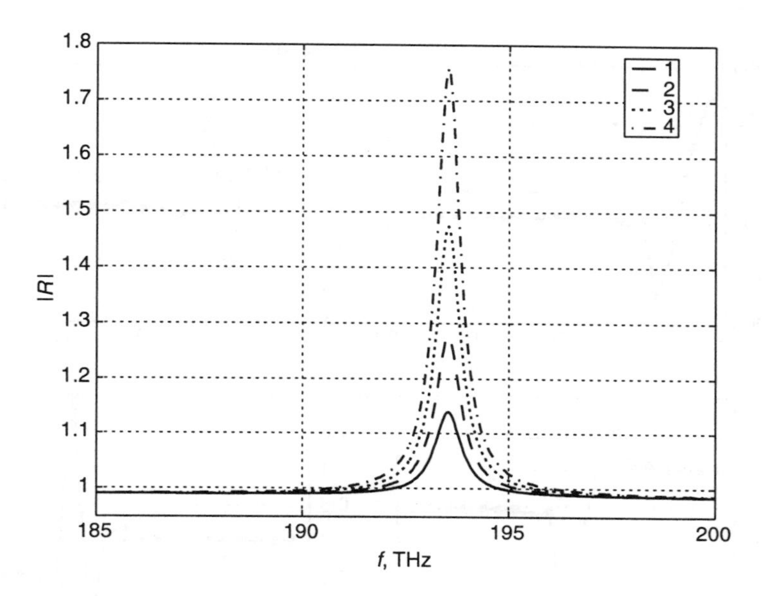

Fig. 6.2 The frequency dependence of modulo of the reflection coefficient $|R|$ for different levels of the gain g_0 of inverted medium in the active substrate, $\sec^{-1}$: (1) $3.7 \cdot 10^9$; (2) $6.56 \cdot 10^9$; (3) $1.03 \cdot 10^{10}$; (4) $1.48 \cdot 10^{10}$. Stack has 5 layers, $\lambda/4$ case.

Table 6.1 Some eigenfrequencies around of the band-gain of the active coated microsphere.

No	a	b	c	d
Re(f), THz	85	150	193.5	231

the image parts of the eigenfrequencies. We analyze four eigenfrequencies (see Table 6.1 and also Fig. 6.1(b,d)). Note that frequency f_c is on the inside of the band-gain resonance, and frequency f_b is in the zone of strong reflection.

In addition we use frequency $f = \omega/2\pi$ instead of ω. In Fig. 6.4 the dependence of Q factor ($Q = \mathrm{Re}(\omega)/2\mathrm{Im}(\omega)$) of eigenfrequencies from Table 6.1 versus the number of layers at various levels for gain-factor g_0 are shown. In Fig. 6.3 the Q factor of eigenfrequency inside of the band-gain $f_c = 193.5\,\mathrm{TGz}$ ($\lambda = 1.55\,\mu\mathrm{m}$) is presented. One can see from Fig. 6.3, while the number of layers in stack is small enough the Q factor is positive

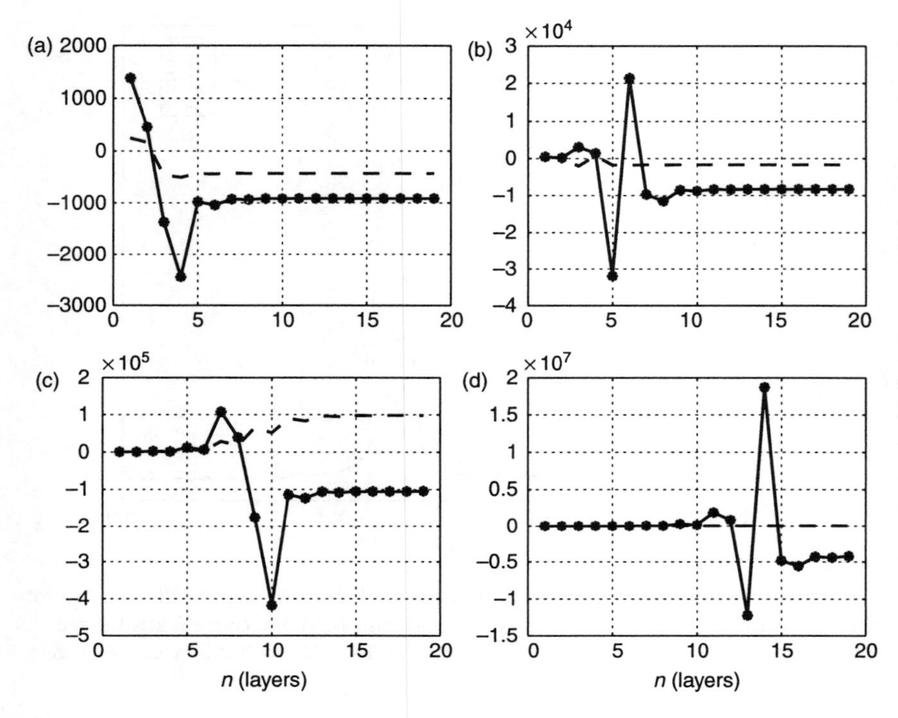

Fig. 6.3 Q factors of eigenfrequency $f = 193.5\,\mu\text{m}$ within a gain-band versus the number of layers in the stack n for the various gain factors, sec^{-1}: (a) $g_0 = 3.7 \cdot 10^9$; (b) $g_0 = 4.1 \cdot 10^8$; (c) $g_0 = 1 \cdot 10^8$; (d) $g_0 = 7.78 \cdot 10^7$. One can see that Q changes sign from positive to negative at large n.

and the system does not generate. It corresponds to a case when the radiative losses of optical oscillations to the surrounding medium are substantial. However with an increasing of number of layers in the stack, the Q factor changes sign and becomes negative (i.e. changes the sign of $\text{Im}(f)$). Due to the exponential factor $\exp(i\,\omega t)$ for the transition of system to the regime of generation of oscillation frequency f_c, one can see that the threshold's number of layers increases while the gain factor g_0 is reduced. Indeed to achieve the generation for $g_0 = 7.78 \cdot 10^7\ \text{sec}^{-1}$ it is necessary to get $\lambda/4$ stack with 12 layers. Our parameters at $g_0 = 7.74 \cdot 10^7\ \text{sec}^{-1}$ (less the change of the sign of Q) was not observed at all. In the case of $g_0 > 7.5 \cdot 10^8\ \text{sec}^{-1}$ and more Q becomes negative for any number of layers in the stack.

Figure 6.4 is similar to Fig. 6.3, but other eigenfrequencies from Table 6.1 is shown. The frequency f_c from the band of the resonance is

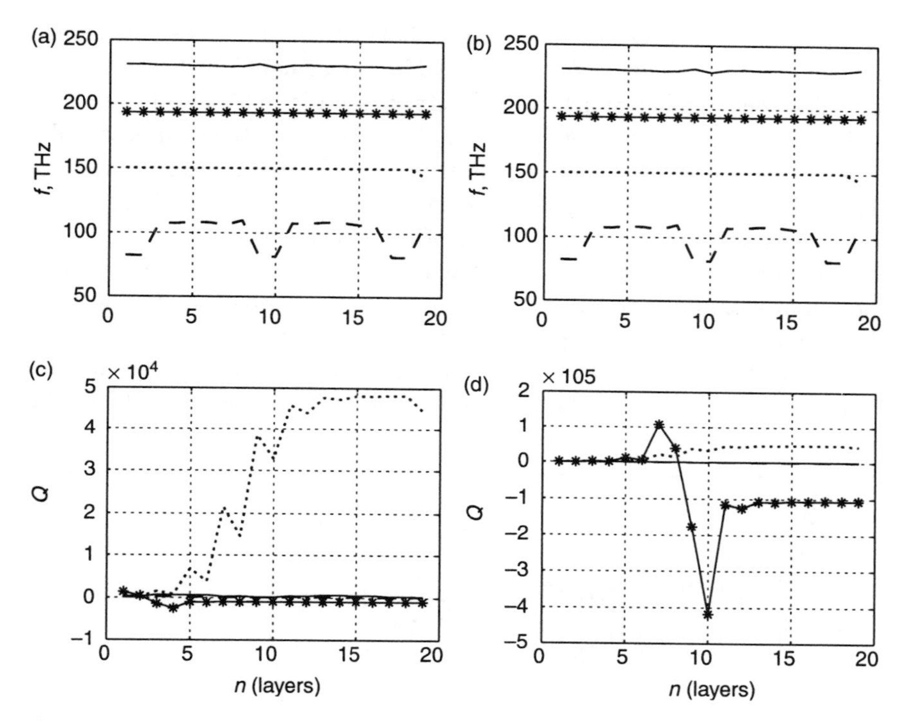

Fig. 6.4 Eigenfrequencies and Q-factors versus number of layers in the stack n. (a, b) eigenfrequencies $f = 82\,\mathrm{THz}$, $150\,\mathrm{THz}$, $193.5\,\mathrm{THz}$, $231\,\mathrm{THz}$; (c, d) corresponding Q factors. The gain factors are, $\sec^{-1}$: (a), (c) $g_0 = 3.7 \cdot 10^9$; (b), (d) $g_0 = 1 \cdot 10^8$. The asterisks signed the eigenfrequency $f = 193.5\,\mathrm{THz}$ and their Q factor finding inside of the gain-band.

marked with sign $*$. One can see from Fig. 6.4(a,b) that the real parts of eigenfrequencies $f_{b,c,d}$ are practically constant. This takes place due to the position of such frequencies in the zone of strong reflection $|R| \approx 1$, what was noted in Chapter 3, see also [Burlak *et al.*, 2001b]. But the real part of frequency f_a, situated in the zone of strong variation of the reflectivity, changes more while the number of layers in the stack varies. Frequencies $f_{a,d}$ have rather low Q factors. Only for frequency f_b that the Q factor increases sharply (exponentially) at the rise of the number of layers. With a large number of layers in the stack such dependence tends towards saturation owing to the material losses both in the layers and the substrate.

Thus in the overcritical regime the generation of the optical radiation with eigenfrequencies in the gain-band can be achieved when the number of layers in the stack is large enough. The radial distribution of the optical field of such eigenfrequency is of interest. We have calculated such dependence for a different number of layers in the stack (see Fig. 6.5). From Fig. 6.5 it is clear that in the active core $r/r_1 < 1$ the structure of the field practically does not change above or below the threshold. However the amplitude of the optical field in the external boundary of the stack $r/r_1 > 1$ (Fig. 6.5b,d) varies substantially while the number of layers changes.

We found that the threshold of generation of the optical eigenfield with the number of the spherical harmonic as $m = 2$ exceeds the threshold for the dipole case of $m = 1$ approximately on one order of magnitude.

We believe that it is due to the radial structure of such eigenmodes that only the $m = 1$ mode has a maximal value of electric field in the

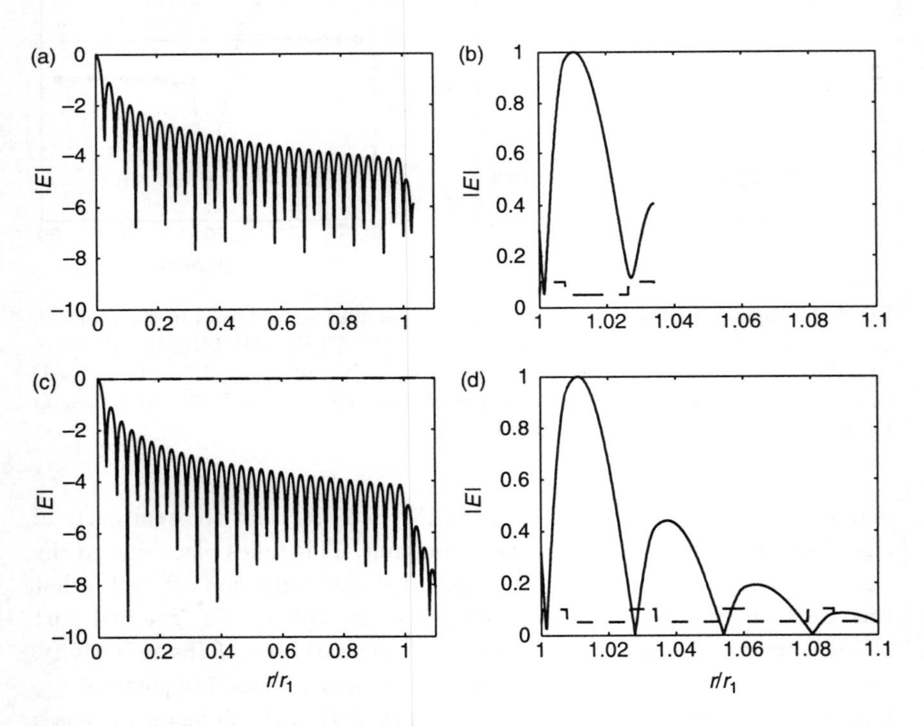

Fig. 6.5 Radial distribution of the normalized amplitude of electric field of the optical radiation in the coated microsphere in the gain-band on frequency 193.5 THz and $g_0 = 3.7 \cdot 10^9\,\text{sec}^{-1}$ versus normalized radius r/r_1 for the various number layers in stack: (a), (b) $n = 3$; (b), (c) $n = 8$. One can see that the distribution of the field in active substrate weakly depends on the number of layers in the stack. r_1 is radius of internal substrate.

center of sphere $r = 0$. For higher modes the eigenfield equals zero in the center of microsphere. Consequently the dipole radiation has optimal overlapping for coupling with the active substrate. We note also that such effect provides the opportunity to use the coated microspheres for effective harmonic generation in the optical range [Hayata & Koshiba, 1992; Centini *et al.*, 1999]. Also an important direction is the use of microspheres for the change of the spontaneous emission of radiation [Klimov & Letokhov, 1999 and references therein].

6.3 Conclusion

In this chapter we have studied the regimes of optical radiation of a compound system: active microsphere with the spherical stack deposited on it. It was supposed that the internal spherical substrate has gain a band of frequencies of the exciton resonance with lengthwave $\lambda = 1.55\,\mu\text{m}$. By means of the matrix method we have derived the eigenfrequencies equation and analyzed some representative cases. Calculation have shown that in the gain regime the coefficient of reflection may exceed unity in the vicinity of the resonance. The Q factor of the eigenfrequencies being in the resonance zone may become negative when the number of layers in the stack is large enough. This occurs due to the compensation of the radiate losses by the gain of the optical field in the active substrate. In such a regime the microsphere generates the optical radiation with the spectrum of eigenfrequencies inside the gain-band. The threshold of generation is least in case of spherical harmonic $m = 1$ (dipole radiation). Let's note that in [Artemyev & Woggon, 2000; Artemyev *et al.*, 2001a,b; Jia *et al.*, 2001] the interaction light with quantum dots was observed for microsphere (without stack) when spectrum optical WGM was registered. We have shown that such effect is rather sensitive to the structure of the spherical stack. This effect may find application in a number of active devices of optoelectronics. In such devices the selection of a few number of modes with opportunities of controlling by parameters of the optical field by means of electric, mechanical or other effects can be achieved.

CHAPTER 7

Transfer Matrix Approach in a Non-Uniform Case

A number of works have been studied on the problems of calculating and application of the various numerical algorithms to study the scattering of the electromagnetic field by a multilayered sphere.

Perelman [1996] studied the problem of scattering by a spherical particle whose refractive index arbitrarily depends on the distance from its center has been solved. A computational scheme for determining the scattering coefficients for the refractive-index profiles given by some piecewise smooth function is constructed, and the models are presented. The simple algorithm for evaluating the scattered and internal field vectors is elaborated. The exact expression for the scattering cross section in terms of the generated Debye partial potentials that are dependent on the given refractive index has been obtained. In [Yang, 2003] an improved recurrence algorithm to calculate the scattering field of a multilayered sphere is developed. The internal and external electromagnetic fields are expressed as a superposition of inward and outward waves. The alternative yet equivalent expansions of fields are proposed by the use of the first kind of Bessel function and the first kind of Hankel function instead of the first and the second kinds of the Bessel function. The final recursive expressions are similar in form to those of the Mie theory for a homogeneous sphere and are proven to be more concise and convenient than the earlier forms. The new algorithm avoids the numerical difficulties, which give rise to significant errors encountered in practice by previous methods, especially for large, highly absorbing thin shells. Various calculations and tests show that this algorithm is efficient, numerically stable, and accurate for a large range of size parameters and refractive indices. In [Wu *et al.*, 1997] an efficient numerical procedure for computing the scattering coefficients of a multilayered sphere is discussed. The stability of the numerical scheme allows us to extend the feasible range of computations, both in size parameter and in the number of layers for a given size, by several orders of magnitude with respect to the previously published algorithms. Exemplifying results, such as scattering diagrams and cross-sectional curves, including the case of Gaussian beam illumination, are provided. Particular attention is paid to the scattering at the rainbow angle for which approaches based on geometrical optics might fail to provide accurate enough results. For early works see [Kai & Massoli, 1994] and reference therein.

In this chapter we develop a simple approach which is suitable to generalize the above developed transfer matrix method for the multilayered microsphere (Sec. 1.4) to a non-uniform layers case. This approach also can be exploited for the calculation of the eigenfrequencies in a non-uniform case.

7.1 Approach to a Non-Uniform Case

In Sec. 1.1.6 we have received the equation for an Debye potential (1.46) in a general case of a inhomogeneous layer in form

$$\varepsilon\frac{\partial}{\partial r}\left(\frac{1}{\varepsilon}\frac{\partial R}{\partial r}\right)+\frac{\omega^2\varepsilon}{c^2}R-\frac{m(m+1)}{r^2}R=0. \tag{7.1}$$

Equation (7.1) is a linear equation with variable coefficients. In the case $\varepsilon(r) = const$, this equation has the form $\frac{d^2R}{dr^2}+\left(\frac{\omega^2}{c^2}\varepsilon-\frac{m(m+1)}{r^2}\right)R=0$ and can easily be solved in the spherical Bessel functions. However in non-uniform layers, where $\varepsilon(r)$ depends on radius r, the Eq. (7.1) becomes essentially complicated and in general it may be solved only by numerical methods.

It is easy to find numerically the partial solution of Eq. (7.1) satisfying to some initial or boundary conditions. An algorithm for solving such an ordinary differential equation numerically can be found, e.g. in [Press *et al.*, 2002]. But for the calculations of fields in each layer by means of the transfer matrix method one requires not the partial but the general solution of (7.1), which are the superposition of direct and reflected waves. However the conditions of reflection can vary. Furthermore in a general case such conditions are determined by the conditions in the external and internal boundaries of the stack. Accordingly the definition of the ratio of amplitudes of the direct and reflected waves in each of the non-uniform layers has to be produced while solving the general problems by a self-consistent way.

To obtain the general solution in a non-uniform layer we will exploit the next way. The idea comes from the general theory of the linear deferential equations of second order for function $y(r)$ in the form

$$\frac{d}{dr}\left(p(r)\frac{dy}{dr}\right) + q(r)y = 0. \tag{7.2}$$

It is well-known [Korn & Korn, 1961; Riley *et al.*, 1998] that if one of the particular solution y_1 of (7.2) is known the general solution of (7.2) is given by

$$y = ay_1(r) + by_2(r), \tag{7.3}$$

where the second linearly independent solution reads

$$y_2(r) = y_1(r)\int_c^r \frac{dr'}{y_1^2(r')p(r')}, \tag{7.4}$$

a, b and c are arbitrary numbers. We admit that $y_1(r)$ satisfies some boundary conditions:

$$y_1(c) = y_{10}, \qquad dy_1(c)/dr = y'_{10}, \tag{7.5}$$

then from (7.4) and (7.5) one can easily find what $y_2(r)$ satisfies the boundary conditions in form

$$y_2(c) = 0, \qquad dy_2(c)/dr = 1/[y_{10}p(c)], \tag{7.6}$$

where values y_{10}, y'_{10} are arbitrary constants. Whereas the Eq. (7.1) is linear, its general solutions does not change by multiplying any constant. Thus it is possible to put $y_{10} = 1$ in (7.5) and (7.6) without any restrictions of generality. As a result we come to the following two-steps algorithm the construction of the general solution of (7.1) in a non-uniform layer with boundary $r = r_k$. Solving the Eq. (7.1) by Runge-Kutta method (or other similar method) we find the solution $y_1(r)$ with boundary conditions $R_1(r_k) = 1$, $dR_1(r_k)/dr = y\prime_{10}$ and, then we find the solution $R_2(r)$ with other boundary conditions: $R_2(r_k) = 0$, $dR_2(r_k)/dr = \varepsilon(r_k)/y_{10}$. Now the general solution in a non-uniform layer can be written down as

$$R_k(r) = a_k R_{k1}(r) + b_k R_{k2}(r), \tag{7.7}$$

where $r_k \leq r \leq r_{k+1}$. In the homogeneous layers, and also in the homogeneous internal and surrounding mediums, we can use the normal solution of the Bessel equation.

Using such approach we can construct a general solution for a Debye potential $R_k(r)$ (7.7) and then obtain the general solution for electromagnetic fields in layers. Now one can apply it to a general theory of a transfer matrix method for the calculation of the electromagnetic fields satisfying the boundary conditions continuity in both homogeneous and inhomogeneous layers of the stack.

To check the accuracy of this approach we have calculated the coefficient of reflection of the metallo-dielectric stack in a homogeneous case by solving (7.7) in a stated approach. The results of comparison with a standard approach are shown in Fig. 7.1. One can see that both curves agree

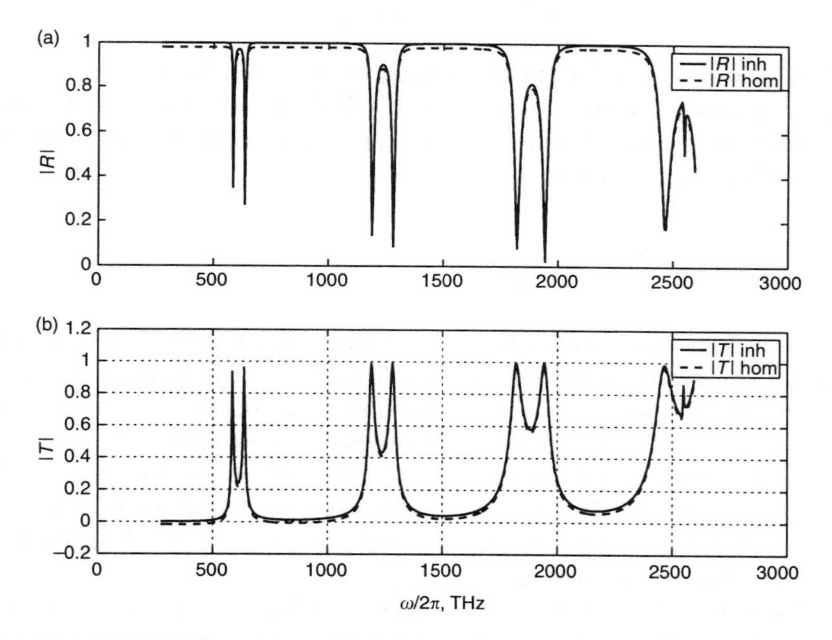

Fig. 7.1 Comparison of the frequency dependence of reflection and transmittance coefficients with the use of the spherical Bessel function and the numerical solving of the Debye potential equation for a homogeneous case. We slightly shifted both curves, but accuracy is fine.

well. In Fig. 7.1 we provided the inhomogeneous curves with a small shift to check for any differences between both lines.

7.2 Example. Non-Uniform Electron's Concentration

We apply this approach to study the propagation electromagnetic waves in non-uniform layers. We use the next simple model. The non-uniformity in dielectric layers has electronic nature and it is due to contact with metallic layers [Sze, 1969]. In the simplest case we neglect the hole's contribution and also the influence of the spherical form layer to a charge distribution. From the equations $\varepsilon\varepsilon_0(d\mathcal{E}/dx) = \rho$, $j = 0$ and $n = n_0 \exp(e\Phi(x)/kT)$ one can obtain the nonlinear Poisson equation for potential Φ ($\mathcal{E} = -\nabla\Phi$) in the form [Sze, 1969]:

$$\frac{d^2\Phi}{dr^2} = \frac{e}{\varepsilon\varepsilon_0}(n - n_0) = \frac{e}{\varepsilon\varepsilon_0} n_0(e^{e\Phi/kT} - 1), \tag{7.8}$$

where $\rho = -en$ is a density of charge, j is the electrical current, n is the electron concentration, k and T are the Boltzmann constants and temperature accordingly. From (7.8) the equation for dimensionless quantity $p = (n/n_0 - 1)$ is given by

$$\frac{d^2}{dx^2} \ln(1 + p) = l^2 p, \tag{7.9}$$

where n_0 is a steady-state concentration of charge, $x = r/d$, $l = d/l_D$, d is the thickness of layer and $l_D = (\varepsilon\varepsilon_0 kT/e^2 n_0)^{1/2}$ is the Debye's length and e is electrical charge of electron. The boundary conditions are $p(0) = A$, $p(1) = B$, where A and B are constants. So far we consider the dielectric layers placed between the metallic layers, the problem becomes symmetrical and $A = B$. Since $p(0) = p(1) = A$ such a two-point problem has solution only if l belongs to the spectrum of eigenvalues. We use the shoot method to search for such a spectrum.

We recall when ordinary differential equation are required to satisfy boundary conditions at more than one value of the independent variable, the resulting problem is called a two-point boundary value problem [Press *et al.*, 2002]. The two-point boundary value problems require considerably more effort to solve than initial value problems.

We rewrite (7.9) in the form of coupled sets of first-order equations:

$$dX_1/dx = X_2, \qquad dX_2/dx = X_2^2/(1+X_1) + p^2 X_1(1+X_1), \qquad (7.10)$$

where $X_1 = p$, $X_2 = dp/dx$. In the shooting method we choose values of the dependent variables $X_1(0) = A$ and $X_2(0) = f(l)$ at the same boundary $x = 0$. Value $X_2(0) = f(l)$ is arranged to depend on parameter l whose values we can initially "randomly" guess. We can then integrate (7.10) by initial value Runge-Kutta methods, arriving at the other boundary $X_2(x=1) = A$. Figure 7.2 shows the electric spatial distribution for $A = B = 0.99$ and $l = 5.97$.

Figure 7.3 shows the reflection coefficients ($|R|^2$ and $|T|^2$) coefficients for such stack with $N = 4$ alternating layers of Si and silver.

In Fig. 7.4. one can see (a) Dielectric permittivity $\varepsilon(r,\omega)$; (b) $\varepsilon(r,\omega)^{-1} d\varepsilon(r,\omega)/dr$ and (c) Plasma frequency $\omega_p(r,\omega)$ of alternating

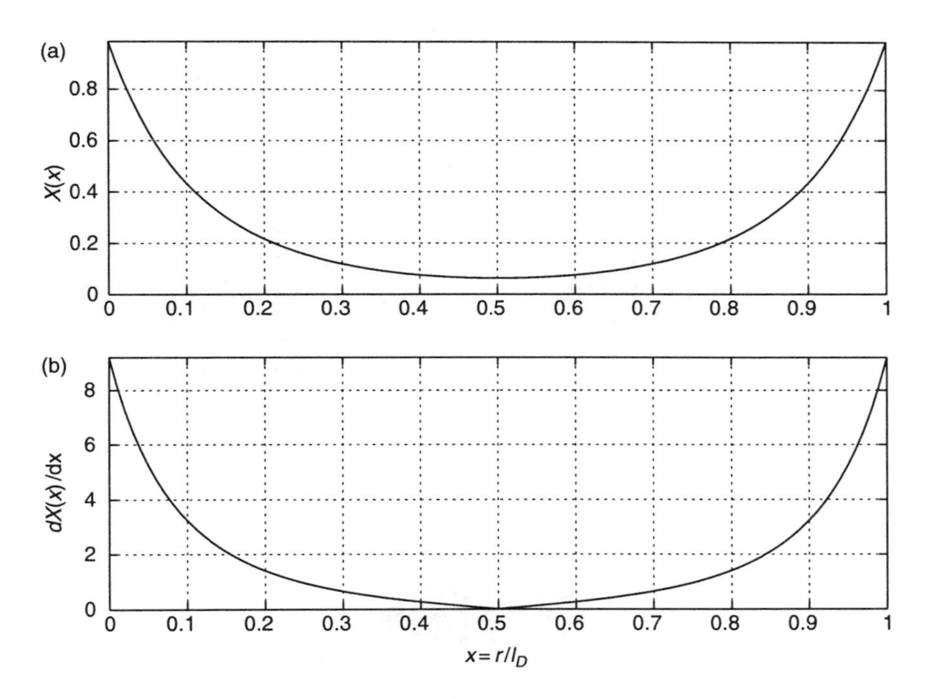

Fig. 7.2 Symmetrical spatial charge distribution in layer, which is governed by nonlinear Poisson equation (7.9) for $A = B = 0.99$, $l = 5.97$: (a) charge $p = X_1$; (b) derivative dp/dx.

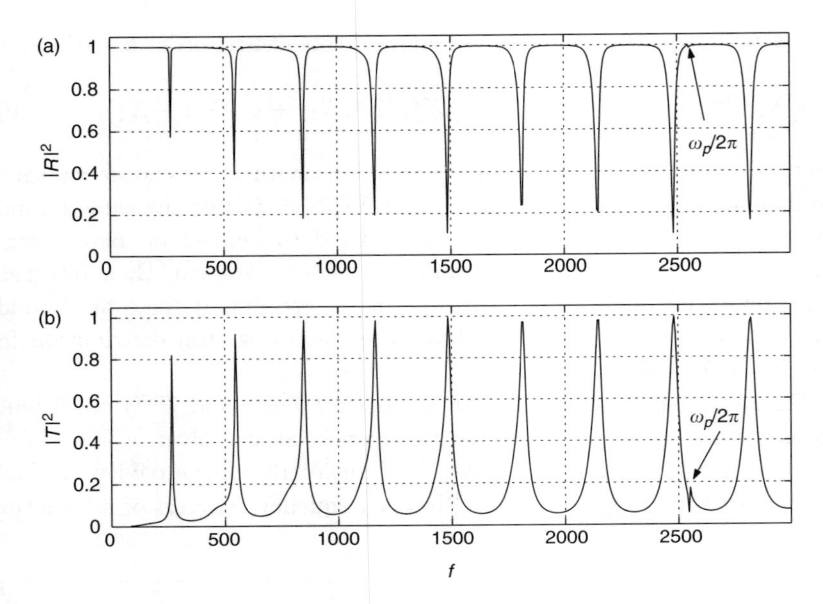

Fig. 7.3 The reflection ($|R|^2$ and $|T|^2$) coefficients for stack with $N = 4$ alternating layers Si and silver.

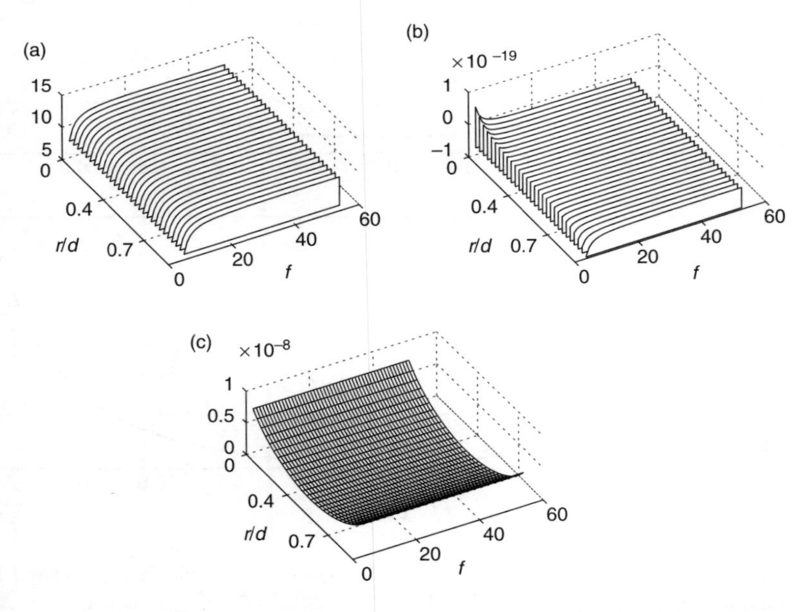

Fig. 7.4 (a) Dielectric permittivity $\varepsilon(r, \omega)$; (b) $\varepsilon(r, \omega)^{-1} d\varepsilon(r, \omega)/dr$; (c) Plasma frequency $\omega_p(r, \omega)$.

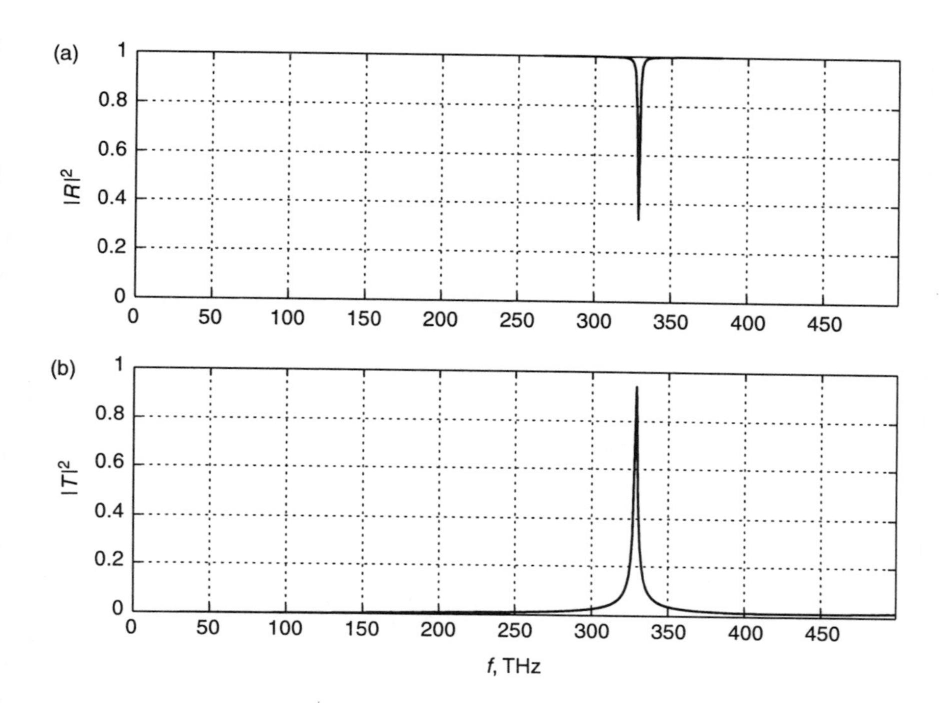

Fig. 7.5 Reflection and transmittance coefficients for non-uniform case.

layers in the stack. Layer 1 is Si and layer 2 is a typical metal similar to silver. The plasmas frequency of metal is $\omega/2\pi \simeq 2600\,\text{THz}$.

Figure 7.5 shows the reflection ($|R|^2$ and $|T|^2$) coefficients for such a stack with $N = 4$ alternating layers for Si and silver and various thickness of layers.

PART II

The Quantum Phenomena in Microspheres

Preface

The investigation and exploitation of the wave properties of atomic matter is of great interest for fundamental as well as applied research and therefore constitutes one of the most active areas in atomic physics and quantum optics. As a consequence of all developments, there is now a vast interest in compact and reliable atom optical setups which not only expand the applicability of atom optics in fundamental research, but also allow for the technological implementation of atom optical measurement systems. A new approach to this challenge lies in the development of miniaturized and integrated atom optical setups based on microfabricated structures [see Birkl *et al.*, 2001 and references therein].

An important line of application of microcavities and microspheres is to use them for the change of the spontaneous emission of radiation by an atom as a result of it being placed into a microcavity [Lai *et al.*, 1988; Bachor, 1998]. Such a problem has fundamental value due to small objects such as microspheres resulting in many important features of the quantum nature of electromagnetic radiation being shown.

As we saw in Part I except WGM a dielectric sphere has a complex spectrum of the electromagnetic low quality eigen oscillations because of leakage in the outer space [Stratton, 1941]. A compound structure: the dielectric sphere coated by an alternative stack possesses more than the manifold properties. The Q factor (quality) of such oscillations strongly depends on the properties of the stack. Q becomes large in the areas of high reflectivity and beyond these zones Q remains small [Brady *et al.*, 1993; Burlak *et al.*, 2001a]. The combination of such factors causes the large variety of optical properties of microspheres with a multilayered stack. Deumie *et al.* produced the metallic coating of the fused-silica microspheres and obtained the artificial optical powders with specific spectral properties. Burlak [2002] found that the Q factor of eigenfrequencies inside the gain area of coated microsphere becomes negative (generations) when the number of layers in the stack is large enough.

It is known [Scully & Zubairy, 1996] that at the interaction of a two-level atom with a single-mode field the phenomena of collapse and revival arise. These phenomena are due to the Rabi oscillations but it is weakly investigated for the multimode case. The important thing is that revivals occur only because of the granular structure of the photon distribution. Revival is thus a pure quantum phenomenon. From the other hand it is well known that a continuum photon distribution would give a collapse, but no revivals [Meystre & III., 1998].

Recent experiments with the quantum dots (QD) introduced into dielectric microsphere, and observed coupling of QD radiation with a spectrum of electromagnetic mode of microsphere (in this case WGM) confirm such a conclusion [Artemyev & Woggon, 2000; Jia *et al.*, 2001; Moller *et al.*, 2002].

For microspheres of sizes of about 1 μm a number of modes can exist in a vicinity of an atom radiation line. Thus the distinction between resonant and non-resonant spontaneous emission no longer appears as pronounced as in the continuum approximation. In case of small-sizes microspheres (emerges) the interplay of quantum and classical fields effects emerges. It defines the fundamental value of studying such processes due to the quantum nature of radiation.

Some new approaches which are valid in the presence of dispersing and absorbing dielectric bodies are developed in [Dung *et al.*, 1998, 2000, 2003; Knoll *et al.*, 2001; Scheel *et al.*, 1999b; Matloob, 2001; Lagendijk, 2003].

Recently a number of groups report on observations of the interplay 3D-confined cavity modes of single microspheres with photons emitted from quantized electronic levels of single semiconductor nanocrystals (the quantum dot states).

The real quantum dots (QDs) are complicated systems, and even in the simplest cases it have a non-trivial energy level structure. We provide some references to various current publications on this topic [Alhassid, 2000; Fiore *et al.*, 2001; Artemyev & Woggon, 2000; Artemyev *et al.*, 2001b; Berg & Mork, 2003; Brun & Wang, 2000; Bryllert *et al.*, 2003; Chen & Brandes, 2003; Costantini *et al.*, 2003; Crooker *et al.*, 2003; Cusack *et al.*, 2001; da Silva *et al.*, 2003; Bimberg *et al.*, 2001; Ferdos *et al.*, 2002; Geller *et al.*, 2003; Hours *et al.*, 2003; Vukocvic & Yamamoto, 2003; Lelong & Lin, 2002; Jacak *et al.*, 2002; Makino *et al.*, 2003; Maksimenko *et al.*, 2000;

Moller *et al.*, 2002; Songmuang *et al.*, 2003; Jia *et al.*, 2001; Slepyan *et al.*, 2002; Wang *et al.*, 2003; LU *et al.*, 2003; Woggon, 1996; Ji *et al.*, 2003].

We therefore approximate the behavior of the real quantum dot (QD) by that of a much simpler quantum system namely a two-level atom (TLA). For our purposes only two energy levels of QD play a significant role in the interactions with the electromagnetic field, so that we represent the QD by a quantum system with only two energy eigenstates. This is the most basic of all quantum systems and it generally simplifies our treatment substantially.

So we assume that a single two-level atom is at rest inside a coated microsphere of sufficiently small dimensions. Under these conditions the TLA effectively interacts with some single electromagnetic mode of microsphere with frequency ω. Our purpose is to study the temporal dynamics of two-level atom and obtain an equation for optical electrical polarization. But first we discuss some more simple models.

First we shall consider the dynamics of a two-level atom interacting with a field, which we assume as classical quantity. It is fair, if the number of field quanta is great $n \gg 1$ and there is no necessity to consider the backward influence of dynamics of the atomic transitions to a field. Furthermore we consider spontaneous emission of atom in microsphere. In that case the field needs to account for the quantum object when the atom and a field form the coupled system. For this case we shall discuss the simple procedure of quantization which takes into account the details of a field in microsphere. The details of dynamics of spontaneous emission are considered in last section.

CHAPTER 8

Coupling of Two-Level Atom with Electromagnetic Field

We consider that a single two-level atom (TLA) is at rest inside a coated microsphere of sufficiently small dimensions. Under these conditions the TLA effectively interacts with some single electromagnetic microsphere mode of frequency ω. We treat the subsystem consisting of the TLA coupled to the electromagnetic microsphere mode system, described by the Hamiltonian $H_s = H_a + H_{af}$, where H_a is the Hamiltonian of TLA, and H_{af} describes the electric dipole coupling between TLA and the microsphere electromagnetic field mode.

Let's consider the following system with Hamiltonian H_a and wave functions $u(q)$

$$H_a(q) \cdot u(q) = Eu(q), \tag{8.1}$$

where H_a is the "unperturbed Hamiltonian" and it is actually independent of time. q is the set of representations of TLA describing a state of system, E is energy. For a two-level TLA we have $E = E_{a,b}$ and $u = u_{a,b}$. General wave function of such a system has the form

$$\psi = au_a(q)\, e^{-i(E_a/\hbar)t-(\gamma_a/2)t} + bu_b(q)\, e^{-i(E_b/\hbar)t-(\gamma_b/2)t}. \tag{8.2}$$

Factors γ_a, γ_b are taking into account phenomenologically some possible mechanisms of decay or transitions from these two levels into some other levels which we do not take into account explicitly. Such wave function (8.2) is a solution of the non-stationary Schrödinger equation

[Maitland & Dunn, 1969 and references therein]

$$i\hbar\frac{\partial\psi}{\partial t} = H_0\psi = \left[H_a - i\hbar\frac{\gamma}{2}\right]\psi, \tag{8.3}$$

where operator γ is defined as:

$$\gamma u_j(q) = \gamma_j u_j, \quad j = a, b. \tag{8.4}$$

We consider the interaction of TLA with an electromagnetic field. To treat such interactions, we add the appropriate interaction energy H_1 to the Hamiltonian, that is

$$i\hbar\frac{\partial\psi}{\partial t} = (H_0 + H_1)\psi, \tag{8.5}$$

where H_0 is Hamiltonian of unperturbed TLA, H_1 defines the interaction of TLA with an electromagnetic field. We search for a solution of (8.5) in form

$$\psi = a(t)u_a(q) + b(t)u_b(q), \tag{8.6}$$

where $a(t), b(t)$ are complex probability amplitudes. We substitute (8.6) in (8.5), multiply $u_a^*(q)$ and $u_b^*(q)$ to it accordingly and integrate with respect to the area of the atom location. Using the orthonormality condition $\int u_i^* u_j dq = \delta_{ij}$ and after some simple calculation, we receive the following equations for $a(t)$ and $b(t)$:

$$i\hbar\frac{da}{dt} = a\left(E_a - i\hbar\frac{\gamma_a}{2}\right) + a(H_1)_{aa} + b(H_1)_{ab}, \tag{8.7}$$

$$i\hbar\frac{db}{dt} = b\left(E_b - i\hbar\frac{\gamma_b}{2}\right) + a(H_1)_{ba} + b(H_1)_{bb}, \tag{8.8}$$

where

$$(H_1)_{ij} = \int u_i^* H_1 u_j\, dq. \tag{8.9}$$

This is the next important circumstance in a damping case. We shall see later the Hamiltonian for a single mode of a quantized radiation field of frequency ω_0 in a cavity is given by $H = \hbar\omega_0 a^+ a$. In the Heisenberg picture (see later), the annihilation and creation operators are given by $a(t) = a \cdot \exp(-i\omega_0 t)$ and $a^+(t) = a^+ \cdot \exp(i\omega_0 t)$, respectively. We assume that we attempt to describe damping in this mode by introducing a phenomenological loss term γ by analogy with a damped oscillator. Since the classical and Heisenberg equations have the same form, the operators would then have the damped solutions $a(t) = a \cdot \exp[-(i\omega_0 + \gamma/2)t]$ and

$a^+(t) = a^+ \cdot \exp[(i\omega_0 - \gamma/2)t]$ and the fields would decay in a time of order γ^{-1}. This result unfortunately violates a fundamental principle of quantum mechanics since the commutator $[a(t), a^+(t)] = \exp[-\gamma t]$ approaches zero. This implies that the uncertainty principle is violated (see Secs. 10.1 and 11.2). But for times short compared with the relaxation time ($t \ll \gamma^{-1}$), the violation is not serious and such a model can be satisfactory [Louisell, 1990].

8.1 Transitions under the Action of the Electromagnetic Field

We now consider the interaction of the medium-assisted electromagnetic field with a point charges e. Applying the minimal-coupling scheme [Cohen-Tannoudji *et al.*, 1989; Scully & Zubairy, 1996; Mandel & Wolf, 1995], we write the complete Hamiltonian in the form

$$H = \frac{1}{2m}\left(\vec{p} + \frac{e}{c}\vec{A}\right)^2 - e\varphi, \tag{8.10}$$

where for scalar φ and vectorial $\vec{A}$ potentials accordingly; we use the Coulomb gauge $\varphi = 0, \nabla\vec{A} = 0$. Neglecting small quantities of the order $\vec{A}^2$ in (8.10), we obtain

$$\hat{H} = \frac{1}{2m}\vec{p}^2 + \frac{e}{mc}(\vec{p} \cdot \vec{A}), \tag{8.11}$$

where $\hat{\vec{p}} = -i\,\hbar\vec{\nabla}$, and the commutator $[\vec{p} \cdot \vec{A}] = 0$. The interaction part has form

$$\hat{H}_1 = -i\frac{e\hbar}{mc}(\vec{A} \cdot \vec{\nabla}). \tag{8.12}$$

We consider a case when the electric dipole coupling is position-independent. The matrix element of the interaction part can be written down as

$$\int u_b^* \hat{H}_1 u_a \, dq = -i\frac{e\hbar}{mc}\vec{A}(t) \int u_b^* \cdot \vec{\nabla} u_a \, dq, \tag{8.13}$$

where we suppose, that $\vec{A}(r,t) \cong \vec{A}(\vec{r}_0, t), \vec{r}_0$ is a position of TLA in microsphere. After some algebra one can get [Landau-Quantum, 1977;

Maitland & Dunn, 1969]

$$\int u_b^* \vec{H}_1 u_j \, dq = -i\frac{e\hbar}{c}\omega_{ab}\vec{A}(t)\int u_b^* \cdot \vec{r} u_a \, dq, \tag{8.14}$$

where $\omega_{ab} = -\omega_{ba} = \frac{1}{\hbar}(E_a - E_b)$.

The expectation value of the dipole moment $\int u_j^* \hat{H}_1 u_j \, dq$ in the lower and upper state must vanish from consideration of symmetry for states of define parity, because the dipole moment has odd parity. Hence we have

$$\vec{D}_{ba} = \int u_b^* \cdot (e\vec{r}) \cdot u_a \, dq = \vec{D}_{ab}^*. \tag{8.15}$$

8.2 The Equations for Probability Amplitudes

Since $\langle q|\widehat{H}_1|q\rangle = 0$ we can rewrite (8.7) and (8.8) in the form

$$i\,\hbar\frac{da}{dt} = a\left(E_a - i\,\hbar\frac{\gamma_a}{2}\right) + b(H_1)_{ab}, \tag{8.16}$$

$$i\,\hbar\frac{db}{dt} = b\left(E_b - i\,\hbar\frac{\gamma_b}{2}\right) + a(H_1)_{ba}. \tag{8.17}$$

We assume $\vec{A} = \vec{A}_0 \cos\omega t$, $\vec{E} = -\frac{1}{c}\partial\vec{A}/\partial t = \vec{A}_0(\omega/c)\sin\omega t = \vec{E}_0 \sin\omega t$, $E_0 = A_o\omega/c$, then in (8.16), (8.17):

$$\begin{aligned}\int u_b^* \widehat{H}_1 u_a \, dq &= -i\frac{e\vec{A}_o\omega}{c}\frac{\sin\omega t}{\omega}\omega_{ab}\frac{1}{e} \cdot \vec{D}_{ba} \\ &= -i(E_0 \cdot \vec{D}_{ba})\frac{\omega_{ab}}{\omega}\sin\omega t,\end{aligned} \tag{8.18}$$

so that

$$\int u_b^* \widehat{H}_1 u_a \, dq = -i(\vec{E}_0 \cdot \vec{D}_{ba})\frac{\omega_{ab}}{\omega}\sin\omega t, \tag{8.19}$$

$$\int u_a^* \widehat{H}_1 u_b \, dq = -i(\vec{E}_0 \cdot \vec{D}_{ab})\frac{\omega_{ba}}{\omega}\sin\omega t, \tag{8.20}$$

and Eqs. (8.16), (8.17) becomes

$$\frac{da}{dt} = -i\,a\left(\omega_a - i\frac{\gamma_a}{2}\right) - b\frac{(\vec{E}_0 \cdot \vec{D}_{ab})}{\hbar}\frac{\omega_{ba}}{\omega}\sin\omega t, \tag{8.21}$$

$$\frac{db}{dt} = -a\frac{(\vec{E}_0 \cdot \vec{D}_{ba})}{\hbar}\frac{\omega_{ab}}{\omega}\sin\omega t - i\,b\left(\omega_b - i\frac{\gamma_b}{2}\right), \tag{8.22}$$

where $\omega_{ab} = -\omega_{ba} = \frac{1}{\hbar}(E_a - E_b)$ and we have to use $\omega_{ab} \approx \omega$ (about equivalence of the $\vec{r}\cdot\vec{E}$ and the $\vec{p}\cdot\vec{A}$ interaction Hamiltonians see discussion in [Scully & Zubairy, 1996, pp. 151, 178]).

Now we write Eqs. (8.21), (8.22) in the following dimensionless form. We use for renormalization a vacuum Rabi frequency in form $\Re = \frac{1}{\hbar}(\vec{E}_0 \cdot \vec{D}_{ab})$ and the next dimensionless variables

$$\tau = |\Re| \cdot t, \quad \Omega = \frac{\omega}{|\Re|}, \quad \alpha_{12} = \frac{\omega_{ba}}{\omega}\frac{\Re}{|\Re|}, \quad \alpha_{21} = \frac{\omega_{ba}}{\omega}\frac{\Re^*}{|\Re|} = \alpha_{12}^*, \tag{8.23}$$

$$\alpha_{11} = \left(\omega_a - i\frac{\gamma_a}{2}\right)\frac{1}{|\Re|}, \quad \alpha_{22} = \frac{\omega_b - i\frac{\gamma_b}{2}}{|\Re|}, \tag{8.24}$$

so that

$$\Omega_{a,b} = \frac{\omega_{a,b}}{|\Re|}, \quad \Gamma_{a,b} = \frac{\gamma_{a,b}}{2|\Re|}, \quad \frac{\omega_{b,a}}{\omega} = \frac{\Omega_{b,a}}{\Omega}, \tag{8.25}$$

$$\alpha_{12} = \frac{\Omega_{ba}}{\Omega}, \quad \alpha_{11} = \Omega_a - i\,\Gamma_a, \tag{8.26}$$

$$\alpha_{22} = \Omega_b - i\,\Gamma_b, \quad \alpha_{21} = \alpha_{12}^*. \tag{8.27}$$

In such new variables one can write down Eqs. (8.21) and (8.22) in the next dimensionless form

$$\frac{da}{d\tau} = -i\,\alpha_{11}a - \alpha_{12}\cdot\sin\Omega\tau\cdot b, \tag{8.28}$$

$$\frac{db}{d\tau} = i\,\alpha_{12}^* a\sin\Omega\tau - i\,\alpha_{22}\cdot b. \tag{8.29}$$

After solving Eqs. (8.28), (8.29), one can represent the electric dipole moment of TLA in the form

$$\vec{P} = e\int \psi^*\cdot\vec{r}\cdot\psi\,dq = a^*b\int u_a^*(e\vec{r})u_b\,dq + ab^*\int u_b^*(e\vec{r})u_a\,dq$$

$$= a^*b\cdot\vec{D}_{ab} + ab^*\cdot\vec{D}_{ba}. \tag{8.30}$$

If the wavefunctions u_a and u_b are real, then both $\vec{D}_{ab} = \vec{D}_{ba}$ and $\Re$ are real quantities. Then the electric dipole moment $\vec{P}$ is given by

$$\vec{P} = (a^*b + ab^*)\vec{D}_{ba}, \tag{8.31}$$

and Eqs. (8.28), (8.29) become following form:

$$\frac{da}{d\tau} = -i\,\alpha_{11}a + V(t)\cdot b, \tag{8.32}$$

$$\frac{db}{d\tau} = -V(t)\cdot a - i\,\alpha_{22}b, \tag{8.33}$$

where $V(t) = \alpha_{12}\sin\Omega\tau$ and $\alpha_{12} = \alpha_{21}$.

8.3 Derivation of the Equation for Polarization of TLA: Dielectric Permittivity

Let's derive the equations for dynamic polarization $P = D_{ba}(a^*b + ab^*)$ based on the equations for amplitudes (8.32) and (8.33). To simplify notation further we use the normalized polarization

$$\widetilde{P} = P/D_{ba} = a^*b + ab^*. \tag{8.34}$$

Proceeding from the Eqs. (8.32) and (8.33) it is easy to obtain the following relations for a and b. Multiplying (8.32) with a complex conjugated equation we obtain the relation for derivation $|a|^2$ in the form

$$a^*\frac{da}{dt} + c.c = \frac{d}{dt}|a|^2 = 2\mathrm{Im}(\alpha_{11})|a|^2 + V\cdot(a^*b + ab^*). \tag{8.35}$$

In a similar way from (8.33) we get the equations for $|b|^2$ as

$$b^*\frac{db}{dt} + c.c = \frac{d}{dt}|b|^2 = 2\mathrm{Im}(\alpha_{22})|b|^2 - V\cdot(a^*b + ab^*). \tag{8.36}$$

Then

$$\frac{d}{dt}(|a|^2 + |b|^2) = 2(\mathrm{Im}(\alpha_{11})|a|^2 + \mathrm{Im}(\alpha_{22})|b|^2), \tag{8.37}$$

and

$$\frac{d}{dt}(|a|^2 - |b|^2) = 2(\mathrm{Im}(\alpha_{11})|a|^2 - \mathrm{Im}(\alpha_{22})|b|^2). \tag{8.38}$$

The first derivative on normalized polarization $\widetilde{P}$ is given by

$$\frac{d\widetilde{P}}{dt} = \frac{d}{dt}(a^*b + ab^*) = -2V\bar{N} + i(-Aab^* + A^*a^*b), \tag{8.39}$$

where $\bar{N} = |a|^2 - |b|^2$ and $A = \alpha_{11} - \alpha_{22}$. Differentiating (8.39) once again, we get

$$\frac{d^2}{dt^2}(a^*b + ab^*) = -2\frac{d}{dt}(\bar{N}V) - [A^2 ab^* + A^{*2}a^*b] + i\bar{N}V(A - A^*), \tag{8.40}$$

which one can rewrite in another form

$$\frac{d^2}{dt^2}\widetilde{P} = -2\frac{d}{dt}(\bar{N}V) - [\mathrm{Re}(A)^2 - \mathrm{Im}(A)^2]\widetilde{P} - 2i\,\mathrm{Re}(A)\mathrm{Im}(A)(ab^* - a^*b) - 2\bar{N}V\mathrm{Im}(A). \tag{8.41}$$

With the help of the above relations Eq. (8.39) can be rewritten as

$$i\,\mathrm{Re}(A)(a^*b - ab^*) = \frac{d}{dt}\widetilde{P} + 2\bar{N}V - \mathrm{Im}(A)\widetilde{P}. \tag{8.42}$$

Substituting it into (8.40). We obtain

$$\frac{d^2}{dt^2}\widetilde{P} - 2\mathrm{Im}(A)^2\frac{d}{dt}\widetilde{P} + (\mathrm{Re}(A)^2 + \mathrm{Im}(A)^2)\widetilde{P} = -2\frac{d}{dt}(\bar{N}V) + 2\bar{N}V\mathrm{Im}(A), \tag{8.43}$$

where $\mathrm{Im}(A) = \frac{1}{i}(\alpha_{11} - \alpha_{22}^*) = \Gamma_a + \Gamma_b$ and $\mathrm{Re}(A)^2 + \mathrm{Im}(A)^2 = \Omega_{ab}^2 + (\Gamma_a + \Gamma_b)^2$. The resulting equation for $\widetilde{P}$ can be written in the final form

$$\frac{d^2\tilde{P}}{d\tau^2} + 2(\Gamma_a + \Gamma_b)\frac{d\tilde{P}}{d\tau} + [\Omega_{ab}^2 + (\Gamma_a + \Gamma_b)^2]\tilde{P} = 2\left[\frac{d}{d\tau}(\bar{N}V) + (\Gamma_a + \Gamma_b)\bar{N}V\right], \tag{8.44}$$

where $\bar{N} = |a|^2 - |b|^2$ is an average number of particles and $\Gamma_{a,b} = \gamma_{a,b}/\wp$, $\Omega_{a,b} = \omega_{a,b}/\wp$, $\Omega = \omega/\wp$ and $V = V(\tau) = \alpha_{11} \cdot \sin\Omega\tau$.

In the simple case $|a|^2 - |b|^2 \cong const = \bar{N}$ Eq. (8.44) becomes

$$\frac{d^2\widetilde{P}}{d\tau^2} + 2\Gamma_{ab}\frac{d\widetilde{P}}{d\tau} + (\Omega_{ab}^2 + \Gamma_{ab}^2)\widetilde{P} = 2\bar{N}\left[\frac{dV(\tau)}{d\tau} + \Gamma_{ab}V(\tau)\right]. \tag{8.45}$$

Since N = *constant* Eq. (8.45) for $\widetilde{P}$ becomes uncoupled from system (8.32), (8.33) and it can now be readily solved by standard methods.

We have

$$\frac{d^2\widetilde{P}}{d\tau^2} + 2\Gamma_{ab}\frac{d\widetilde{P}}{d\tau} + \tilde{\Omega}_{ab}^2\widetilde{P} = 2\bar{N}\alpha_{12}[\Omega\cos\Omega\tau + \Gamma_{ab}\sin\Omega\tau], \tag{8.46}$$

so that Eq. (8.46) becomes

$$\frac{d^2\widetilde{P}}{d\tau^2} + 2\Gamma_{ab}\frac{d\widetilde{P}}{d\tau} + \tilde{\Omega}_{ab}^2\widetilde{P} = N\alpha_{12}(\Omega - i\,\Gamma_{ab})\,e^{i\Omega\tau} + k.c., \tag{8.47}$$

where $\Gamma_{ab} = \Gamma_a + \Gamma_b$ and $\tilde{\Omega}_{ab}^2 = \tilde{\Omega}_{ab}^2 + \Gamma_{ab}^2$.

The solution for Eq. (8.47) has the form

$$\widetilde{P} = \widetilde{P}_0\,e^{i\Omega\tau}, \tag{8.48}$$

where

$$\widetilde{P}_0 = -\frac{\bar{N}\alpha_{12}\cdot(\Omega - i\,\Gamma_{ab})}{\Omega^2 - \tilde{\Omega}^2 - 2i\,\Gamma_{ab}\Omega}. \tag{8.49}$$

We consider in (8.49) $|\Gamma_{ab}|^2 \ll \Omega_{ab}^2$ and $\Omega^2 \cong \tilde{\Omega}_{ab}^2$, and after rewriting (8.49) in the initial dimensional variables, we get

$$P_0 = -\frac{\bar{N}\omega_{ab}\cdot\left(1 - i\frac{\gamma_{ab}}{\omega}\right)\left(\frac{D_{ba}^2}{\hbar}\right)}{\omega^2 - \omega_{ab}^2 - 2i\,\gamma_{ab}\omega}E_o. \tag{8.50}$$

Since the dielectric induction has the form $D = \varepsilon_0 E + P$ one can write the dielectric susceptibility in the form

$$\chi(\omega) = \frac{P_0}{E_0} = \frac{-G}{\omega^2 - \omega_{ab}^2 - 2i\,\gamma_{ab}\omega}, \tag{8.51}$$

where

$$G = \frac{N\omega_{ab}\cdot D_{ba}^2\cdot\left(1 - i\frac{\gamma_{ab}}{\omega}\right)}{\varepsilon_0\hbar}. \tag{8.52}$$

The dielectric permittivity of environment $\varepsilon(\omega) = \varepsilon_h + \chi(\omega)$ is given by

$$\varepsilon(\omega) = \varepsilon_h - \frac{G}{\omega^2 - \omega_{ab}^2 - 2i\,\gamma_{ab}\omega}, \tag{8.53}$$

where $\omega_{ab} = \omega_a - \omega_b = (E_a - E_b)/\hbar > 0$, $\bar{N} = |a|^2 - |b|^2 = |a_0|^2 - |b_0|^2 = constant$. ε_h is a dielectric permittivity of a host medium. Later we use formula (8.53) for the calculation of electromagnetic field in a coated microsphere.

While $\bar{N} = |a|^2 - |b|^2 > 0$ and $G > 0$, such a system dwells in the equilibrium steady-state. But while population is inverted in $\bar{N} = |a|^2 - |b|^2 < 0$, we get $G < 0$. In this case TLA is in exited state and it can amplify a resonance electromagnetic field. In the zone of a resonance where $\omega - \omega_{ab} = \delta$, $|\delta| \ll \omega, \omega_{ab}$ and in a case of weak dissipation $|\gamma_{ab}/\omega_{ab}| \ll 1$ the dielectric permittivity becomes a well-known form

$$\varepsilon(\omega) = \varepsilon_h - \frac{\bar{N} \cdot \left(\frac{D_{ba}^2}{2\hbar}\right)}{\omega - \omega_{ab} - i\,\gamma_{ab}}. \tag{8.54}$$

One can see again that the properties of such a system are strongly depending on a sign of the quantity $\bar{N} = |a|^2 - |b|^2$.

8.4 Temporal Dynamics of Polarization and the Probability Amplitudes

The analysis of a temporal dynamics becomes simpler if one writes Eqs. (8.32), (8.33) for $a(\tau)$ and $b(\tau)$ in another form. We change the variables as $a \to \bar{a}$ and $b \to \bar{b}$, where

$$a(\tau) = \bar{a}(\tau)\, e^{-i\alpha_{11}\tau}, \quad b(\tau) = \bar{b}(\tau)\, e^{-i\alpha_{22}\tau}. \tag{8.55}$$

After some algebra manipulations the equations for $\bar{a}, \bar{b}$ become

$$\frac{d\bar{a}}{d\tau} = V(t) \cdot b e^{iB\tau}, \tag{8.56}$$

$$\frac{d\bar{b}}{d\tau} = -V(t) \cdot \bar{a} \cdot e^{-iB\tau}, \tag{8.57}$$

and the normalized polarization $\widetilde{P} = P/D_{ba}$ in (8.31) is now given by

$$\tilde{P} = \bar{a}^* \bar{b}\, e^{-iA^*\tau} + \bar{a}\bar{b}^*\, e^{iA^*\tau}, \tag{8.58}$$

where $V(\tau) = \frac{\Omega_{ba}}{\Omega} \sin \Omega\tau$ and

$$\begin{aligned} B &= -\alpha_{11} + \alpha_{22} = \Omega_{ba} + i(\Gamma_a - \Gamma_b), \\ A &= -\alpha_{11} + \alpha_{22}^* = \Omega_{ba} + i(\Gamma_a + \Gamma_b), \\ |a|^2 \pm |b|^2 &= |\bar{a}|^2\, e^{-2\Gamma_a\tau} \pm |\bar{b}|\, e^{-2\Gamma_b\tau}. \end{aligned} \tag{8.59}$$

Now we study the temporal dynamics the $\tilde{P}$ and the probability density $|C_{a,b}(r)|^2$ of the lower and upper atomic state due to of the external time-depended electromagnetic field. Also we explore a frequency response

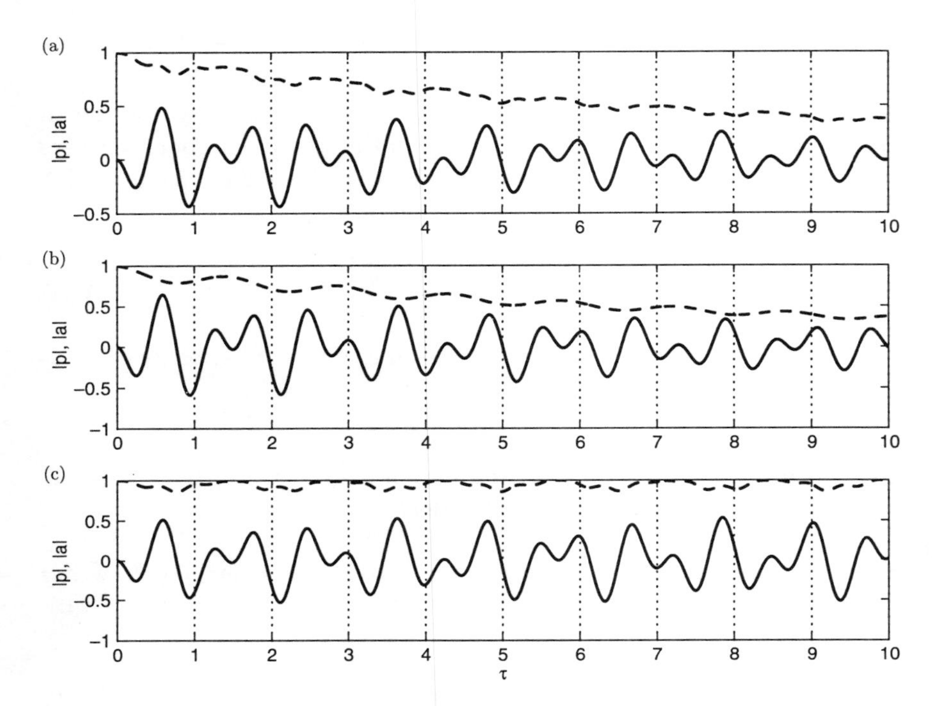

Fig. 8.1 Non-resonant case with $\Omega_{ab} = 10$, $\Omega = 6$. Solid line is a modulo of dynamical polarization $p \equiv \widetilde{P}$ and the dash line is the expectation value $|a|^2$. (a) $\Gamma_a = 0.05$ and $\Gamma_b = 0.04$ without RWA; (b) is the same as case (a) but with RWA; (c) is the same as case (a) but with $\Gamma_a = 0$ and $\Gamma_b = 0$.

of $\widetilde{P}$. It is clear from (8.58), that in general the polarization is a nonlinear function of the field. Both A and B in (8.59) are complex quantities, besides Eqs. (8.56) and (8.57) have the variable coefficients. Therefore is rather difficult to find the exact solution of Eqs. (8.56), (8.57). Therefore in addition we study system (8.56), (8.57) numerically. We use the initial conditions $\overline{a}(\tau = 0) = a_0, \overline{b}(\tau = 0) = b_0$ and we calculate the amplitudes $\overline{a}$, $\overline{b}$ and the polarization of TLA $\tilde{P}$ by Runge–Kutta method. After that we calculate the frequency spectrum by a fast Fourier transformation (FFT). The results are shown in Figs. 8.1–8.4. We used the following parameters. Parameters of absorption are $\Gamma_a = 0.05$ and $\Gamma_b = 0.04$. In Fig. 8.1 is shown a non-resonant case when both systems and fields frequencies are higher than Rabi frequency. Figure 8.2 shows corresponding Fouries spectrum.

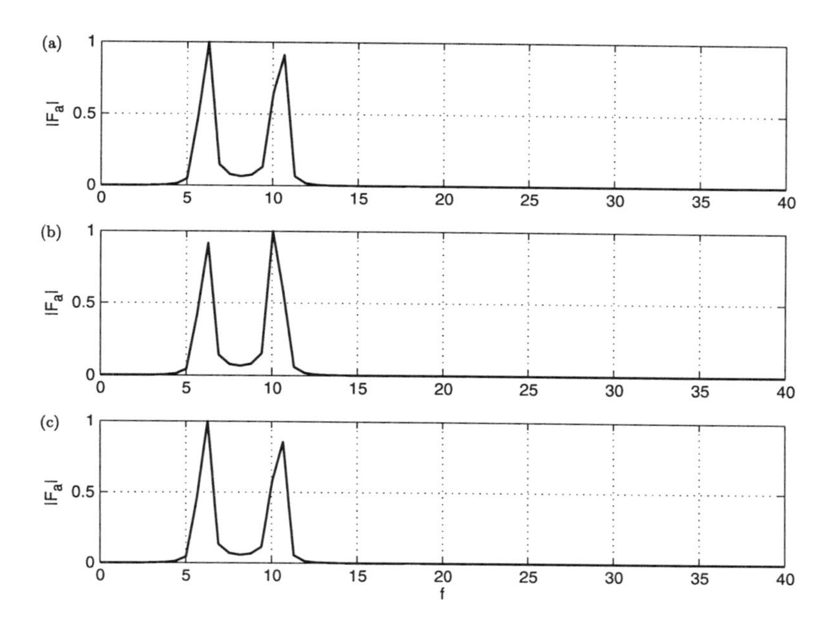

Fig. 8.2 Fourier spectrum of modulo of dynamical polarization $|\tilde{P}|$ for no resonant case (see Fig. 8.1).

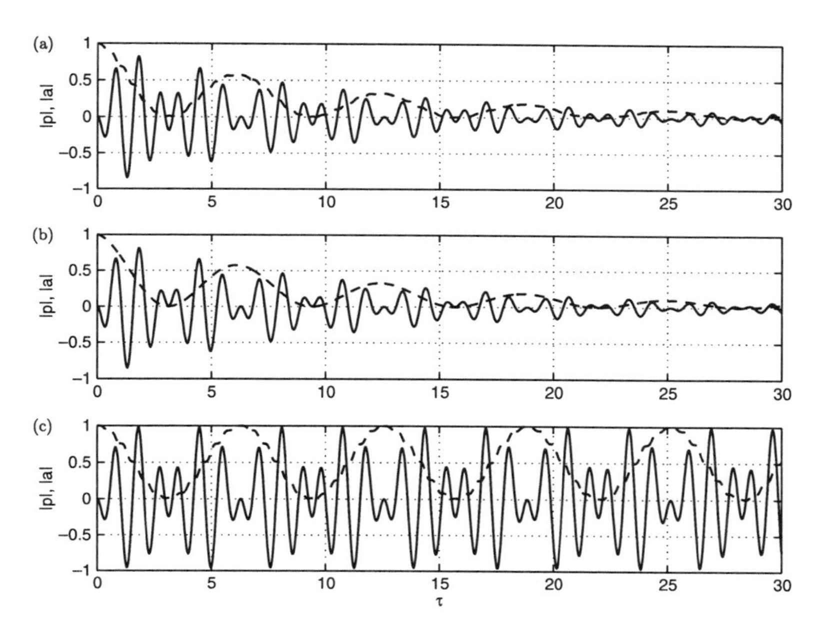

Fig. 8.3 Resonant case with $\Omega_{ab} = 6$, $\Omega = 6$. Solid line is a modulo of the dynamical polarization $p \equiv \widetilde{P}$ and the dash line is the expectation value $|a|^2$. (a) $\Gamma_a = 0.05$ and $\Gamma_b = 0.04$ without RWA; (b) is the same as case (a) but with RWA; (c) is the same as case (a) but with $\Gamma_a = 0$ and $\Gamma_b = 0$.

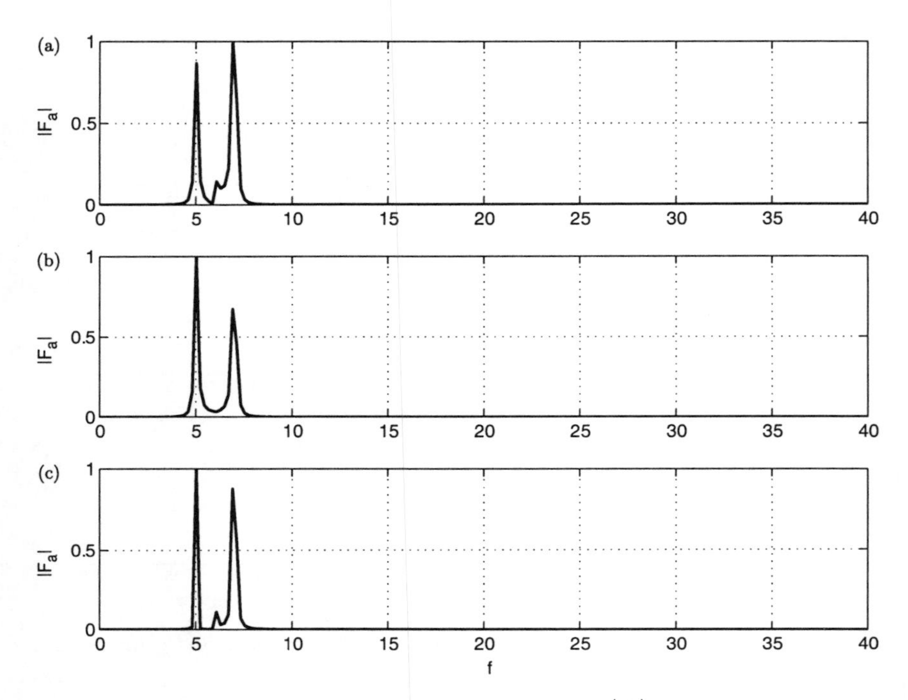

Fig. 8.4 Fourier spectrum of dynamical polarization $\left|\widetilde{P}\right|$ for resonant case (see Fig. 8.3).

Both TLA transition line and line of polarization are well separated and well coincide with the initial values.

Figures 8.3 and 8.4 show the resonant case. One can see from these figures that repulsion both lines from the initial values now occur. The frequency spectrum consists of a doublet of lines split by the vacuum Rabi frequency from initial values. This spliting equals 2 for our renormalization.

CHAPTER 9

Classical Field

9.1 Schrödinger Equation

We shall rewrite the basic formulas of the theory, stated in the previous sections, in the Dirac notations. It allows us to proceed to the matrix formulation of the problem and then to the discussion of the procedure of quantization of the field that is necessary for the study of spontaneous emission. In [Louisell, 1964a, Meyste, 1998a] one can find much more about the Dirac notations. Typically we consider the interaction of atoms with electromagnetic fields. To treat such interactions, we add the appropriate interaction energy H_1 to the unperturbed Hamiltonian H_0, that is

$$i\,\hbar\frac{\partial}{\partial t}|\psi\rangle = (H_0 + H_1)|\psi\rangle, \tag{9.1}$$

where $|\psi\rangle$ is an unknown state vector. If we expand (9.1) the state vector $|\psi\rangle$ in terms of the eigenfunctions $|u_m\rangle$ of the "unperturbed Hamiltonian" H_0, rather than those of the total Hamiltonian H the probability amplitudes $C_m(t)$ change in time. So

$$|\psi\rangle = \sum_m C_m(t)|u_m\rangle, \tag{9.2}$$

where $|u_m\rangle$ are state vectors of the non-perturbed Hamiltonian H_0 so $H_0|u_m\rangle = E_m|u_m\rangle$, and E_m represent a corresponding energy spectrum and $C_m(t)$ is the unknown time-dependent factors that we have to find. C_m are complex probability amplitudes, which completely determine the wave function $|\psi\rangle$. $|C_m|^2$ can be interpreted as the probability that the system is in the mth energy state. In particular if we are only concerned about how a system (such as an atom) absorbs energy from a light field, this development is completely described by the changes in the C_m. Substituting (9.2) in (9.1), we obtain

$$i\hbar \sum_m \frac{\partial C_m(t)}{\partial t} |u_m\rangle = \sum_m C_m(t) H_0 |u_m\rangle + \sum_m H_1 C_m(t) |u_m\rangle. \tag{9.3}$$

Let's multiply the right side of Eq. (9.3) by $\langle u_k|$ and sum on m with the account, that $\langle u_k|u_m\rangle = \delta_{km}$, we obtain

$$i\hbar \frac{\partial C_k(t)}{\partial t} = E_k C_k(t) + \sum_m \langle u_k|H_1|u_m\rangle C_m(t). \tag{9.4}$$

Now we consider

$$H_1 = -e(\vec{r} \cdot \vec{E}), \tag{9.5}$$

where $\vec{E} = \vec{E}(\vec{r}, t)$. Furthermore we write $|u_m\rangle$ simply as $|m\rangle$. We assume a two-level atom is located at point $\vec{r_0}$. In the dipole approximation $\vec{r} = \vec{r}_0$, so $\vec{E} = \vec{E}(\vec{r}_0, t) = \vec{E}_0 F(t)$. Taking into account the next operational identity, which is valid for a general operator L

$$L = \sum_{m,n} |m\rangle\langle m|L|n\rangle\langle n|,$$

we write (9.5) as follows:

$$H_1 = -e\vec{E}(t) \cdot \sum_{i,j} |i\rangle\langle i|\vec{r}|j\rangle\langle j| = -\vec{E} \cdot \sum_{i,j} |i\rangle \vec{\wp}_{ij} \langle j|, \tag{9.6}$$

where $\vec{\wp}_{ij} = e\langle i|\vec{r}|j\rangle$.

9.2 Matrix Form for Two-Level Atom

Now we assume a two-level atom, where level a is an excited state, and b is a ground state and one can neglect the influence of the other states. Such assumption does not follow from the theory and can be justified only by a comparison with an experiment. For such an atom we have

$\vec{\wp}_{ij} = \vec{\wp}_{ab}\delta_{ia}\delta_{jb} + \vec{\wp}^*_{ba}\delta_{ja}\delta_{ib}$, where $i, j = a, b$, $\vec{\wp}_{ab} = \vec{\wp}_{ab} \equiv \vec{\wp}$ and $\vec{\wp}_{aa} = \vec{\wp}_{bb} = 0$. Then (9.6) becomes

$$H_1 = -\hbar F(t)\Re \cdot (|a\rangle\langle b| + |b\rangle\langle a|), \tag{9.7}$$

where $\vec{E} = \vec{E}_0 F(t)$, $\Re = (\vec{E}_0 \cdot \vec{\wp})/\hbar$. The parameter $\Re$ is known as the Rabi frequency. For such two-level atom Eq. (9.4) becomes

$$\begin{aligned} i\hbar\frac{\partial C_a(t)}{\partial t} &= E_a C_a(t) + \hbar F(t)\Re C_b(t), \\ i\hbar\frac{\partial C_b(t)}{\partial t} &= E_b C_b(t) + \hbar F(t)\Re C_a(t). \end{aligned} \tag{9.8}$$

If we use the matrix notation, the Eq. (9.8) one can rewrite in the form

$$i\hbar\frac{\partial \vec{C}}{\partial t} = \begin{bmatrix} E_a & 0 \\ 0 & E_b \end{bmatrix} \vec{C} + \hbar F(t)\Re \begin{bmatrix} 0 & 1 \\ 1 & 0 \end{bmatrix} \vec{C}, \tag{9.9}$$

where the vector $\vec{C} = \{C_a(t), C_b(t)\}$. We can reorganize (9.9) in other useful form. We write $E_a = E_0 + \Delta E$ and $E_b = E_0 - \Delta E$, where $E_0 = (E_a + E_b)/2$ and $\Delta E = (E_a - E_b)/2$. Then

$$\begin{bmatrix} E_a & 0 \\ 0 & E_b \end{bmatrix} = E_0 \begin{bmatrix} 1 & 0 \\ 0 & 1 \end{bmatrix} + \Delta E \begin{bmatrix} 1 & 0 \\ 0 & -1 \end{bmatrix}.$$

As the origin of energy can be chosen by an arbitrary way, we can set $E_0 = 0$, $\Delta E = (E_a - E_b)/2 = \hbar(\omega_a - \omega_b)/2 = \hbar\omega_{ab}/2$, then Eq. (9.8) is written as

$$i\hbar\frac{\partial \vec{C}}{\partial t} = \frac{\hbar\omega_{ab}}{2}\sigma_z\vec{C} + \hbar F(t)\Re\sigma_1\vec{C}, \tag{9.10}$$

where 2d matrixes σ_z and σ_1 are

$$\sigma_z = \begin{bmatrix} 1 & 0 \\ 0 & -1 \end{bmatrix} \quad \text{and} \quad \sigma_1 = \begin{bmatrix} 0 & 1 \\ 1 & 0 \end{bmatrix}. \tag{9.11}$$

From (9.10) we can see that for two-level atom the Schrödinger equation reads

$$i\hbar\frac{\partial \vec{C}}{\partial t} = H\vec{C} = (H_a + H_1)\vec{C}, \tag{9.12}$$

where

$$H_a = (\hbar\omega_{ab}/2)\sigma_z \tag{9.13}$$

is the Hamiltonian of atom and

$$H_1 = (\vec{E}(t) \cdot \vec{\wp})\sigma_1 = \hbar F(t)\Re\sigma_1 \tag{9.14}$$

is the field-atom interaction part in full Hamiltonian H. With new variables $\bar{C}_{a,b}(t) = C_{a,b}(t)\, e^{i\omega_{a,b}t}$ Eqs. (9.9) become

$$\frac{\partial \bar{C}_a(t)}{\partial t} = -i\, F(t) Re^{-i\omega_{ab}t}\bar{C}_b(t), \qquad \frac{\partial \bar{C}_b(t)}{\partial t} = -i\, F(t) Re^{i\omega_{ab}t}\bar{C}_a(t). \tag{9.15}$$

For important case of oscillatory field $F(t) = \cos(\omega t)$ the Eqs. (9.15) become

$$\frac{\partial \bar{C}_a(t)}{\partial t} = -i\Re\cos(\omega t)\, e^{-i\omega_{ab}t}\bar{C}_b(t), \qquad \frac{\partial \bar{C}_b(t)}{\partial t} = -i\,\Re\cos(\omega t)\, e^{i\omega_{ab}t}\bar{C}_a(t). \tag{9.16}$$

Very often on the right hand sides of (9.16) only the nearly resonant terms with $\omega \approx \omega_{ab}$ are important, then

$$\cos(\omega t)\, e^{\pm i\omega_{ab}t} = \frac{1}{2}(e^{i(\omega\pm\omega_{ab})t} + e^{i(-\omega\pm\omega_{ab})t}) \approx \frac{1}{2}\begin{cases} e^{i(\omega-\omega_{ab})t}, & \text{if+} \\ e^{-i(\omega-\omega_{ab})t}, & \text{if}- \end{cases}.$$

Such approximation named RWA (rotative-wave approximation) is used very often on studying the dynamics of interaction of two-level atom with a field. In the RWA case Eqs. (9.16) become

$$\frac{\partial \bar{C}_a(t)}{\partial t} = -i\frac{\Re}{2}\, e^{i(\omega-\omega_{ab})t}\bar{C}_b(t), \qquad \frac{\partial \bar{C}_b(t)}{\partial t} = -i\frac{\Re}{2}\, e^{-i(\omega-\omega_{ab})t}\bar{C}_a(t). \tag{9.17}$$

Equations (9.17) can be simplified if we were to use the unidimensional variable $\tau = \Re t$, then Eqs. (9.16) acquire the form

$$\frac{\partial \bar{C}_a}{\partial \tau} = -i\cos(\nu\tau)\, e^{-i\nu_{ab}\tau}\bar{C}_b, \qquad \frac{\partial \bar{C}_b}{\partial \tau} = -i\cos(\nu\tau)\, e^{i\nu_{ab}\tau}\bar{C}_a, \tag{9.18}$$

where $\nu = \omega/\Re, \nu_{ab} = \omega_{ab}/\Re$.

CHAPTER 10

Quantization of Electromagnetic Field

In previous chapters we have used the quantum expressions for the description of the atomic transitions but the classical expressions for electromagnetic fields. One would naturally ask a question, How valid is such a mixed description of the atom-fields interaction? Whether is it always true? At the atomic transitions an atom absorbs or radiates the energy of the electromagnetic field by fixed portions, which are multiples of the energy of quantum $\hbar\omega$, where ω is the frequency of the field and $\hbar$ is the Planck's constant. Therefore one can say that the radiated field represents the set of quanta, thus, the quantum is the least particle of field. Such particles are referred to as photons. The latter allows one to get a fundamental conclusion that the quantum description of a field is more general than the classical one. The classical description represents only the asymptotic case of the quantum description when the number of photons n is great, i.e. $n \gg 1$.

Naturally the following question arises. Maxwell equations (1.1)–(1.4) do not contain the fundamental constant of quantum physics namely in the Planck's constant $\hbar$. Then in what way the quantum aspect appears in the theory of the electromagnetic field? The advanced answer can be found in textbooks on Quantum mechanics and Quantum optics [Basdevant & Dalibard, 2002; Cohen-Tannoudji *et al.*, 1989; D. & M., 1977; Liboff & L., 1998; Louisell, 1964; Mandel & Wolf, 1995; Meystre & III., 1998; Scully & Zubairy, 1996]. In this chapter we only briefly depict the logic scheme of the answer. Various approaches to the quantization of electromagnetic field were proposed in the last few years. Nowadays the main efforts are concentrated in the development of a quantization scheme for the radiation field in the dispersive and absorptive dielectrics, which can be applied to both homogeneous materials and multilayer dielectric structures. Various approaches to the radiation-field quantization for homogeneous and inhomogeneous Kramers-Kronig dielectrics have been developed in [Crenshaw, 2003; Glauber & Lewenstein, 1991; Gruner & Welsch, 1996; Dung *et al.*, 1998, 2002; Knoll *et al.*, 2001; Lagendijk, 2003; Matloob, 2001; Slepyan *et al.*, 2002; Scheel *et al.*, 1998].

In this section we discuss a simple approach to fields' quantization in non-uniform spherical structures. We will calculate the amplitude of vacuum field (field per photon) in spherical geometry.

10.1 Energy of Field

We recall the Heisenberg uncertainty principle of $\Delta p \Delta q \geq \hbar/2$, where Δp and Δq are uncertainties of the measurements of a momentum p and position q of a particle accordingly. This form of uncertainty principle comes from the commutation relations of operators momentum and position of particle

$$[p, p] = 0, \quad [q, q] = 0 \quad \text{and} \quad [p, q] = -i\,\hbar, \tag{10.1}$$

where commutators are $[a, b] = ab - ba$. Such relations for ensemble of particles are given by

$$[p_k, p_m] = 0, \quad [q_k, q_m] = 0 \quad \text{and} \quad [p_k, q_m] = -i\,\hbar\delta_{km}. \tag{10.2}$$

It is well-known [Landau & Lifshits, 1975; Jackson, 1975; Loudon, 1994; Scully & Zubairy, 1996; Meystre & III., 1998] that the energy of electromagnetic field

$$w = \tfrac{1}{2}(\varepsilon_0 \varepsilon E^2 + \mu_0 H^2) \tag{10.3}$$

can be represented in the form of energy of unit-mass oscillator (quantum oscillator)

$$w = \tfrac{1}{2}(p^2 + q^2\omega^2),$$

where p is momentum, q is position and ω is the oscillatory eigenfrequency. This represents the equivalence of the electromagnetic field to the ensemble of oscillators (fields oscillators). Also well-known is (see textbooks on quantum mechanics [D. & M., 1977; Liboff & L., 1998]) that the energy of quantum oscillator can be changed only in the discrete portions $n\hbar\omega$ where

n is an integer. Operators p and q of momentum and position of quantum oscillators satisfies to commutation relations (10.1), where $\hbar$ is written. The electric and magnetic fields E and H can be expressed linearly with p and q of the field's quantum oscillators, therefore both electric and magnetic fields become operators. One can say that due to the commutator relations $[q,p] = i\,\hbar$, the quantity $\hbar$ arrives into the field's equations. As a result of such procedure, named the quantization of field, the Planck's constant $\hbar$ comes to the description of fields and the Maxwell equations acquire the fundamental quantum aspect.

After such a brief comment we turn to the description of quantization of a free electromagnetic field in microspheres. For simplicity we deal with a one-mode field.

The procedure of quantization of the electromagnetic field in a plane-waves case is well known [Meystre & III., 1998; Scully & Zubairy, 1996]. In a plane-waves case the following expression for an electric field turns $E = \mathcal{E}_0(a^+e^{i\theta} + ae^{-i\theta})$, where the quantity $\mathcal{E}_0 = (\hbar\omega/2\varepsilon_0 V)^{1/2}$ has the dimensions of electric field, V is a volume, $\theta = \omega t - \overrightarrow{k}\,\overrightarrow{r}$, a^+ and a are the creation and annihilation operators of photon on a corresponding mode respectively. Such a field is included into the atom-field interactions part of Schrödinger equation $H_{af} = -e(\overrightarrow{r} \cdot \overrightarrow{E})$. For exited atom this interaction occurs even for vacuum state $|0\rangle$ and can stimulate the atom to emit spontaneously.

It is well-known that the expecting value of linear polarized field vanish $\langle n|E|n\rangle = 0$ for a number state $|n\rangle$, but the expectation value of the intensity operator E^2 is given by $\langle n|E^2|n\rangle = (\mathcal{E}_0)^2(2n+1)$. The non-zero fluctuations take place even for a vacuum state $|0\rangle$ i.e. $\langle 0|E^2|0\rangle = (\mathcal{E}_0)^2$. From the latter the physical sense of a quantity $\mathcal{E}_0$ as the field's vacuum amplitude (or field per photon) can be recognized.

In plane-waves case the expectation value $\langle 0|E^2|0\rangle$ depends only on the eigenfrequency of cavity mode as $\omega^{1/2}$. But as we show in this chapter, such simple behavior is modified in microspheres and becomes more intricate: a field vacuum amplitude acquires more complex dependence on the radial structure. This peculiarity becomes important due to a recent progress in studying the various fundamental effects in microspheres [Moller *et al.*, 2002; Deumie *et al.*, 2002] as well as ensembles of microspheres [Furukawa & Tenjimbayashi, 2001], where such quantum effects can become important following a field-atom coupling.

Furthermore we study a case of true microsphere having a number of the well-separated field modes of a spherical multilayered structure placed in a quantized sphere with radius a_0. Figure 10.1 shows the geometry of system.

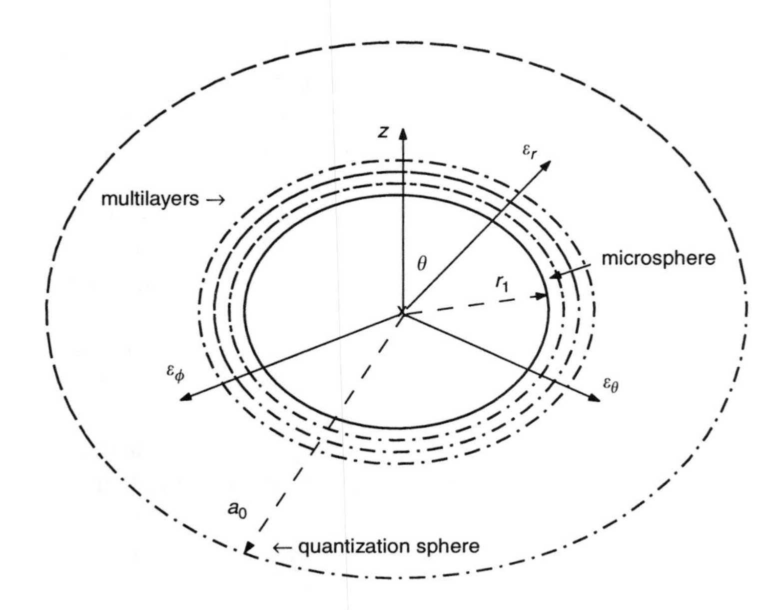

Fig. 10.1 Geometry of system.

Due to spherical geometry of the problem we will characterize the state of electromagnetic field in a microsphere with indexes ν, m and j (radial, spherical and azimuthal quantum number) accordingly. In this case for a photon state we write $|n_{\vec{k}}\rangle \to |n_{\nu m}\rangle, \omega_{\vec{k}} \to \omega_{\nu m}$ (the eigenfrequencies of sphere do not depend on azimuthal quantum number [Stratton, 1941]), $E_{\vec{k}} \to E_{\nu m j}$.

Quantization of the electromagnetic field is accomplished by choosing a and a^+ to be mutually adjoin operators. In Sec. 1.1.7 was derived the next expression for the field energy

$$W = \frac{\varepsilon_0}{8}\overline{\beta}^2[(A + A^*)^2 - (A - A^*)^2] = \frac{\varepsilon_0}{4}\overline{\beta}^2(AA^* + A^*A). \tag{10.4}$$

Now in treating (10.4) we represent the amplitudes A and A^* as

$$A = \sqrt{2}\mathcal{E}e^{i\varsigma}a, \qquad A^* = \sqrt{2}\mathcal{E}e^{-i\varsigma}a^+, \tag{10.5}$$

where $\mathcal{E}$ is a normalized constant which has a unit of field, $\mathcal{E}$ and ς are real numbers, $(\varsigma = wt)$ $\mathcal{E} > 0, a^+$ and a are creation and annihilations fields operators in the appropriate field modes with frequency ω. Quantity $\mathcal{E}$ will be defined later from the relations with energy of quantum field. Since the particle of fields (photons) are bosons the appropriate commutation relations must be applied for the operators of a and a^+ (the boson commutation relations). For the case of one particle we have the commutation relations in the form

$$[a, a] = [a^+, a^+] = 0, \quad [a, a^+] = 1, \tag{10.6}$$

and the rules for the action of a and a^+ to the number state vector $|n\rangle$ are given by

$$a|n\rangle = \sqrt{n}|n-1\rangle, \quad a^+|n\rangle = \sqrt{n+1}|n+1\rangle. \tag{10.7}$$

At the use of pair operators a^+ and a one of the most important relation reads

$$a^+a|n\rangle = a^+\sqrt{n}|n-1\rangle = \sqrt{n}\sqrt{n}|n\rangle = n|n\rangle. \tag{10.8}$$

This means that the eigenvalue of a^+a is the number of particles in state $|n\rangle$. Other useful relations are

$$a|0\rangle = 0, \quad |n\rangle = \frac{(a^+)^n}{\sqrt{n!}}|0\rangle. \tag{10.9}$$

In the case of ensemble of particles (Fock's state) $|\{n\}\rangle = |n_1\rangle|n_2\rangle\cdots|n_N\rangle = \prod_{m=1}^{N}|n_m\rangle$ we can write the similar relations in form

$$[a_k, a_m] = [a_k^+, a_m^+] = 0, \quad [a_m, a_k^+] = \delta_{km}, \tag{10.10}$$

$$a_k|\{n\}\rangle = \sqrt{n_k}|n_1\rangle|n_2\rangle\cdots|n_{k-1}\rangle|n_k-1\rangle|n_{k+1}\rangle\cdots|n_N\rangle, \tag{10.11}$$

and

$$a_m^+|\{n\}\rangle = \sqrt{n_m+1}|n_1\rangle|n_2\rangle\cdots|n_{m-1}\rangle|n_m+1\rangle|n_{m+1}\rangle\cdots|n_N\rangle. \tag{10.12}$$

Also

$$a_m^+a_k|\{n\}\rangle = a_m^+\sqrt{n_k}|n_1\rangle|n_2\rangle\cdots|n_k-1\rangle\cdots|n_N\rangle \tag{10.13}$$

$$= \sqrt{n_k}\sqrt{n_m+1}|n_1\rangle|n_2\rangle\cdots|n_k-1\rangle|n_m+1\rangle\cdots|n_N\rangle(1-\delta_{km})$$
$$+ n_k|\{n\}\rangle\delta_{km}. \tag{10.14}$$

After substitution (10.5) in (10.4) we obtain the fields Hamiltonian in the form

$$H_f = \varepsilon_0 a^3 \mathcal{E}^2 \beta^2 \left(a^+a + \tfrac{1}{2}\right) = \hbar\omega\left(a^+a + \tfrac{1}{2}\right),$$

with expectation value (energy)

$$\langle n|H_f|n\rangle = 2a_0^3\mathcal{E}^2\overline{\beta}^2\frac{\varepsilon_0}{4}\left(n+\tfrac{1}{2}\right) = \hbar\omega\left(n+\tfrac{1}{2}\right), \tag{10.15}$$

where n is integer, and

$$\overline{\beta}^2 = \overline{\beta}^2(\alpha) = m(m+1)I_m(\alpha),$$

where m is the spherical quantum number, α the eigenfrequency and quantity $I_m(\alpha)$ defined by the radial structure of a field and is given by

$$I_m(\alpha) = \frac{1}{\alpha^3}\int_0^\alpha R_m^2(y)\,dy. \tag{10.16}$$

10.2 Structure of Vacuum Field

We can use (10.15) as a normalizing condition and define $\mathcal{E}^2$ as

$$\mathcal{E}^2 = \mathcal{E}(\alpha)^2 = \frac{\hbar\omega}{2a_0^3\overline{\beta}^2\frac{\varepsilon_0}{4}}. \tag{10.17}$$

Then we can rewrite the fields amplitudes in the next form. Substituting formulas from Sec. 1.1.7 in (10.5) we obtain

$$E_s = \frac{\mathcal{E}}{\sqrt{2}}(e^{i\varsigma}a + e^{-i\varsigma}a^+)F_s, \tag{10.18}$$

$$H_s = i\frac{\mathcal{E}}{\sqrt{2}}(e^{i\varsigma}a - e^{-i\varsigma}a^+)f_s, \tag{10.19}$$

where $F_s = F_s(r,\theta,\varphi)$, $f_s = f_s(r,\theta,\varphi)$, $s = r,\theta,\varphi$ are written in Sec. 1.1.7. Let us calculate the expected value of field E_s for a state $|n\rangle$. We have

$$\begin{aligned}
\langle n|E_s|n\rangle &= \langle n|\frac{\mathcal{E}}{\sqrt{2}}(e^{i\varsigma}a + e^{-i\varsigma}a^+)F_s|n\rangle \\
&= \frac{\mathcal{E}}{\sqrt{2}}F_s(e^{i\varsigma}\langle n|a|n\rangle + e^{-i\varsigma}\langle n|a^+|n\rangle) \\
&= \frac{\mathcal{E}}{\sqrt{2}}F_s(e^{i\varsigma}\sqrt{n}\langle n||n-1\rangle + e^{-i\varsigma}\sqrt{n+1}\langle n|n+1\rangle = 0,
\end{aligned} \tag{10.20}$$

where $\langle n_k|n_m\rangle = \delta_{km}$. Thus the expecting value of field vanish $\langle n|E_s|n\rangle = 0$ for a number state $|n\rangle$.

The expectation value of the intensity operator E_s^2 is given by

$$\begin{aligned}\langle n|E_s^2|n\rangle &= \langle n|\tfrac{1}{2}\mathcal{E}^2F_s^2(e^{2i\varsigma}a^2 + e^{-2i\varsigma}(a^+)^{2i\varsigma} + a^+a + aa^+)|n\rangle \\ &= \tfrac{1}{2}\mathcal{E}^2F_s^2(\langle n|2a^+a+1|n\rangle) = \tfrac{1}{2}\mathcal{E}^2F_s^2(2n+1). \end{aligned} \qquad (10.21)$$

As far as for fluctuation of a variable z we have $\langle z^2 - \langle z\rangle^2\rangle = \langle z^2 - 2z\langle z\rangle + \langle z\rangle^2\rangle = \langle z^2\rangle - \langle z\rangle^2$, the fluctuation of field is given by $\langle E^2\rangle - \langle E\rangle^2$. Since $\langle 0|E^2|0\rangle = \frac{1}{2}\mathcal{E}^2F_s^2$, one can conclude: the nonzero fluctuations of field take place even for a vacuum state $|0\rangle$. From the latter the physical sense of quantity $\mathcal{E}$ as the field's vacuum amplitude (or field per photon) can be recognized.

From (10.15) the amplitude $\mathcal{E}$ for vacuum state $|0\rangle$ can be written in the form

$$\mathcal{E} = \mathcal{E}(\alpha) = \mathcal{E}_0\Delta_m(\alpha), \quad \Delta_m(\alpha) = \left(\frac{16\pi}{3\overline{\beta}^2}\right)^{1/2} = \left(\frac{16\pi\alpha^3}{3m(m+1)I_m(\alpha)}\right)^{1/2}, \qquad (10.22)$$

where $\mathcal{E}_0 = (\hbar\omega/(2\varepsilon_0 V))^{1/2}$ is a well-known amplitude of the vacuum field (field per photon) for a plane geometry case [Scully & Zubairy, 1996], which does not depend on the structure of field, and $V = (4\pi a_0^3)/3$ is the volume of sphere. The quantity $\Delta_m(\alpha)$ in (10.22) defines the correction of such amplitude due to the spherical geometry. Equation (10.22) provides the solution of the vacuum field problem for a microsphere case. One can see that the amplitude $\mathcal{E}(\alpha)$ in the spherical geometry depends on the spherical m quantum number and also on the radial quantum number ν through the eigenfrequencies $\alpha_{\nu m}$.

CHAPTER 11

Schrödinger and Interaction Pictures

11.1 Equations for the State Vectors

The Schrödinger equation becomes a more simple form in the interaction picture. We start from Schrödinger picture in the form

$$i\hbar\frac{\partial}{\partial t}|\Psi_S\rangle = H_S|\Psi_S\rangle, \tag{11.1}$$

where $H_S = H_0 + H_S^1$, H_0 is the unperturbed Hamiltonian, which is not depending on time t, H_S^1 is a part corresponding to interaction in the full Hamiltonian H_S. We search for the solution (11.1) in form

$$|\Psi_S\rangle = \sum_n C_n(t)e^{-(i/\hbar)\mathcal{E}_n t}|n\rangle, \tag{11.2}$$

where $|n\rangle$ and $\mathcal{E}_n$ are the set of known stationary eigenvectors and eigenvalues of Hamiltonian H_0, $H_0|n\rangle = \mathcal{E}_n|n\rangle$, and $C_n(t)$ are the time-dependent unknown amplitudes subjected to definition.

In a general new representation we write down the transformation of a wave function $|\Psi_I\rangle$ accordingly to the formula

$$|\Psi_I\rangle = K(t)|\Psi_S\rangle, \tag{11.3}$$

where $K(t)$ is some operator, which for interaction picture has the form

$$K = K(t) = \exp(i\,H_0 t/\hbar) = \exp(\alpha H_0), \quad K^{-1} = \exp(-\alpha H_0), \alpha = i\,t/\hbar. \tag{11.4}$$

Furthermore we shall supply both operators and wavefunctions with the bottom indexes S or I to separate the Schrödinger pictures (S) or the interaction pictures (I). In accordance to the general rules, at such transformations an operator G_S will be transformed according to $G_I = KG_SK^{-1}$. For finding-out the sense of use of operational expression in the exponents, it is useful to derive the next important formula. For a function $u(t)$ for time t one can write down the following expansion in the Taylor's series:

$$
\begin{aligned}
u(t+t_0) &= u(t_0) + t\frac{d}{dt_0}u(t_0) + \frac{t^2}{2!}\frac{d^2}{dt_0^2}u(t_0) + \cdots = e^{td/dt_0}u(t_0) \\
&= e^{-(i/\hbar)H_S t}u(t_0), \qquad (11.5)
\end{aligned}
$$

where we have used the expression $d/dt = -(i/\hbar)H_S$, which follows from the Schrödinger Equation (11.1). Using (11.3) and (11.5) in (11.3), we obtain

$$
|\Psi_I\rangle = K(t)|\Psi_S\rangle = \sum_n C_n(t)e^{-(i/\hbar)\mathcal{E}_n t}e^{\alpha H_0}|n\rangle = \sum_n C_n(t)|n\rangle. \qquad (11.6)
$$

Expression $G_I = KG_SK^{-1}$ applied to a case $G_S = H_S$ gives

$$
\begin{aligned}
H_I &= e^{\alpha H_0}H_S e^{-\alpha H_0} = e^{\alpha H_0}(H_0 + H_S^1)e^{-\alpha H_0} \\
&= H_0 + e^{\alpha H_0}H_S^1 e^{-\alpha H_0}. \qquad (11.7)
\end{aligned}
$$

Here we considered the known formula [Scully & Zubairy, 1996] for operators A and B, α is number, as follows

$$
e^{\alpha A}Be^{-\alpha A} = B + \alpha[AB] + \frac{\alpha^2}{2!}[A[AB]] + \cdots, \qquad (11.8)
$$

where commutator $[AB] = AB - BA$. It is easy to see that applying the formula (11.8) for a case $A = B = H_0$ gives us

$$
e^{\alpha H_0}H_0 e^{-\alpha H_0} = H_0. \qquad (11.9)
$$

Now we can obtain Schrödinger equation in the interaction representation. For this purpose we substitute derived formulas into Eq. (11.1), that for $|\Psi_S\rangle = K^{-1}|\Psi_I\rangle$ gives us

$$i\hbar\frac{\partial}{\partial t}[K^{-1}\,|\Psi_I\rangle] = i\hbar\left[-\frac{i}{\hbar}H_0|\Psi_I\rangle + \frac{\partial}{\partial t}|\Psi_I\rangle\right]K^{-1}$$
$$= \left(H_0 + H_S^1\right)K^{-1}|\Psi_I\rangle. \tag{11.10}$$

So

$$i\hbar\frac{\partial}{\partial t}|\Psi_I\rangle = -H_0|\Psi_I\rangle + K\left(H_0 + H_S^1\right)K^{-1}|\Psi_I\rangle$$
$$= -H_0|\Psi_I\rangle + \left[H_0 + (KH_S^1K^{-1})\right]|\Psi_I\rangle = H_I^1|\Psi_I\rangle. \tag{11.11}$$

Thus, the Schrödinger equation in interaction picture is given by

$$i\hbar\frac{\partial}{\partial t}|\Psi_I\rangle = H_I^1|\Psi_I\rangle, \tag{11.12}$$

where H_I^1 is the interactions part in full Hamiltonian and has the form

$$H_I^1 = \left(KH_S^1K^{-1}\right) = e^{iH_0t/\hbar}H_S^1e^{iH_0t/\hbar}. \tag{11.13}$$

Substituting (11.2) in (11.12), and after simple algebra one obtains

$$i\hbar\frac{\partial}{\partial t}\langle m|\sum_n C_n(t)|n\rangle = i\hbar\frac{\partial}{\partial t}C_m(t) = \sum_n C_n(t)\langle m|H_I^1|n\rangle, \tag{11.14}$$

where $\langle m|n\rangle = \delta_{mn}$. Thus in the interaction picture the equation for amplitudes $C_n(t)$ has the form

$$i\hbar\frac{\partial C_m(t)}{\partial t} = \sum_n C_n(t)\langle m|H_I^1|n\rangle \tag{11.15}$$

that we often will use further.

11.2 Equations for Operators

Now we can derive the motion equations for some operator $A_I(t)$ in the interaction picture. Let us calculate a temporal derivative

$$\begin{aligned}\frac{dA_I(t)}{dt} &= \frac{d}{dt}[e^{iH_0t/\hbar}A_Se^{iH_0t/\hbar}]\\ &= \frac{i}{\hbar}H_0e^{iH_0t/\hbar}A_Se^{iH_0t/\hbar} + e^{iH_0t/\hbar}\left(\frac{dA_S}{dt}\right)e^{iH_0t/\hbar}\\ &-\frac{i}{\hbar}e^{iH_0t/\hbar}A_Se^{iH_0t/\hbar}H_0\\ &= \frac{i}{\hbar}[A_IH_0 - H_0A_I] + e^{iH_0t/\hbar}\left(\frac{dA_S}{dt}\right)e^{iH_0t/\hbar}. \end{aligned} \tag{11.16}$$

If in the Schrödinger picture the operator A_S does not depend on time, Eq. (11.16) acquires a more simple form

$$i\hbar\frac{dA_I(t)}{dt} = [A_IH_0], \tag{11.17}$$

where on the right there is a commutator with the unperturbed Hamiltonian $[A_IH_0] = [A_IH_0 - H_0A_I]$.

As an example we apply (11.17) for the annihilation operator a in the interaction picture. In this case $H_0 = \hbar\omega(a^+a + 1/2)$. As follows from the aforesaid, at such transformation H_0 does not change and we can simply write down $H_0 = \hbar\omega(a_I^+a_I + 1/2)$. After that, using (11.17), we obtain

$$\begin{aligned}i\hbar\frac{da_I(t)}{dt} &= a_IH_0 - H_0a_I = \hbar\omega\left[a_I\left(a_I^+a_I + \tfrac{1}{2}\right) - \left(a_I^+a_I + \tfrac{1}{2}\right)a\right]\\ &= \hbar\omega[a_Ia_I^+a_I - a_I^+a_Ia] = \hbar\omega[(1 + a_Ia_I^+)a_I - a_I^+a_Ia]\\ &= \hbar\omega a_I. \end{aligned} \tag{11.18}$$

On deriving (11.18), we have used the commutator $a_Ia_I^+ - a_I^+a_I = 1$. Equation (11.18) can be solved easily, and as a result we obtain

$$a_I(t) = a_I(0)e^{-i\omega t} = a_Se^{-i\omega t}, \tag{11.19}$$

where we considered, that $a_I(0) = a_S$ for $t = 0$. Thus in the interaction picture (and in the Heisenberg picture) the annihilation operator acquires a negative frequency part $e^{-i\omega t}$. The solution of the similar equation for the creation operator gives us

$$a_I^+(t) = a_S^+ e^{i\omega t}, \tag{11.20}$$

and a^+ gets a positive frequency part $e^{i\omega t}$. Such positive or negative frequency parts at creation or annihilation operators have a deep physical sense in a context of the description of creation or annihilation of a photon with energy $\hbar\omega$.

11.2.1 *Operator's calculations*

Some useful formulas for calculations of the operators equations are written below. We use the next matrixes $\sigma_1 = \sigma_+ + \sigma_-$

$$\sigma_z = \begin{pmatrix} 1 & 0 \\ 0 & -1 \end{pmatrix}, \quad \sigma_+ = \begin{pmatrix} 0 & 1 \\ 0 & 0 \end{pmatrix}, \quad \sigma_- = \begin{pmatrix} 0 & 0 \\ 1 & 0 \end{pmatrix}, \tag{11.21}$$

for which the next commutators are valid

$$[\sigma_z \sigma_+] = 2\sigma_+, \quad [\sigma_z \sigma_-] = -2\sigma_-. \tag{11.22}$$

For calculation the exponential operators can use the next few relations:

$$e^{\alpha A} B e^{-\alpha A} = B + \alpha[AB] + \frac{\alpha^2}{2!}[A[AB]] + \frac{\alpha^3}{3!}[A[A[AB]]] + \cdots, \tag{11.23}$$

$$e^{\alpha\sigma_z}\sigma_- e^{-\alpha\sigma_z} = e^{-2\alpha}\sigma_-, \quad e^{\alpha\sigma_z}\sigma_+ e^{-\alpha\sigma_z} = e^{2\alpha}\sigma_+, \tag{11.24}$$

$$e^{\alpha a^+ a} a^+ e^{-\alpha a^+ a} = e^{\alpha} a^+, \quad e^{\alpha a^+ a} a e^{-\alpha a^+ a} = e^{-\alpha} a. \tag{11.25}$$

$$e^{A+B} = e^A e^B e^{-1/2[A,B]} = e^B e^A e^{1/2[A,B]} \tag{11.26}$$

CHAPTER 12

Two-Level Atom (The Matrix Approach, a Quantized Field)

As was already derived before, the Hamiltonian of atom H_a reads

$$H_a = \sum_i E_i |i\rangle \langle i| , \tag{12.1}$$

and

$$H_a|k\rangle = \sum_i E_i|i\rangle\langle i|k\rangle = E_i|k\rangle. \tag{12.2}$$

For two-level atoms i, j accept values of $i, j = a, b$, then $H_a = E_a|a\rangle\langle a| + E_b|b\rangle\langle b| = E_a\sigma_{aa} + E_b\sigma_{bb}$, where $\sigma_{ij} = |i\rangle\langle j|$. We enter $E = (E_a + E_b)$ and $\Delta E = E_a - E_b$, then $E_{a,b} = (E \pm \Delta E)/2$ and H_a becomes

$$\begin{aligned} H_a &= \frac{E + \Delta E}{2}\sigma_{aa} + \frac{E - \Delta E}{2}\sigma_{bb} \\ &= \frac{E}{2}(\sigma_{aa} + \sigma_{bb}) + \frac{\Delta E}{2}(\sigma_{aa} - \sigma_{bb}), \end{aligned} \tag{12.3}$$

where $\omega_{ab} = (E_a - E_b)/\hbar$. The energy level E can be chosen by an arbitrary way and we put $E = 0$, then the atom Hamiltonian (12.3) is reduced to

$$H_a = \hbar\frac{\omega_{ab}}{2}(\sigma_{aa} - \sigma_{bb}) = \hbar\frac{\omega_{ab}}{2}\sigma_z. \tag{12.4}$$

Expression (12.4) is received from the following calculations. Since $\langle a|b\rangle = \delta_{ab}$, (12.4) follows as:

$$H_a|a\rangle = \hbar\frac{\omega_{ab}}{2}|a\rangle\langle a|a\rangle = \hbar\frac{\omega_{ab}}{2}|a\rangle \quad \text{and} \quad H_a|b\rangle = -\hbar\frac{\omega_{ab}}{2}|b\rangle. \tag{12.5}$$

If we write for the state vector $|\Psi\rangle$ as

$$|\Psi\rangle = C_a|a\rangle + C_b|b\rangle = \begin{bmatrix} C_a \\ C_b \end{bmatrix}, \tag{12.6}$$

we obtain

$$H_a|\Psi\rangle = H_a \begin{pmatrix} C_a \\ C_b \end{pmatrix} = \hbar\frac{\omega_{ab}}{2}\begin{bmatrix} 1 & 0 \\ 0 & -1 \end{bmatrix}\begin{pmatrix} C_a \\ C_b \end{pmatrix} = \hbar\frac{\omega_{ab}}{2}\sigma_z \begin{pmatrix} C_a \\ C_b \end{pmatrix}, \tag{12.7}$$

where matrix σ_z is given by

$$\sigma_z = \begin{pmatrix} 1 & 0 \\ 0 & -1 \end{pmatrix}.$$

The field's Hamiltonian has the form

$$H_f = \sum_k \frac{\hbar\omega_k}{2}\left(a_k^+ a_k + a_k a_k^+\right) = \sum_k \hbar\omega_k \left(a_k^+ a + \frac{1}{2}\right), \tag{12.8}$$

where $a^+a + aa^+ = 2a^+a + 1$. Then

$$H_f|\{n\}\rangle = \sum_k \omega_k \hbar \left(a_k^+ a + \frac{1}{2}\right)|n_1\rangle|n_2\rangle \cdots = \sum_k \hbar\omega_k \left(n_k + \frac{1}{2}\right)|\{n\}\rangle, \tag{12.9}$$

where $|\{n\}\rangle = |n_1\rangle|n_2\rangle \cdots = \prod_k |n_k\rangle$ is the Fock's state for field.

$$H_f|\{n\}\rangle = \sum_k \hbar\omega_k \left(a_k^+ a + \frac{1}{2}\right)\begin{bmatrix} n_1 \\ n_2 \\ \dots \end{bmatrix} = \begin{bmatrix} \hbar\omega_1\left(n_1 + \frac{1}{2}\right) \\ \hbar\omega_2\left(n_2 + \frac{1}{2}\right) \\ \dots \end{bmatrix}. \tag{12.10}$$

Unpeturbated Hamiltonian for free atom and field is given by

$$H_0 = H_a + H_f. \tag{12.11}$$

To calculate the electro-dipole transition matrix element one can write

$$e\vec{r} = \sum_{ij} e|i\rangle\langle i|\vec{r}|j\rangle\langle j| = \sum_{ij} \vec{\wp}_{ij}|i\rangle\langle j| = \sum_{ij} \vec{\wp}_{ij}\sigma_{ij}, \tag{12.12}$$

where $\vec{\wp}_{ij} = \langle i|e\vec{r}|j\rangle$ and $\sigma_{ij} = |i\rangle\langle j|$. In the two-level atom case we have $i, j = a, b$, $\vec{\wp}_{ab} = \langle a|\, e\vec{r}\, |b\rangle$, $\vec{\wp}_{aa} = \vec{\wp}_{bb} = 0$, $\vec{\wp}_{ab} = \vec{\wp}_{ba} = \vec{\wp}$ and one can rewrite $e\vec{r}$ in the form $\vec{\wp}\sigma_1$, where matrixes $\sigma_1 = \sigma_+ + \sigma_-$ are written in (11.21),(9.11). Since the interaction atom-field's part in the Hamiltonian is given by $H_1 = H_{af} = -e\left(\vec{r}\cdot\vec{E}\right)$, $\vec{E} = \sum_k \vec{E}_k\widehat{a}_k + H.c.$ one can rewrite H_1 in the form

$$H_1 = -e(\vec{r}\cdot\vec{E}) = -(\sigma_+ + \sigma_-)\cdot\left[\sum_k \widetilde{\mathcal{E}}_k\widehat{a}_k + H.c.\right], \tag{12.13}$$

where $\widetilde{\mathcal{E}}_k = (\vec{E}_k\cdot\vec{\wp})$, and we sum over all the modes of field.

Now we can use the above written technique to pass from the Schrödinger picture to the interaction picture. In this case operator K is given by $K = K_a\cdot K_f$, where $K_a = \exp(iH_at/\hbar)$ and $K_f = \exp(i\,H_ft/\hbar)$, while operator $H_S^1 = H_{Sa}^1H_{Sf}^1$, as it follow from (12.13). Since commutator $[H_aH_f] = 0$ one has to rewrite Eq. (11.13) in the form

$$\begin{aligned} V \equiv H_I^1 &= (K_aK_f)\left(H_{Sa}^1H_{Sf}^1\right)(K_aK_f)^{-1} \\ &= \left(K_aH_{Sa}^1K_a^{-1}\right)\left(K_fH_{Sf}^1K_f^{-1}\right), \end{aligned} \tag{12.14}$$

or

$$\begin{aligned} V = -\left[e^{\alpha\sigma_z}(\sigma_+ + \sigma_-)\,e^{-\alpha\sigma_z}\right]\cdot\Bigg[&e^{\sum_k\beta_k\left(a_k^+a_k+\frac{1}{2}\right)}\cdot\sum_k\widetilde{\mathcal{E}}_k\left(\widehat{a}_k + \widehat{a}_k^+\right) \\ &\cdot\, e^{-\sum_k\beta_k\left(a_k^+a_k+\frac{1}{2}\right)}\Bigg], \end{aligned} \tag{12.15}$$

where $\alpha = i\,(\omega_{ab}/2)t$; $\beta_k = (it/\hbar)\hbar\omega_k$. Taking into account Eqs. (11.22)–(11.25) one can rewrite V in (12.15) in the form

$$V = -\left[e^{i\omega_{ab}t}\sigma_+ + e^{-i\omega_{ab}t}\sigma_-\right]\sum_k\widetilde{\mathcal{E}}_k\left(e^{i\nu_kt}\widehat{a}_k^+ + e^{-i\nu_kt}\widehat{a}_k\right). \tag{12.16}$$

Now the Schrödinger equation in the interaction picture becomes

$$i\hbar\frac{\partial}{\partial t}|\Psi\rangle = V|\Psi\rangle. \tag{12.17}$$

The solution of Schrödinger equation in (12.17) can be written in the form

$$|\Psi\rangle = C_a(t)|a\rangle|n_1\rangle|n_2\rangle\cdots + C_b(t)|b\rangle|n_1\rangle|n_2\rangle\cdots \tag{12.18}$$

Now we consider simplest case when field can be only in one of two states: the vacuum state $|\{0\}\rangle$ or in the one-photon state $|\{1_{\vec{k}}\}\rangle$. Such assumption is normally treated as the simplest model of the spontaneous emission. In this case from (12.18) we have

$$|\Psi\rangle = C_a(t)|\{0\}\rangle + \sum_{\vec{k}=1} C_b(t)\{\cdots 1_{\vec{k}} \cdots\}, \tag{12.19}$$

where $\{0\} = |0\rangle|0\rangle \cdots |0\rangle$ is the vacuum state of the field, and $\{1_{\vec{k}}\} = |n_{\vec{k}}\rangle|0\rangle|0\rangle \cdots |0\rangle$ are all possible states of the field with one photon in the $\vec{k}$th mode.

As it follows from (12.12) and (11.24) the operator of the dynamic dipole moment of two-level system in the interaction picture can be written in the form

$$\vec{P} = -\vec{\wp} \cdot \left[e^{i\omega_{ab}}\sigma_+ + e^{-i\omega_{ab}}\sigma_-\right]. \tag{12.20}$$

Let us calculate the expectation value of polarization $\vec{P}$, with $|\Psi\rangle$ given by (12.19). We have

$$\langle \vec{P} \rangle = \langle \Psi | \vec{P} | \Psi \rangle = -\vec{\wp}_{ab} \langle \Psi | e^{i\omega_{ab}t}\sigma_+ + e^{-i\omega_{ab}t}\sigma_- | \Psi \rangle. \tag{12.21}$$

In (12.21) we first calculate

$$\left(e^{i\omega_{ab}t}\sigma_+ + e^{-i\omega_{ab}t}\sigma_-\right)|\Psi\rangle = \begin{bmatrix} e^{i\omega_{ab}} \sum_{\vec{k}} C_b \{1_{\vec{k}}\} \\ e^{-i\omega_{ab}} C_a\{0\} \end{bmatrix}, \tag{12.22}$$

then we calculate the scalar product

$$\begin{bmatrix} C_a^*\{0\} \sum_{\vec{k}} C_b^*\{1_{\vec{k}}\} \end{bmatrix} \cdot \begin{bmatrix} e^{i\omega_{ab}t} \sum_{\vec{k}} C_b\{1_{\vec{k}}\} \\ e^{-i\omega_{ab}t} C_a\{0\} \end{bmatrix}$$

$$= e^{i\omega_{ab}t} C_a^*\{0\} \sum_{\vec{k}} C_b\{1_{\vec{k}}\} + \text{c.c.} \tag{12.23}$$

So, the expectation value of the dynamic dipole moment of atom $\langle \vec{P} \rangle$ is given by

$$\langle \vec{P} \rangle = -\vec{\wp}_{ab} \cdot \left\{ [C_a\{0\}e^{-i\omega_{ab}t}]^* \sum_{\vec{k}} C_b\{1_{\vec{k}}\} + \text{c.c.} \right\}. \tag{12.24}$$

Now we write down the equations for amplitudes $C_{a,b}$ in the Schrödinger equation (12.17). Substituting (12.18) into (12.17) we obtain

$$i\hbar\frac{d}{dt}|\Psi\rangle = i\hbar\frac{d}{dt}\begin{bmatrix} C_a\{0\} \\ \sum_{\vec{k}} C_b\{1_{\vec{k}}\} \end{bmatrix} = \widehat{V}|\Psi\rangle. \tag{12.25}$$

First we calculate in the right of (12.25):

$$\widehat{V}|\Psi\rangle = -[e^{i\omega_{ab}t}\sigma_+ + e^{-i\omega_{ab}t}\sigma_-]\cdot\sum_k \widetilde{\mathcal{E}}_k\left[e^{i\nu_k t}\widehat{a}_k^+ + e^{-i\nu_k t}\widehat{a}_k\right]\cdot\begin{bmatrix} C_a\{0\} \\ \sum_{\vec{k}} C_b\{1_{\vec{k}}\} \end{bmatrix} \tag{12.26}$$

$$= -\sum_k \widetilde{\mathcal{E}}_k\left[e^{i\nu_k t}a_k^+ + e^{-i\nu_k t}a_k\right]\cdot\begin{bmatrix} e^{i\omega_{ab}t}\sum_{\vec{m}} C_b\{1_{\vec{m}}\} \\ -e^{-i\omega_{ab}t}C_a\{0\} \end{bmatrix}$$

$$= -\begin{bmatrix} e^{i\omega_{ab}t}\left\{\sum_k \widetilde{\mathcal{E}}_k\left[e^{i\nu_k t}\sum_{\vec{m}} a_k^+ C_b\{1_{\vec{m}}\} + e^{-i\nu_k t}\sum_{\vec{m}} a_k C_b\{1_{\vec{m}}\}\right]\right\} \\ -e^{-i\omega_{ab}t}\left\{\sum_k \widetilde{\mathcal{E}}_k[e^{i\nu_k t a_k^+}C_a\{0\} + e^{-i\nu_k t}a_k C_a\{0\}]\right\} \end{bmatrix}, \tag{12.27}$$

where we have taken into account the usual relations $a_k C_a|\{0\}\rangle = 0$, $a_m C_{bj}|\{1_j\}\rangle = C_{bj}\delta_{mj}|\{1_j\}\rangle$ and also

$$a_k^+|1_m\rangle = \begin{Bmatrix} \sqrt{1_m+1}\,|2_m\rangle & \text{if } k=m \\ \sqrt{0+1}\,|1_k\rangle & \text{if } k\neq m \end{Bmatrix}. \tag{12.28}$$

Now the Schrödinger equation (or the equations for amplitudes) become

$$i\hbar\frac{dC_{a\{0\}}}{dt} = -\sum_k \widetilde{\mathcal{E}}_k\left[e^{i(\omega_{ab}+\nu_k)t}Q_k + e^{i(\omega_{ab}-\nu_k)t}C_{b\{1_k\}}\right], \tag{12.29}$$

and

$$i\hbar\frac{dC_{b\{1_{\vec{k}}\}}}{dt} = -\widetilde{\mathcal{E}}_k[e^{-i(\omega_{ab}-\nu_k)t}]C_{a\{0\}}, \tag{12.30}$$

where quantity Q_k is given by

$$Q_k = \sum_m \begin{Bmatrix} \sqrt{1_m+1} & \text{if } (m=k) \\ 1 & \text{if } (m\neq k) \end{Bmatrix} \begin{matrix} C_b\{2_m\} \\ C_b\{1_m+1\} \end{matrix}. \tag{12.31}$$

One can see that quantity Q_k answers to the transitions, when atom rises in exited level and photons are emitted. We drop such non-conserving states in spirit of the rotating-wave approximation (RWA) [Scully & Zubairy, 1996]. Now Eqs. (12.32) and (12.30) become

$$\frac{dC_{a\{0\}}}{dt} = -\frac{1}{i\hbar}\sum_{\vec{k}} \widetilde{\mathcal{E}}_{\vec{k}} e^{i(\omega_{ab}-\nu_k)t} C_{b\{1_k\}}, \tag{12.32}$$

$$\frac{dC_{b\{1_{\vec{k}}\}}}{dt} = -\frac{1}{i\hbar}\widetilde{\mathcal{E}}_{\vec{k}}[e^{-i(\omega_{ab}-\nu_k)t}]C_{a\{0\}}, \tag{12.33}$$

where $\widetilde{\mathcal{E}}_k = (\vec{E}_k \cdot \vec{\wp})$, $\vec{\wp} = \langle a|e\vec{r}|b\rangle$ and further we will rewrite $C_a\{0\} \rightarrow C_a$ and $C_{b\{1_{\vec{k}}\}} \rightarrow C_{b\vec{k}}$.

12.1 Equations for Probability Amplitudes in Spherical Coordinates

We have obtained the motion Eqs. (12.32) and (12.33) for the probability amplitudes C_a and $C_{b,\vec{k}}$. Now due to a spherical geometry of a problem we will characterize the state of the electromagnetic field in microsphere with indexes ν, m and j (radial, spherical and azimuthal quantum number) accordingly. In this case for a photon state we write $|1_{\vec{k}}\rangle \rightarrow |1_{\nu mj}\rangle$, $\omega_{\vec{k}} \rightarrow \omega_{\nu m}$ (the eigenfrequencies of metallized sphere do not depend on azimuthal quantum number [Stratton, 1941]), $E_{\vec{k}} \rightarrow E_{\nu mj}$ and $\sum_{\vec{k}} = \sum_{n=1}^{\infty}\sum_{m=1}^{\infty}\sum_{j=-m}^{m}$. Since the eigenfrequencies spectrum of the microsphere $\omega_{\nu m}$ is known, it is convenient to use the dimensionless spatial variable $\alpha_{\nu m}z$, where $z = r/a_0$, $\alpha_{nm} = \omega_{\nu m}a_0/c$, and a_0 is the radius of the microsphere. The electrical field in the spherical coordinate frame (ρ, θ, φ) can be calculated by means of the Debye potentials Π (see Chapter 1) and has the form

$$\vec{E}(y,\theta,\varphi) = \sum_{\nu,m,j} \vec{E}_{\nu mj}(y,\theta,\varphi), \tag{12.34}$$

$$\vec{E}_{\nu mj}(y,\theta,\varphi) = \mathcal{E}_{\nu m}[\varepsilon_r(y,\theta,\varphi)\widehat{e}_r + \varepsilon_\theta(y,\theta,\varphi)\widehat{e}_\theta + \varepsilon_\varphi(y,\theta,\varphi)\widehat{e}_\varphi,$$

where $\mathcal{E}_{\nu m}$ has the dimension of a field, and ε_s, $s = r,\theta,\varphi$ for TM waves are given by

$$\varepsilon_r = \frac{m(m+1)}{(\alpha_{mn}z)^2}\Pi; \quad \varepsilon_\theta = \frac{1}{\alpha_{mn}^2 z}\frac{\partial^2\Pi}{\partial z\partial\theta}; \quad \varepsilon_\varphi = \frac{1}{\alpha_{mn}^2 z}\frac{\partial^2\Pi}{\partial z\partial\varphi}, \tag{12.35}$$

$$\Pi = \Pi(r, \theta, \varphi) = R_m(\alpha_{mn} z) Y_m^j(\theta, \varphi). \tag{12.36}$$

We rewrite Eqs. (12.32), (12.33) in the form

$$\frac{dC_0}{d\tau} = i \sum_{\nu m j} W_{\nu m j} e^{i(1-\nu_{\nu m})\tau} C_{\nu m j}, \tag{12.37}$$

$$\frac{dC_{\nu m j}}{d\tau} = i W^*_{\nu m j} e^{-i(1-\nu_{\nu m})\tau} C_0, \tag{12.38}$$

where $\nu_{\nu m} = \omega_{\nu m}/\omega_{ab}$, $W_{\nu m j} = (\vec{\wp}_{ab} \cdot \vec{E}_{\nu m j})$. Let us calculate $|\Psi|^2$ for (12.18)

$$|\Psi|^2 = |C_0|^2 + \sum_{\nu m j} |C_{\nu m j}|^2, \tag{12.39}$$

and (12.37), (12.38). We obtain

$$C_0^* \frac{d}{d\tau} C_0 + C_0 \frac{d}{d\tau} C_0^* = \frac{d}{d\tau} |C_0|^2 = i \left(\sum_{\nu m j} W_{\nu m j} e^{i(1-\nu_{\nu m})\tau} C_0^* C_{\nu m j} \right) + \text{c.c.}, \tag{12.40}$$

and

$$\sum_{\nu m j} \left(C^*_{\nu m j} \frac{d}{d\tau} C_{\nu m j} + \text{c.c.} \right) = \frac{d}{d\tau} \sum_{\nu m j} |C_{\nu m j}|^2$$
$$= i \left(\sum_{\nu m j} W^*_{\nu m j} e^{-i(1-\nu_{\nu m})\tau} C^*_{\nu m j} C_0 \right) + \text{c.c.}$$

So

$$\frac{d}{d\tau} \left[|C_0|^2 + \sum_{\nu m j} |C_{\nu m j}|^2 \right] = 0, \tag{12.41}$$

and

$$|C_0(\tau)|^2 + \sum_{\nu m j} |C_{\nu m j}(\tau)|^2 = C^2 = \text{constant}. \tag{12.42}$$

Since $|C_0(0)|^2 = 1$ and $|C_{\nu m j}(0)|^2 = 0$ we have $C^2 = 1$ and (12.42) becomes the normalize condition in the form

$$|\Psi|^2 = |C_0(\tau)|^2 + \sum_{\nu m j} |C_{\nu m j}(\tau)|^2 = 1. \tag{12.43}$$

This condition is certainly nothing but only the condition of normalization of a state vector $|\Psi\rangle$. It is necessary to note the following: eigenfrequencies of the spherical resonator derived from the dispersion equations (for instance, metallized microsphere) do not depend on the azimuth index j, and depend only on indexes ν and m. For every m there exist $2m+1$ various angular functions $R_{\nu m}(r)Y_m^j(\theta,\varphi) = R_{\nu m}(r)P_m^j(\cos\theta)e^{\pm j\varphi}$ (see Chapter 1, (1.109)) where each eigenfrequency $\omega_{\nu m}$ has the degeneracy rate of $2m+1$. This degeneracy is caused by a high degree of symmetry of sphere. The choice of an axis z ($\theta = 0$) for sphere is arbitrary, at other choice the new field's oscillations become the linear combinations of former oscillations with the same eigenfrequency [Vainstein, 1988]. Equations (12.37) and (12.38) have a large dimension $\nu m(2m+1)$, but its can be simplified by the next way. Since $\omega_{\nu m}$ and $\nu_{\nu m}$ do not depend on index j it is convenient in (12.37), (12.38) to introduce new variables $\widetilde{C}_{\nu m}$ in the form

$$\widetilde{C}_{\nu m} = \frac{1}{\beta_{\nu m}}\left(\sum_{j=-m}^{m} W_{\nu mj}C_{\nu mj}\right), \tag{12.44}$$

where $\beta_{\nu m}$ is some normalized quantity. Then Eqs. (12.37), (12.38) become

$$\frac{dC_0}{d\tau} = i\sum_{\nu m}\beta_{\nu m}e^{i(1-\nu_{\nu m})\tau}\frac{\sum_j W_{\nu mj}C_{\nu mj}}{\beta_{\nu m}}, \tag{12.45}$$

$$\frac{1}{\beta_{\nu m}}\sum_{j=-m}^{m} W_{\nu mj}\frac{dC_{\nu mj}}{d\tau} = i\,\beta_{\nu m}^{-1}C_0 e^{-i(1-\nu_{\nu m})\tau}\sum_{j=-m}^{m} W_{\nu mj}W_{\nu mj}^*. \tag{12.46}$$

As a result Eqs. (12.37) and (12.38) can be reduced to the next form

$$\frac{dC_0}{d\tau} = i\sum_{\nu m}\beta_{\nu m}e^{i(1-\nu_{\nu m})\tau}\widetilde{C}_{\nu m}, \tag{12.47}$$

$$\frac{d\widetilde{C}_{\nu m}}{d\tau} = i\frac{1}{\beta_{\nu m}}e^{-i(1-\nu_{\nu m})\tau}C_0\sum_{j=-m}^{m}|W_{\nu mj}|^2. \tag{12.48}$$

Equations (12.47) and (12.48) have dimensions νm only. Let us calculate the norm of the state vector $|\Psi\rangle$ in such new variables. We have

$$C_0^*\frac{dC_0}{d\tau} + C_0\frac{dC_0^*}{d\tau} = \frac{d}{d\tau}|C_0|^2 = iC_0^*\sum_{\nu m}\beta_{\nu m}e^{i(1-\nu_{\nu m})\tau}\widetilde{C}_{\nu m} + \text{c.c.}, \tag{12.49}$$

and

$$\sum_{\nu m}\left[\widetilde{C}^*_{\nu m}\frac{d\widetilde{C}_{\nu m}}{d\tau}+\widetilde{C}_{\nu m}\frac{d\widetilde{C}^*_{\nu m}}{d\tau}\right]=\frac{d}{d\tau}\sum_{\nu m}|\widetilde{C}_{\nu m}|^2 \tag{12.50}$$

$$=iC_0\sum_{\nu m}\widetilde{C}^*_{\nu m}\beta^{-1}_{\nu m}e^{-(1-\nu_{\nu m})\tau}\sum_{j=-m}^{m}|W_{\nu mj}|^2+\text{c.c.} \tag{12.51}$$

Therefore

$$\frac{d}{d\tau}\left[|C_0|^2+\sum_{\nu m}|\widetilde{C}_{\nu m}|^2\right]$$
$$=iC_0^*\sum_{\nu m}\frac{1}{\beta^*_{\nu m}}\left[|\beta_{\nu m}|^2-\sum_{j=-m}^{m}|W_{\nu mj}|^2\right]e^{i(1-\nu_{\nu m})\tau}\widetilde{C}^*_{\nu m}+\text{c.c.} \tag{12.52}$$

Since quantities $\beta_{\nu m}$ are an undefined variables we can choose $\beta_{\nu m}$ in the form

$$|\beta_{\nu m}|=\left[\sum_j|W_{\nu mj}|^2\right]^{1/2}. \tag{12.53}$$

Then (12.52) is given by

$$\frac{d}{d\tau}\left[|C_0|^2+\sum_{\nu}m|\widetilde{C}_{\nu m}|^2\right]=0, \tag{12.54}$$

and

$$|C_0|^2+\sum_{\nu m}|\widetilde{C}_{\nu m}|^2=C^2=\text{constant}. \tag{12.55}$$

Since initial conditions $|C_0(0)|^2=1$, $|C_{\nu mj}(0)|^2=0$ and $|\widetilde{C}_{\nu m}(0)|^2=0$ we have $C^2=1$ and (12.55) becomes

$$|\Psi|^2=|C_0(\tau)|^2+\sum_{\nu m}|\widetilde{C}_{\nu m}(\tau)|^2=1. \tag{12.56}$$

We have obtained and the normalize condition for (12.47), (12.48).

CHAPTER 13

Dynamics of Spontaneous Emission of Two-Level Atom in Microspheres: Direct Calculation

13.1 Introduction

A variety of aspects of spontaneous emission was studied in a number of works, see [Rahman & Bryant, 2002; Beige *et al.*, 2002; Barut & Dowling, 1987; Boone *et al.*, 2003; Bunkin & Oraevsky, 1959; Burlak *et al.*, 2003; Hooijer *et al.*, 2001; Datsyuk, 2002b; Yablonovich, 1987; Kien *et al.*, 2000; Gacuteerard *et al.*, 1998; Juzeliunas, 1997; Schniepp & Sandoghdar, 2002; Dung & Ujihara, 1999; Dung *et al.*, 2000; Kapale *et al.*, 2003; Khodjasteh & Lidar, 2003; Ujihara & Dung, 2002; Klimov *et al.*, 1996; Klimov & Letokhov, 1999; Rogobete *et al.*, 2003; Lewis *et al.*, 2002; Lewenstein *et al.*, 1988; Macovei & Keitel, 2003; Fleischhauer, 1999; Oraevskii, 1994; Sanchez-Mondragon *et al.*, 1983; Scheel *et al.*, 1999a; Tomas & Lenac, 1999; Bondarev *et al.*, 2002; Fan *et al.*, 2001; Lee & Yamanishi, 1995; Xu *et al.*, 2000b; Xu *et al.*, 2000a; Li *et al.*, 2000 and reference therein]. Various new effects were predicted and different theoretical models were proposed [Yokoyama & Ujihara, 1995].

It is well known that the placing of the excited atom in a cavity strongly modifies the dynamics of spontaneous emission of such an atom [Purcell, 1946a]. At the corresponding description spontaneous emission in cavities usually use qualitatively two approaches. In the first case a spontaneous emission is studied at an exact resonance of frequency of the atom transition with an eigenmode of a cavity. In this case the decay rate appears more in comparison with a vacuum case (the enhancement of spontaneous emission). If the frequency of transition is not in a resonance the decay rate appears less than the vacuum value (the forbidden spontaneous emission). Such conclusions follow from the simple reasons based on the updating of the density states of cavity modes and they are well confirmed experimentally [Scully & Zubairy, 1996; Meystre & III., 1998; Bachor, 1998 and references therein].

However the correctness of the use of quantity of the mode density is based on the assumption that eigenfrequencies of a cavity are located so closely that the continuum mode approximation is valid. Mathematically it is expressed in the replacement of operation of summation over all states of a field by integration,

$$\sum_{\vec{k}} \rightarrow \frac{V}{(2\pi)^3} \int d^3k. \tag{13.1}$$

The accuracy of such mathematical approach can be estimated from the Euler–Maclaurin's [Korn & Korn, 1961] formula:

$$\sum_{k=0}^{n} \rho(k) \approx \int_0^n \rho(x)dx + \frac{1}{2}[\rho(n) + \rho(0)]$$
$$+ \sum_{k=1}^{m} \frac{B_{2k}}{(2k)!}[\rho^{(2k-1)}(n) - \rho^{(2k-1)}(0)],$$

where B_{2k} are Bernoulli numbers, $B_2 = 1/6$, $B_4 = -1/30$, $B_6 = 1/42, \ldots$ If the states density ρ is a rather smooth function the error of such approximation is small. But such approximation has large error if ρ is not smooth and has large derivations $\rho^{(2k-1)}$.

Therefore such approach is not valid enough for true microspheres with the sizes about the wavelength of an atom transition. In this case the distance between modes has the order of magnitude of the microsphere sizes, so the continuum modes approximation is no longer valid. It means that when studying the atom-field interaction one has to take into account the spatial structure of a variable optical field of such a microsphere. It means that the number of modes involved in the interaction and also the dynamics of spontaneous emission can differ for the atoms located near and far from the center of microsphere.

Thus, in microspheres factors such as no-zero intermodes separation, and the spatial inhomogeneity of field become important. Due to mentioned factors the simple models of spontaneous emission cease to be analytically solvable.

In such models (a continuum mode case) for a proper account of the atomic decay, corresponding to a quantization cavity one usually makes the next approximations. First is the assumption that the modes of the field are closely spaced in frequency $\Delta\omega \to 0$ and one can replace the summation over $\vec{k}$ by an integration. The second is that the photon frequency ω varies little around the frequency of atoms transition ω_{ab} and one can therefore replace ω by ω_{ab}. The resulting continuum mode decay constant has the form [Scully & Zubairy, 1996; Meystre & III., 1998]

$$\Gamma_v = \omega_{ab}^3 |\wp|^2 / 3\pi\varepsilon_0 \hbar c^3.$$

But mentioned assumptions are not valid enough for true microspheres, where $\Delta\omega \sim \omega_0 = c/a_0$ (a_0 is radius of microsphere, c is the light velocity in vacuum). For instance if $a_0 = 1\,\mu\text{m}$ one has $\Delta\omega \sim 3 \cdot 10^{14} c^{-1}$. So converting a sum into integral becomes inexact. In this case one has to solve the complete set of differential equations for probabilities amplitudes.

For detailed studying of spontaneous emission we use the numerical methods by means of the direct solution the equations for probability amplitudes of the excited and ground states of atom interacting with the electromagnetic field of microsphere. Such a general approach allows us to study the dynamics of spontaneous emission in the uniform frameworks, without separation into resonant and non-resonant cases.

In this chapter we mainly concentrate on the dynamics of spontaneous emission in the lossless metallized microsphere. Due to simplicity such a model can serve as a fundamental model at studying the details of spontaneous emission. A few good recognized factors influence the temporal dynamics in this model. The spectrum of eigenfrequencies of such system is well known. The quantum volume of the field fits the volume of the microsphere since the boundary conditions of equality to zero of the tangential fields on boundary are satisfied. Nevertheless even in such a simple model a variety of dynamic properties can be recognized.

In more complex dielectric microspheres the high-quality electromagnetic whispering gallery modes (WGM) oscillations exist. The contribution of spectrum WGM in such a system demands for the consideration of all modes with various values of spherical quantum numbers. In this case the radiating losses begin to depend on the frequency of oscillations. In a bare dielectric microsphere the photon modes having small spherical quantum numbers have low Q factor. They can have a great value of Q factor only for microspheres with the superficial structures similar to an alternative stack, deposited on dielectric microsphere [Brady *et al.*, 1993; Sullivan, 1994; Burlak *et al.*, 2000]. However the dynamics of spontaneous emission in this a case appears quite sensitive to the concrete model of boundary conditions on a surface of microsphere.

13.2 Basic Equations

Consider a metallized hollow microsphere of radius a_0 and a two-level atom located in point $\vec{r_0}$ in microsphere. Figure 13.1 shows the geometry of the system. Due to the perfectly conducting walls the transverse component of the electromagnetic field obeys boundary conditions of equality to zero and the field volume in the microsphere fits the volume of microsphere.

We assume that at time $t = 0$ the atom is in the excited state $|a\rangle$ and the field modes are in the vacuum state $|\{0\}\rangle$. The ground state is $|b\rangle$ and $|\{1_{\vec{k}}\}\rangle$ is a photons state with one photon in some $\vec{k}$ state. The state vector

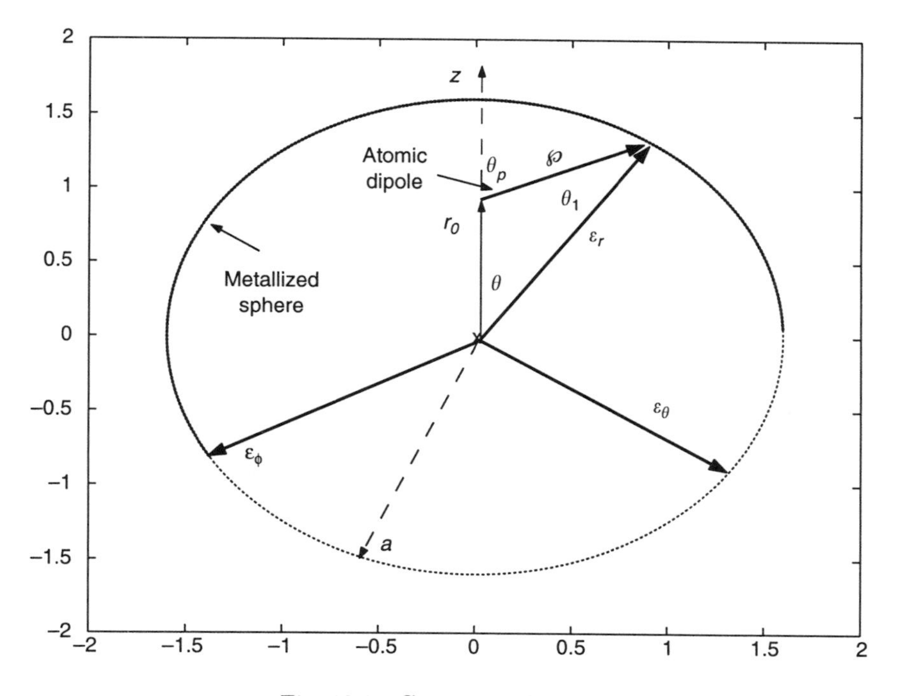

Fig. 13.1 Geometry of system.

is therefore

$$|\psi(\tau)\rangle = C_a(\tau)|a,\{0\}\rangle + \sum_{\vec{k}} C_{b\vec{k}}(\tau)|b,\{1_{\vec{k}}\}\rangle \tag{13.2}$$

with initial conditions $C_a(0) = 1$ and $C_{b\vec{k}}(0) = 0$ of probability amplitudes for the upper exited level $C_a(t)$ and the ground levels $C_{b\vec{k}}(t)$ accordingly. We want to determine the state of the atom and the state of the radiation field at some later time when the atom begins to emit photons. The Hamiltonian of system in the dipole approximation has the form [Scully & Zubairy, 1996]

$$H = H_a + H_f + H_{af},$$

where

$$H_a = (\hbar\omega_{ab}/2)\sigma_z, \quad H_f = \sum_{\vec{k}} \hbar\omega_{\vec{k}} \left[a^+_{\vec{k}} a_{\vec{k}} + \tfrac{1}{2}\right],$$

$$H_f = -e\vec{r}\cdot\vec{E}, \quad \omega_{ab} = (E_a - E_b)/\hbar > 0$$

is a transition frequency of an atom, e is the electron charge, $\vec{r}$ is the place of atom in microsphere, and $\vec{E}$ is a variable electric field. The Schrödinger equation in the interaction picture is given by

$$i\,\hbar\frac{\partial}{\partial t}|\psi(\tau)\rangle = \mathcal{V}|\psi(\tau)\rangle, \tag{13.3}$$

where

$$\mathcal{V} = -\wp\sum_{\vec{k}}[\vec{E}_{\vec{k}}\cdot\sigma_+ e^{i(\nu_{\vec{k}}-\omega_{ab})t}a_{\vec{k}} + H.c.] \tag{13.4}$$

is the interaction part of the Hamiltonian, $\vec{\wp} = \langle a|e\vec{r}|b\rangle$ is a dipole matrix element (we assume it to be real), $\omega_{\vec{k}}$ is a frequency of field, $\vec{k}$ is a wave vector, $a^+_{\vec{k}}$ and $a_{\vec{k}}$ are creation and annihilation operators for the radiation field, $\sigma_\pm = (\sigma_x \pm i\,\sigma_y)/2$, $\sigma_x, \sigma_y, \sigma_z$ are Pauli spin matrices. In the dipole approximation $\vec{r} = \vec{r}_0$, we take into account the usual relations $a_k C_a|\{0\}\rangle = 0$, $a_m C_{bj}|\{1_j\}\rangle = C_{bj}\delta_{mj}|\{1_j\}\rangle$ and we drop the non-conserving two-photon states accordingly to the rotating-wave approximation (RWA) [Scully & Zubairy, 1996].

The apparently simple model (13.3)–(13.4) is not exactly solvable unless the rotating-wave approximation is done, but even in this latter case the solutions are known for two asymptotic approaches: single-mode field case (the Jaynes-Cummings model) and the continuum modes case (Weisskopf-Wigner approach) [Scully & Zubairy, 1996; Meystre & III., 1998]. We study an intermediate case of true microsphere having a large number of the well-separated fields modes.

From (13.2)–(13.4) we receive the motion equations for the probability amplitudes C_a and $C_{b,\vec{k}}$. Due to a spherical geometry of problem we will characterize the state of electromagnetic field in microsphere with indexes n, m and j (radial, spherical and azimuthal quantum number) accordingly. In this case for a photon state we write $|1_{\vec{k}}\rangle \to |1_{nmj}\rangle$, $\omega_{\vec{k}} \to \omega_{nm}$ (the eigenfrequencies of metallized sphere do not depend on azimuthal quantum number [Stratton, 1941]), $E_{\vec{k}} \to E_{nmj}$.

Since the eigenfrequencies spectrum of the microsphere ω_{nm} is known, it is convenient to use the dimensionless spatial variable $\alpha_{nm}z$, where $z = r/a$, $\alpha_{nm} = \omega_{nm}a/c$, a is radius of microsphere. The electrical field in the spherical coordinate frame (ρ, θ, φ) can be calculated by means of scalar functions called the Debye potentials (see Sec. 1.1.6), and has form

$$\vec{E}(y,\theta,\varphi) = \sum_{n,m,j} \vec{E}_{nmj}(y,\theta,\varphi), \tag{13.5}$$

$$\vec{E}_{nmj}(y,\theta,\varphi) = \mathcal{E}_{nm}[\varepsilon(y,\theta,\varphi)\widehat{e}_r + \varepsilon_\theta(y,\theta,\varphi)\widehat{e}_\theta + \varepsilon_\varphi(y,\theta,\varphi)\widehat{e}_\varphi,$$

We focus our attention on the *TM* wave case, as the *TE* wave case can be studied in much the same way. Using the Debye potential approach [Stratton, 1941] the fields components in (13.5) can be written in the next form [Stratton, 1941]

$$\begin{aligned} \varepsilon(y,\theta,\varphi) &= \frac{m(m+1)}{y^2}u(y,\theta,\varphi), \quad \varepsilon_\theta(y,\theta,\varphi) = \frac{1}{y}\frac{\partial^2 u(y,\theta,\varphi)}{\partial y \partial\theta} \\ \varepsilon_\varphi(y,\theta,\varphi) &= \frac{1}{y}\frac{\partial^2 u(y,\theta,\varphi)}{\partial y\partial\varphi}, \end{aligned} \tag{13.6}$$

where $u(y,\theta,\varphi)$ is the Debye potential for *TM* waves, which is given by $u(y,\theta,\varphi) = R_m(y)Y_m^j(\theta,\varphi)$, and for our model metallized microsphere $R_m(y) = (2y/\pi)^{1/2}J_{m+1/2}(y)$. $J_m(y)$ is the Bessel function, $Y_m^j(\theta,\varphi) = C \cdot P_m^{(j)}(\cos\theta)e^{ij\varphi}$ is a spherical function, $P_m^{(j)}$ is the associate Legendre function, C is a constant (see Appendix F), $y = \alpha_{nm}z$, $\mathcal{E}_{\vec{k}} = \left(\frac{\hbar\omega_{\vec{k}}}{2\varepsilon_0 V}\right)^{1/2} \to \mathcal{E}_{nm} = \left(\frac{\hbar\omega_{nm}}{2\varepsilon_0 V}\right)^{1/2}$ is a field per photon, V is a volume of microsphere, $\widehat{e}_{r,\theta,\varphi}$ are the basis set for spherical coordinates. We consider the dipole orientation as $\vec{\wp} = \wp\{0, \sin\theta_p, \cos\theta_p\}$(see Fig. 13.1). In a spherical cavity the probability amplitudes C_a and $C_{b,\vec{k}}$ do not depend on angles. The scalar product $(\vec{\wp}\cdot\vec{E}_{nmj})$ in (13.4) after integration with respect to the solid angle $d\Omega = \sin(\theta)d\theta d\varphi$ becomes (see Appendix 5 for details)

$$\int d\Omega(\vec{\wp}\cdot\vec{E}_{nmj}) = \wp\mathcal{E}_{nm}2\pi F_m(y_0)\delta_{j0}, \tag{13.7}$$

where

$$F_m(y) = \frac{\cos\theta_p}{3}[A_m(y_0) + 2B_m(y_0)]\delta_{m1} + \sin\theta_p[A_m(y_0)\widetilde{A}_m + B_m(y_0)\widetilde{B}_m]. \tag{13.8}$$

In (13.8) $A_m(y_0) = \frac{m(m+1)}{y_0^2}R_m(y_0)$, $B_m(y_0) = \frac{1}{y_0^2}\frac{dR_m(y_0)}{dy}$, $y_0 = \alpha_{nm}z_0$, $z_0 = r_0/a \le 1$, r_0 is a point of location atom in the microsphere, quantities $\widetilde{A}_m$ and $\widetilde{B}_m$ are written in Appendix 5. The requirement that the tangential electric field components of the TM mode should vanish on the inside surface of the microsphere $\varepsilon_\theta(\alpha_{nm},\theta,\varphi) = \varepsilon_\theta(\alpha_{nm},\theta,\varphi) = 0$ at $r = a$ gives rise to the following the eigenfrequencies equation $B_m(\alpha_{nm}) = 0$, whose solution is α_{nm} or $\omega_{nm} = \alpha_{nm}c/a$. Note that E_{nmj} and other quantities in (13.5)–(13.8) depend on n through $\omega_{nm}r/c$. One can see from (13.7) and (13.8) that in case $\theta_p = 0$ (longitudinal dipole) only the field dipole mode $m = 1$ contributes in the dynamics of the spontaneous emission. Higher spherical modes $m > 1$ give contribution in a general case $\theta_p \neq 0$ (see Appendix 5). Using (13.4)–(13.7) we obtain the following equations for probability amplitudes in the dimensionless form

$$\begin{aligned} \frac{dC_a}{d\tau} &= -i\,K\sum_n^N\sum_m^M F_{nm}\nu_{nm}^{1/2}e^{i(1-\nu_{nm})\tau}C_{b,nm}, \\ \frac{dC_{b,nm}}{d\tau} &= i\,KF_{nm}\nu_{nm}^{1/2}e^{-i(1-\nu_{nm})\tau}C_a, \end{aligned} \tag{13.9}$$

where $N \equiv n_{\max}$, $M \equiv m_{\max}$, $\nu_{nm} = \omega_{nm}/\omega_{ab}$, $\tau = \omega_{ab}t$, $F_{nm} = F_m(\alpha_{nm}z_0)$, $K = (\wp^2/4\hbar\omega_{ab}\varepsilon_0 V)^{1/2}$.

The set of Eqs. (13.9) contains all the necessary physical information regarding the dynamics and strength of the spontaneous emission. For the study of such a dynamics process one has to solve the set of Eqs. (13.9) for set of eigenfrequencies ω_{nm}.

We solve such a system for a various number of microsphere modes (number of the equations), up to $NM \sim 3600$. The algorithms for the integration of a system of ordinary differential equations numerically can be found, e.g. in [Press *et al.*, 2002].

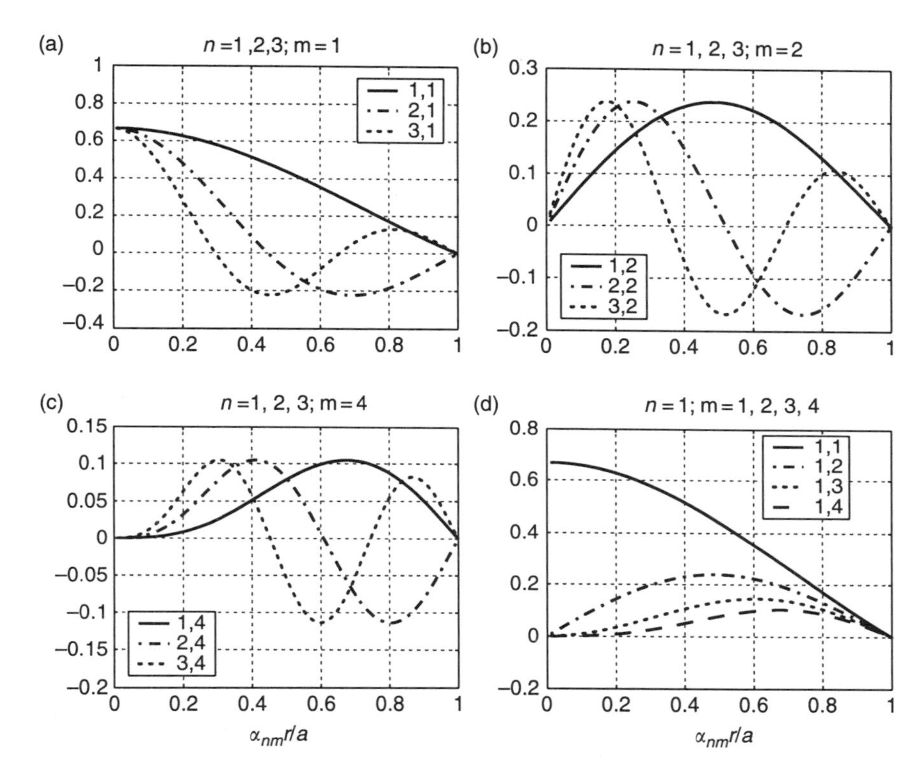

Fig. 13.2 Radial distribution of the electric fields for a number of first spherical eigenmodes. α_{nm} are the eigenfrequencies, a is the radius of microsphere.

The physical factor determining the values of both radial and spherical numbers $N = n_{Max}$ and $M = m_{Max}$ is the number of the fields eigenmodes, which influence an atom-field interaction in (13.9). One can see the eigenfrequencies α_{nm} are included into denominators of $A_m(\alpha_{nm} z_0)$ and $B_m(\alpha_{nm} z_0)$ in Eq. (13.8). Moreover the quantities $\widetilde{A}_m$ and $\widetilde{B}_m$ are decreasing since the growth of the number m. Therefore the contribution of the highest eigenfrequencies α_{nm} is decreasing on increasing N and M. Due to the complexity of the spatial structure of the fields it is difficult to establish some analytical criterion for N and M. While performing calculations we increased the quantities of N and M till the sum in (13.9) did not cease to change essentially. Usually it did not exceed $N \leq 50$ and $M \leq 10$.

Table 13.1 Eigenfrequencies α_{nm} for TM waves.

m,n	1	2	3	4	5	6
1	2.74	6.12	9.32	12.5	15.6	18.8
2	3.87	7.44	10.7	13.9	17.1	20.3
3	4.97	8.72	12.1	15.3	18.5	21.7
4	6.06	9.97	13.4	16.7	19.9	23.1

13.3 Results and Discussions

First we calculated the radial structure of the electromagnetic eigenmodes metallized microsphere for some first eigenmodes, which can interact with an atom. In Fig. 13.2 such modes $\varepsilon_\theta(\alpha_{nm}z)$ are shown for the TM case. Corresponding values α_{nm} are written in Table 13.1.

The asymptotic formula for α_{nm} has the form $\alpha_{nm} = \frac{\pi}{2}(m+2n)$ and is valid for $n > m \gg 1$ [Vainstein, 1969, 1988]. One can see from Fig. 13.2(a) for a case $m = 1$ and $n \geq 1$ the fields have a maximum at the center of $r = 0$. Thus the atom-field interaction in this case must be most effective near the center of the microsphere. Modes with spherical quantum numbers $m > 1$ have zeros in the center (see Fig. 13.2(b,c)), consequently one has to expect the effective interaction beyond the center. Figure 13.2(d) shows that the modes with radial quantum numbers $n = 1$ and $m > 1$ does not have zero volume (except $r = 0$), only in boundary $r = a$. Since m increases the maximum of field migrates from the center. If $m =$ constant and n increases, the fields have the oscillatory structure: zeros separate the maximums of fields.

In Fig. 13.3 the temporal dynamics of the probability density of the excited state $|C_a(\tau)|^2$ under various number of radial and spherical modes N and M is shown. The almost tangential dipole orientation is used $\theta_p = 1.57$. From Fig. 13.3(a) one can see that at a small number of interacting modes the density of probability of the excited state $|C_a(\tau)|^2$ weakly oscillates near 1 and it practically does not decay. However since growth the radial quantum number n and the oscillator amplitude becomes deeper. One can see from Fig. 13.3(b) that since $N = 7$ the dynamics takes a peak-like form. Such a tendency becomes more pronounced at an increased N. The peak-like picture arises since spherical number $M = 2$ and it weakly changes if number M increases. From Fig. 13.3(d) one can see that for large N such dynamics has a form of a clear sequence of alternative collapses and revivals of atom to the excited state. We have found that in a large time such dynamics acquires more of a noise component. Similar behavior takes place in a single-mode fields model [Scully & Zubairy, 1996; Meystre & III., 1998]. In the case of small $|\theta_p| < 1$ no revivals were found.

One can see from Fig. 13.3(d) that at small time the exited state decays quickly, however, after some transitions the oscillating dynamics occurs. Our calculations show that the tendency of damping of such the oscillatory dynamics is weakly pronounced at a small time and it became clear only at a later time. The norm of wave function (13.2) is preserved (a dash-point line in Fig. 13.3(a,b)).

We found that the distance between peaks in Fig. 13.3(d) is sensitive to a value of parameter ω_0/ω_{ab}, which reflects the number of photon states in

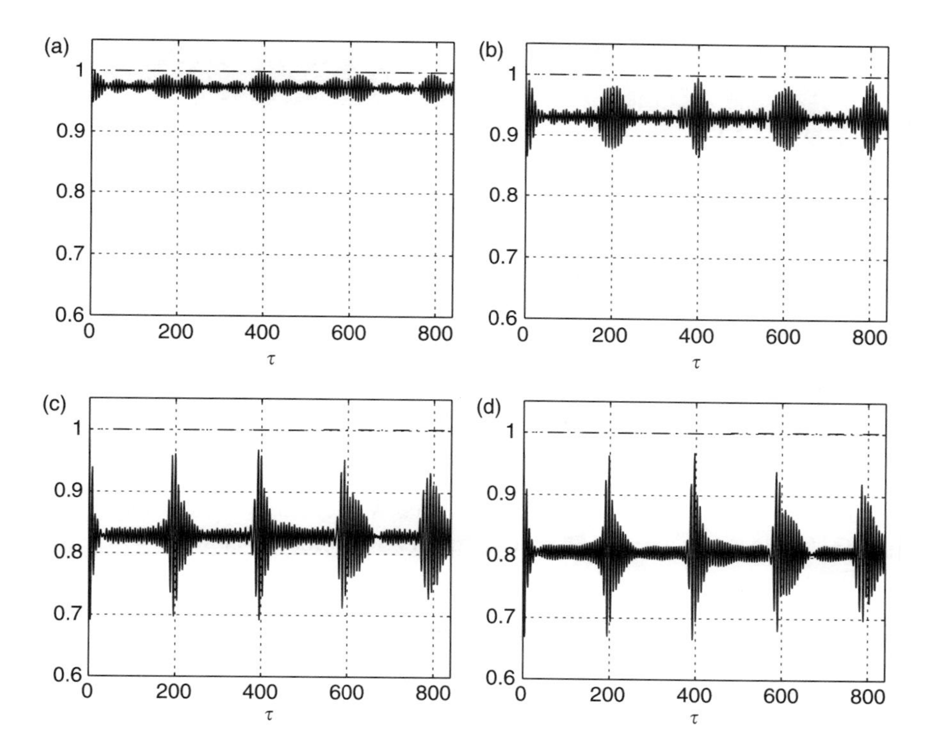

Fig. 13.3 Temporal dynamics of the probability density for the excited state $|C_a(\tau)|^2$ on various number of modes in (13.9). (a) $N = 4$, $M = 2$; (b) $N = 7$, $M = 4$; (c) $N = 15$, $M = 4$; (d) $N = 30$, $M = 4$; $\omega/\omega_{ab} = 0.01$, $K = 8$, $\theta_p = 1.57$, $z_0 = 0.1$. The dash-dot line shows $|\psi(\tau)|^2$ (13.2).

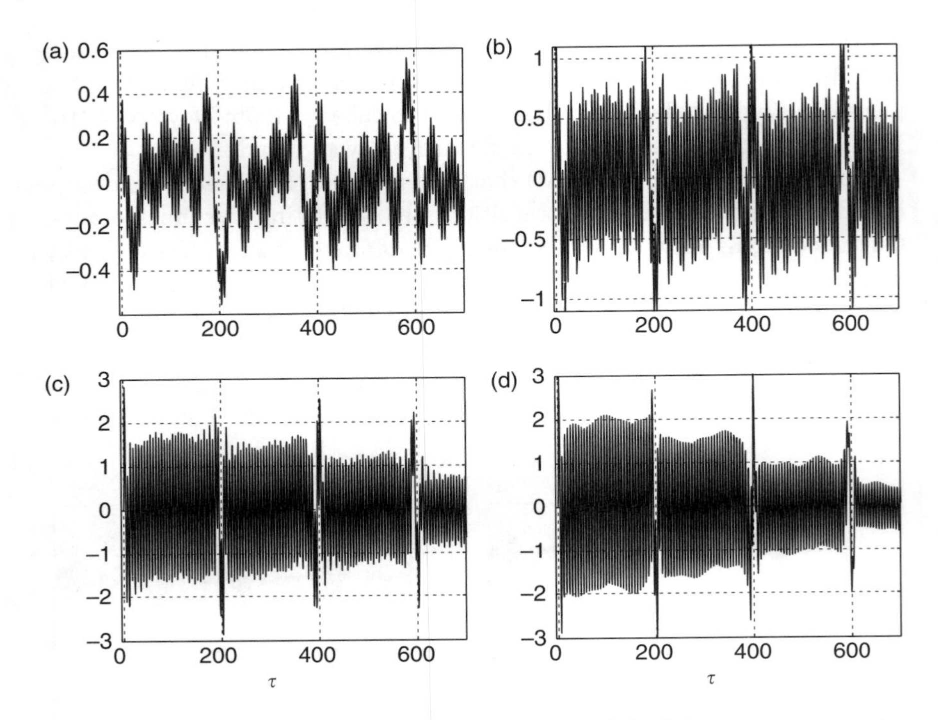

Fig. 13.4 Temporal dynamics of the polarization $\mathcal{P}(\tau)$ while spontaneous emission for the same parameters as in Fig. 13.3.

the vicinity of the atom transition frequency. We believe that in this case the influence of the non-resonant part of the photon modes and the various Rabi effects (oscillations and splitting) [Sanchez-Mondragon *et al.*, 1983; Zhu *et al.*, 1990] are not accounted for in the continuum mode case.

Following Eqs. (13.3)–(13.9) the normalized dynamical electrical polarization of two-level atom $\mathcal{P}$ in the interaction picture can be written in the form

$$\mathcal{P} = e^{-i\tau} C_a \sum_{n,m} C^*_{b,nm} + \text{c.c.}, \tag{13.10}$$

where for normalizing $\wp$ is used. One can see that in (13.10) both probability amplitude of the excited state, and the probability amplitudes of multimode fields of the ground state contribute significantly. Therefore we can refer to $\mathcal{P}$ as a natural collective quantity reflecting the internal dynamics of spontaneous emission in a multimode case. In Fig. 13.4 the temporal dynamics of $\mathcal{P}$ (while spontaneous emission is happening) is shown for the same parameters as in Fig. 13.3 . One can see from Fig. 13.4 that upon the quite smooth curve $\mathcal{P}$ are imposed the sharp peaks, placed similarly to peaks in Fig. 13.3. From Fig. 13.4(d) one can see the well-pronounced tendency for large N to decay at a longer time. However the revival peaks were well-observed at our calculations. We believe that such a temporal behavior is due to the Rabi oscillations and it can be physically understood by the following points. At the initial time, $t = 0$, the atom is prepared in a definite state and therefore all periodical motions are correlated. As time increases the Rabi oscillations associated with the different multimode excitations have different frequencies and therefore they become uncorrelated. This leads to a collapse. As time is further increased, the correlation is restored and revival occurs. However one can see from Fig. 13.4(d) that in a multimode field case the correlation is only partly restored. Such dynamics generate a damped sequence of revivals and after a few peak-like revivals the system begins to transit into an uncorrelated state. Similarly to a single-mode case the revival occurs only because of the granular structure of the photon distribution and it is a pure quantum phenomenon.

In Fig. 13.5 the variation of the first peak of revival (see Fig. 13.3(d)) at various dipole angles θ_p is shown. Our calculations found that the location of such a peak practically does not change while the number of modes increases. However the form of peak becomes sharper, and we found the strong dependence of such dynamics on a value of angle θ_p. One can see from Fig. 13.5(a) that in a case of almost tangential dipole $\theta_p \sim \pi/2$ the revivals are expressed clearly. However this dynamics acquire an oscillatory structure since there is little deviations from $\theta_p = \pi/2$. Thus we can conclude that both collapses and revivals in $|C_a(\tau)|^2$ are mainly due to the presence of the tangential dipoles component.

We are interest in the temporal dynamics of the probability amplitudes in both the excited and ground states, as well as the distribution of the probability density for various principal n and spherical m quantum numbers of the emitted photons. The resonance selects from the complete spectrum only when the mode for which prove to be $\omega_{\nu m} \approx \omega_{ab}$, i.e. $\nu_{\nu m} \approx 1$, that fixes the numbers n and m.

It would be obvious to assume that such a resonant mode has to have the greatest probability density and its temporal-space parameters will determine the details both in the dynamics and the atom-field system, as well as the dynamic polarization $\mathcal{P}(\tau)$. However the quantity $F_{\nu m}$, as we already

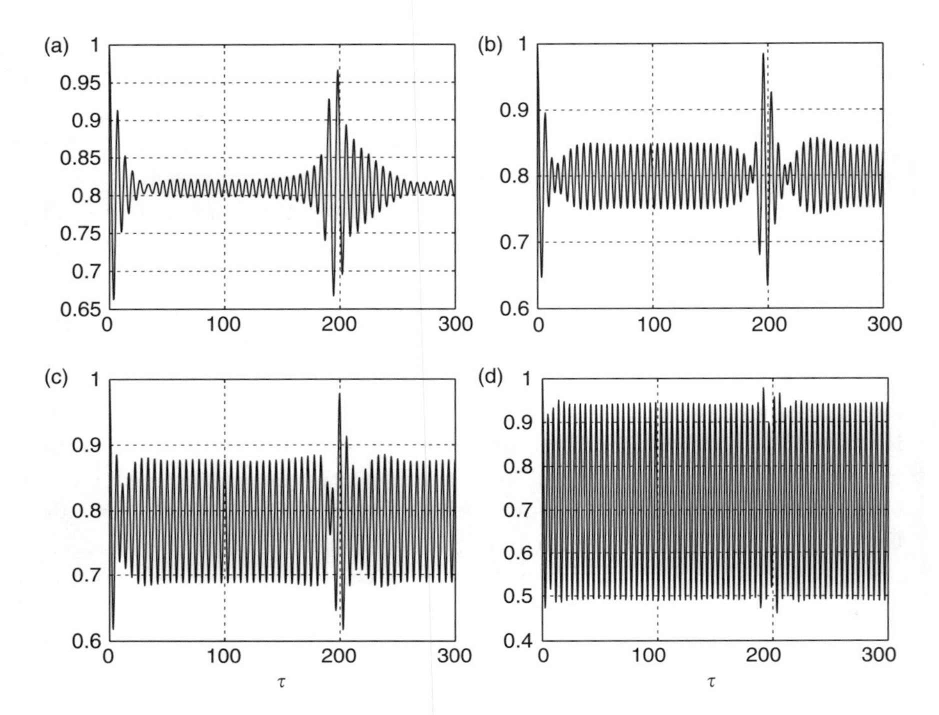

Fig. 13.5 Temporal dynamics of the probability density of the excited state $|C_a(\tau)|^2$ at various atomic dipole angles θ_p: (a) 1.57; (b) 1.569; (c) 1.568; (d) 1.565.

have mentioned, begins to play an essential role and results in the change of such behavior. The corresponding dynamics is shown in Figs. 13.6 and 13.8.

Figure 13.6 shows the temporal dynamics of the probability density of the excited state $|C_a(\tau)|^2$ and the dynamic polarization $\mathcal{P}(\tau)$ (13.10) for parameters $N = 15$, $M = 2$, $\nu = 0.093$ when the dipole is close to tangential orientation $\theta_p \sim \pi/2$. The atom transition frequency is in resonance with the field mode $n = 3$, $m = 2$, so $\omega_{32}/\omega_{ab} = 1$. The spectrum of normalized eigenfrequencies $\alpha_{\nu m}\nu = \omega_{\nu m}/\omega_{ab}$ is written in Table 13.2.

Since step $\Delta(\alpha_{\nu m}\nu) \leq 0.3$ there are quite large number of eigenfrequencies near the resonance in this case. In Fig. 13.6(a,b) the dynamics of $|C_a|^2$ and $\mathcal{P}(\tau)$ is shown for dipole angle $\theta_p = 1.557$, and in Fig. 13.6(c,d) the same is shown but for $\theta_p = 1.57$. One can see that the temporal dynamics of the probability density $|C_a|^2$ in Fig. 13.6(a) has a periodic character. However the dynamics of $\mathcal{P}$ in Fig. 13.6(b) seems to be quasi-stochastic. One can see from formula (13.9) that such dynamics can take place while a large

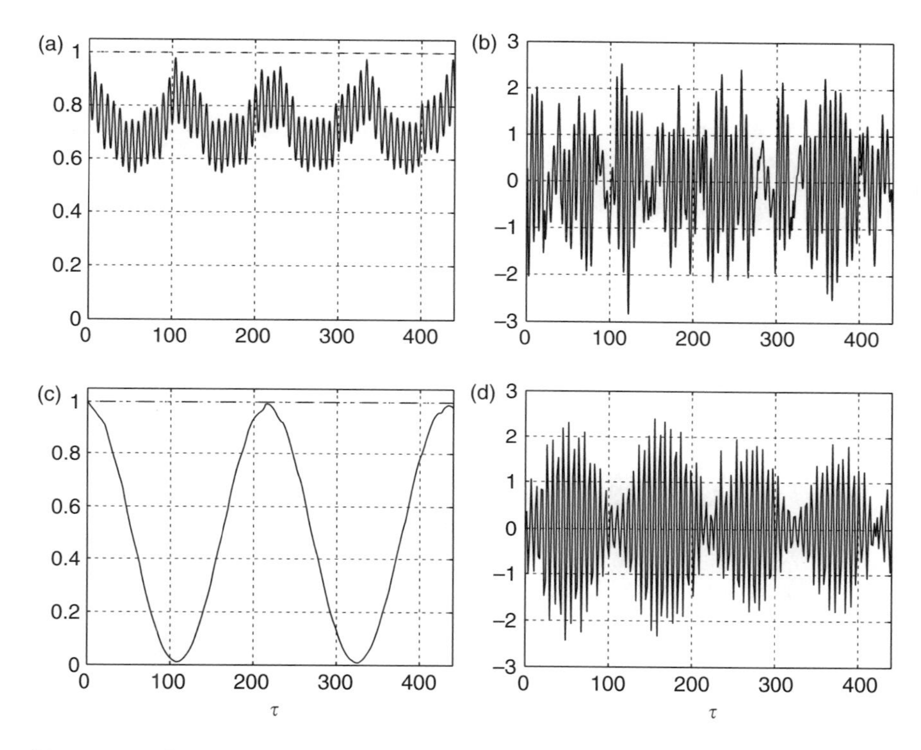

Fig. 13.6 The temporal dynamics of the probability density of the excited state $|C_a|^2$ (a,c) and polarization $\mathcal{P}$ (b,d) for parameters $N = 15$, $M = 2$. The atom transition frequency is in resonance with a field mode $n = 3$, $m = 2$. (a) and (b) $\theta_p = 1.557$; (c) and (d) $\theta_p = 1.57$.

number of excited eigenmodes with closely placed frequencies is involved in interaction. In Fig. 13.6(c,d) the dynamics both $|C_a|^2$ and of $\mathcal{P}$ have a periodical structure.

Figure 13.7 shows the temporal dynamics of probability densities of the ground states $|C_{b,nm}|^2$ associated with the radiated photons modes $1_{\nu m}$. For a clear graphical representation of the results of calculations it is convenient to perform the following renumeration of the probability amplitudes. The idea is to collect in the same picture the dynamics of the probability densities of ground states $|C_{b,nm}|^2$ for different principal and spherical quantum numbers simultaneously. We use the next vectorial designations. The amplitude of probability of excited state C_a is supplied

Table 13.2 Eigenfrequencies $\alpha_{\nu m}\nu$ for TM waves.

m,n	1	2	3	4	5	6	7	8
1	0.25	0.57	0.87	1.17	1.46	1.75	2.05	2.34
2	0.36	0.70	1	1.30	1.60	1.89	2.19	2.48
m,n	9	10	11	12	13	14	15	
1	2.64	2.93	3.22	3.52	3.81	4.10	4.40	
2	2.78	3.07	3.36	3.66	3.95	4.25	4.54	

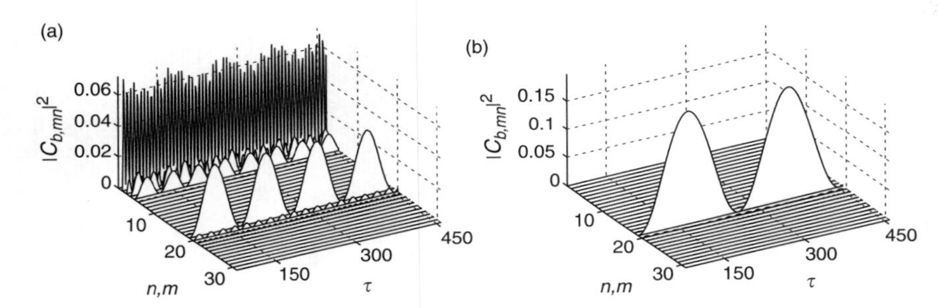

Fig. 13.7 The temporal dynamics of probability densities of the ground states $|C_{b,nm}|^2$ for parameters $N = 15$, $M = 2$. The atom transition frequency is in resonance with field mode $n = 3$, $m = 2$. (a) $\theta_p = 1.557$; (b) $\theta_p = 1.57$. See text for details.

with index 1 (C_1) and it is shown in Fig. 13.6(a,c). The probability amplitudes of ground states $C_{b,nm}$ we provide with single index β (C_β), where $\beta = 2, 3, 4, \ldots, n + 1, n + 2, \ldots, nm + 1$. For instance, modes with any n and $m = 1$ have C_β, $\beta = 2, 3, 4, \ldots, n + 1$, and for $m = 2$ one has $\beta = n + 1, n + 2, \ldots$. Such quantities we present in a separate 3D plot (see Fig. 13.7).

In Fig. 13.7(a) the density of probabilities from $|C_2|^2$ to $|C_{16}|^2$ answer to the $m = 1$ mode and $n = 1, \ldots, 15$, quantities $|C_{17}|^2 \div |C_{31}|^2$ answer to the $m = 2$ mode, etc. The resonant mode $|C_{b,32}|^2$ meets $|C_{19}|^2$. Figure 13.7(a) shows that for $\theta_p = 1.557$ such resonant mode has quite a large amplitude. However, the non-resonant background from several first modes with $m = 1$ and $n = 1, 2, 3$ has the same amplitudes as well and it is

Table 13.3 Eigen frequencies $\alpha_{\nu m}\nu$ for TM waves.

m,n	1	2	3	4	5	6	7
1	0.19	0.44	0.67	0.90	1.12	1.35	1.58
2	0.28	0.54	0.77	1	1.23	1.46	1.68
3	0.36	0.62	0.87	1.1	1.33	1.56	1.79
4	0.44	0.71	0.96	1.2	1.43	1.66	1.89

mixed up to the dynamics of $\mathcal{P}$ (13.10). The latter non-resonant background produces a high-frequency modulation of the probability density of $|C_a|^2$ in Fig. 13.6(a). The sum of such terms, having close frequencies, gives rise to the quasi-stochastic dynamics of dynamic polarization on Fig. 13.6(b). Figures 13.6(c,d) and 13.7(b) is similar to Figs. 13.6(a,b) and Fig. 13.7(a) accordingly, but for the dipole angle $\theta_p = 1.57$. One can see that the contribution of non-resonant background sharply decreases since the dipole polarization θ_p is very close to $\pi/2$. Figure 13.6(c) shows that the giant periodic energy exchange between electromagnetic field and the atom occurs.

In addition we investigate the frequency spectrum of the dynamic polarization $\mathcal{P}$ at a resonance conditions. It is well known that due to such a resonance the Rabi splitting of the radiation lines arises [Scully & Zubairy, 1996], which was observed experimentally [Zhu *et al.*, 1990]. We investigate such spectrum for the case $N = 7$, $M = 4$, $r_0/a = 0.1$ and $\nu = 0.078$. The spectrum of corresponding eigenfrequencies $\alpha_{\nu m}\nu$ is written in Table 13.3.

Figure 13.8 shows the frequency spectrum of polarization $\mathcal{P}$ (13.10) at the resonance with fields mode $n = 4$, $m = 2$ for various values of the parameter K. One can see from Fig. 13.8(a) that at $K = 1$ the well expressed doublet (the first-order peak), displaced concerning the center on $\pm 2\,\Omega$ (Ω see below). Figure 13.8(b) shows that for $K = 3$ the distance between peaks increases as well, and in the bottom the second-order peaks arise. A further increase K in Fig. 13.8(c) results in the evolution of the first and second-order peaks and the occurrence of the higher-order peaks. At a large value of $K = 10$ (see Fig. 13.8(d)) the dynamic polarization $\mathcal{P}$ generates a rather complex frequency spectrum. The doublets of higher-orders peaks lose symmetry concerning the center of atom transition frequency ω_{ab}. Figure 13.9 shows the corresponding temporal dynamics of the probability density of

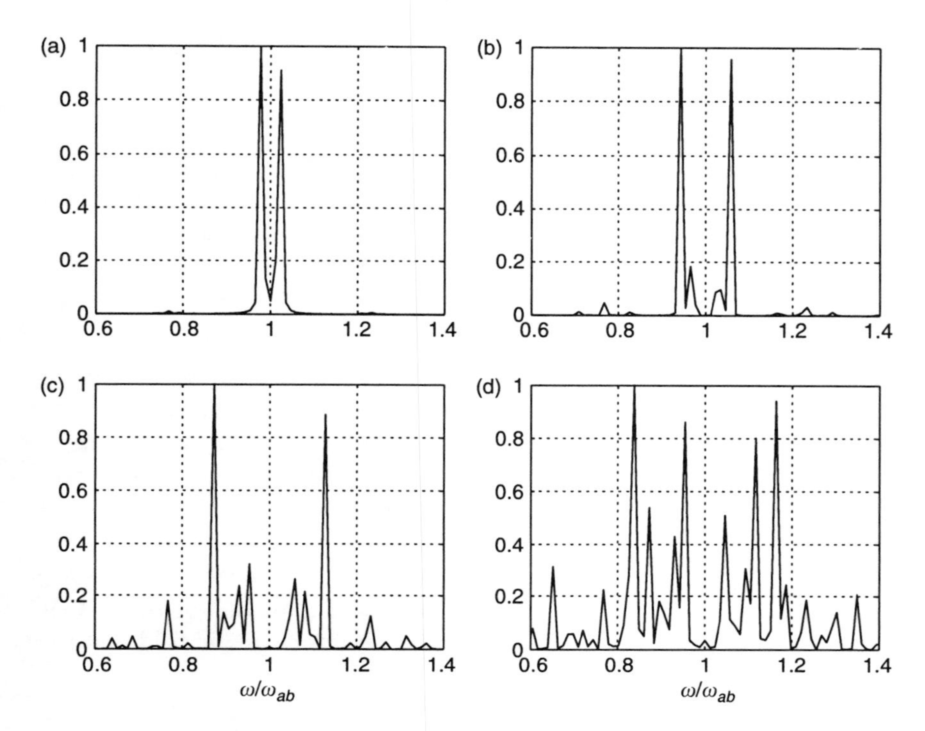

Fig. 13.8 Frequency spectrum of polarization $\mathcal{P}$ for parameters $N = 15$, $M = 2$. At resonances with mode $n = 4$, $m = 2$ for $\omega/\omega_{ab} = 0.0718$ and various values of parameter K. (a) $K = 1$; (b) $K = 3$; (c) $K = 5$; (d) $K = 10$.

the ground states $|C_{b,nm}|^2$. Figure 13.9(a,b) correspond to Fig. 13.8(a,d) accordingly. One can see from Fig. 13.9(a) that at $K = 1$ the single resonance with mode $n = 4$, $m = 2$ occurs, but the expected values of the other modes are small.

Such behavior can be explained by the following. Near to the resonance frequency $\nu_{\nu m}$ one can neglect by the non-resonant parts in (13.9). Then Eqs. (13.9) reduce to the simple form

$$\frac{dC_a}{d\tau} = -i\widetilde{K}e^{i\alpha\tau}C_b, \quad \frac{dC_b}{d\tau} = -i\widetilde{K}e^{-i\alpha\tau}C_a, \tag{13.11}$$

where $\widetilde{K} = KF_{\nu m}\nu_{nm}^{1/2}$, $\alpha = 1 - \nu_{\nu m}$ is a small mismatch. The solution of the Eqs. (13.11) has the form

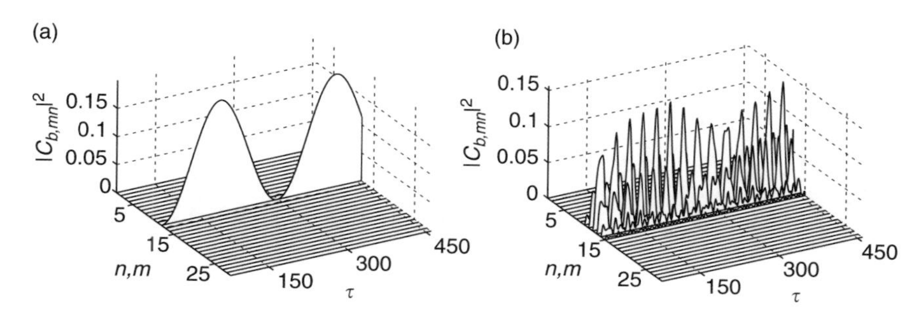

Fig. 13.9 The temporal dynamics of probability densities of the ground states $|C_{b,nm}|^2$ for parameters similar to Fig. 13.8 (a) $K = 1$; (b) $K = 10$.

$$C_a = \frac{1}{s_1 - s_2}[-s_2 e^{is_1\tau} + s_1 e^{is_2\tau}], \quad C_b = \frac{e^{-i\alpha\tau}}{\widetilde{K}} \frac{s_1 s_2}{s_1 - s_2}\left[e^{is_1\tau} - e^{is_2\tau}\right], \tag{13.12}$$

where $s_{1,2} = \frac{\alpha}{2} \pm \widetilde{K}$, $|\alpha| \ll 2\widetilde{K}$. Substituting (13.12) in (13.10) gives

$$\mathcal{P} = \frac{e^{i(1-\alpha)\tau}}{4\widetilde{K}^3}[\alpha - s_2 e^{2i\widetilde{K}\tau} - s_1 e^{-2i\widetilde{K}\tau}] + \text{c.c.}, \tag{13.13}$$

i.e. the spectrum of dynamic polarization $\mathcal{P}$ consists of oscillations with frequencies $1 - \alpha$, $1 - \alpha \pm 2\widetilde{K}$, shifted by α from center 1. For the exact resonate case $(\alpha = 0)$ $\mathcal{P}$ becomes

$$\mathcal{P} = \frac{e^{i\tau}}{4\widetilde{K}^2}[e^{2i\widetilde{K}\tau} - e^{-2i\widetilde{K}\tau}] + \text{c.c.} \tag{13.14}$$

i.e. for exact resonance in the $\mathcal{P}$ spectrum, only oscillations with frequencies $1 \pm 2\widetilde{K}$ is present. Such case is shown in Fig. 13.8(a). However at large values of K such as $K = 10$ (see Fig. 13.9(b)) some non-resonant photon modes with $|\alpha| \ll 1$ provide the non-zero contribution in $\mathcal{P}$. In such a case the simple expression (13.14) is no longer valid. Such case answers to Fig. 13.8(d). As a result the $\mathcal{P}$ frequency spectrum acquires a number of peaks corresponding to the background oscillations and it becomes more indented. We note that one can see from Figs. 13.7 and 13.9 that the contribution of highest values of n,m-field modes is very small and therefore in the calculations we can restrict ourselves by $N = 15$ and $M = 4$ numbers.

13.4 Triple photon state

13.4.1 *Basic equations*

Other example is how to create the photon's triple state.* Fig. 13.10 shows the geometry of system.

Due to the spherical geometry of the problem, we will characterize the state of electromagnetic field in microsphere with indexes ν, m and j (radial, spherical and azimuthal quantum numbers). In this case for a photon state we write $|n_{\vec{k}}\rangle \rightarrow |n_{\nu m}\rangle$, $\omega_{\vec{k}} \rightarrow \omega_{\nu m}$ (the eigenfrequencies of sphere do not depend on azimuthal quantum number [Stratton, 1941; Vainshtein, 1969], but the field do), $E_{\vec{k}} \rightarrow E_{\nu m j}$. In this chapter we study how to use such

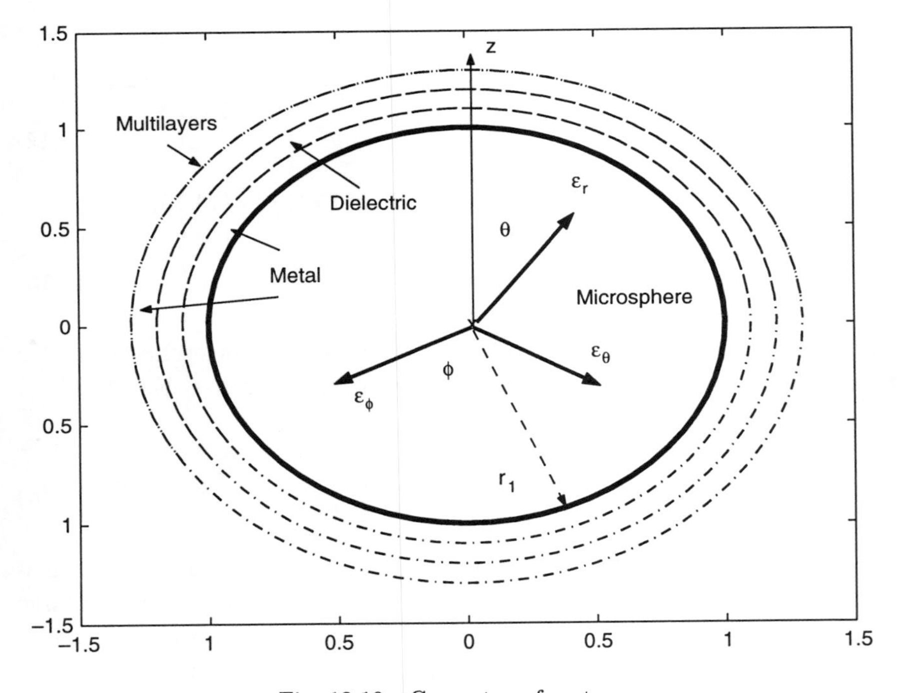

Fig. 13.10 Geometry of system.

*We note the relation of this state with the qutrit objects, which are now under intensive studying, see e.g. A. Galindo *et al.*, *Rev. Mod. Phys.* **74**, 347 (2002); Alipasha Vaziri *et al.*, *Phys. Rev. Lett.* **91**, 227902 (2003) and references therein.

optical oscillations in microspheres to produce the triple photon states. The idea comes from the next paragraphs.

Eigenoscillations of field in the spherical symmetry systems by natural quantization possess both angular spherical m and azimuthal j quantum numbers assuming that the integer values are $|j| <= m$. Due to high symmetry the eigenfrequencies of oscillations do not depend on azimuthal number and they have $2m + 1$ rates of degeneracy. In whispering gallery mode (WGM) regime (see [Buck & Kimble, 2003; Ilchenko *et al.*, 1998] and references therein) when $m >> 1$ the eigenfrequencies spectrum has huge degrees of degeneracy. So our proposition is to use the oscillations with a small spherical number m. In microsphere the fundamental oscillations with spherical number $m = 1$ allow only three independent field states $j = -1$, $j = 0$ and $j = 1$, which have the same frequency $\omega_{\nu 1}$ and thus cannot be separated. We suppose also that such a photon is in a coherent state, whose eigenvalues are α, where $|\alpha| = \sqrt{n}$, n is average photons number in mode $\omega_{\nu 1}$ [Scully & Zubairy, 1996]. We write such a state as $|j, \alpha\rangle$. We consider the superposition of such states in the form $|\psi\rangle = q_1| - 1, \alpha\rangle + q_2|0, -\alpha\rangle + q_3| + 1, \alpha\rangle$, where the complex coefficients q_1, q_2, q_3 amount to six real parameters. In the simplest case, which we suppose here to be $|q_i| = r$, writing them as $q_i = r\ e^{i\phi_i}$, $i = 1, 2, 3$, factoring out a global irrelevant phase, and imposing $|\psi\rangle$ to be of unit norm, we can write down

$$|\psi\rangle = r(e^{i\Phi_1}\,|{-1}, \alpha\rangle + e^{i\Phi_2}\,|0, -\alpha\rangle + |{+1}, \alpha\rangle), \tag{13.15}$$

where r is a normalization constant and $\Phi_1 = \phi_1 - \phi_3$, $\Phi_2 = \phi_2 - \phi_3$. The decoherence phenomenon can transform this superposition (13.15) into a statistical mixture [Zurek, 2003], but its lifetime may be rather long if the Q-factor of corresponding photons mode has a great enough value. Let us evaluate the decoherence time of such a state in a microsphere. Following [Zurek, 2003] the decoherence time is given by $\tau_D = \tau_R(\hbar^2/2m_e k_B T_0 \Delta x^2)$, where m_e is the electron mass, T_0 is a temperature, Δx is a typical spatial scale, $\tau_R \sim \gamma^{-1}$, γ is a relaxation rate, in our case $\gamma = \mathrm{Im}(f) = \mathrm{Re}(f)/2Q$, f is a photons mode frequency. For parameters $T_0 = 27\,\mathrm{K}$, $Q = 10^9$, $dx = 50\,\mathrm{nm}$, $f = 500\,\mathrm{THz}$ ($\lambda = 600\,\mathrm{nm}$) one obtains $\tau_D = 0.23\tau_R$, which is in good agreement with the experiment involved Rydberg atoms interacting with photon coherent field [Brune *et al.*, 1996]. If $|{-1}, \alpha\rangle$, $|0, \alpha\rangle$ and $|{+1}, \alpha\rangle$ are an orthonormal basis one has to write Eq. (13.15) in the form

$$|\psi\rangle = p_1 e^{i\Phi_1}\,|{-1}, \alpha\rangle + p_2 e^{i\Phi_2}\,|0, -\alpha\rangle + p_3\,|{+1}, \alpha\rangle), \tag{13.16}$$

where pure state $|\psi\rangle$ is a superposition of the states $|j, \alpha\rangle$ and due to the normalization $\langle\psi|\psi\rangle = 1$ we have $|p_1|^2 + |p_2|^2 + |p_3|^2 = 1$. One can satisfy this if he were to write the amplitudes p_i in the form $p_1 = \sin\theta\cos\varphi$, $p_2 = \sin\theta\sin\varphi$ and $p_3 = \cos\theta$, where θ and φ are spherical angles, $0 \leq \theta$, $\varphi \leq \pi/2$, $0 \leq \Phi_{1,2} \leq 2\pi$. So (13.16) has four independent parameters $(\theta, \varphi, \Phi_{1,2})$. In this a case state (13.16) is qutrit, which is a system whose states live in a three-dimensional Hilbert space [Arvind & Mukunda, 1997; Caves & Milbum, 2000]. We suppose also that (13.16) is prepared by an external classical source from a vacuum state, and therefore is in a coherent state. In which way one can prepare the latter?

We refer to Chapter 4, where is shown that in a case of spherical metallo-dielectric stack the narrow peaks almost complete transmittance arise in a frequency range of opacity of metallic layers ($\varepsilon < 0$) [Burlak *et al.*, 2002]. The details of peaks are defined by the structure of such a stack.

Due to the high Q factor of oscillations in coated microspheres such a state can be well isolated from an environment. Further we study the transmittance peaks in a layered microsphere coated by metallo-dielectric stack to support such a field state (see Fig. 13.10).

The following parameters have been used in calculations: The geometry of a system is $SLNLV$ where letters S, L, N, V indicate the materials in the system, the radius of the internal substrate of microsphere is $r_1 = 4\,\mu\text{m}$. We have used the next parameters of materials [Hodgson & Weber, 1997] S (spherical substrate): glass $n_S = 1.5$, L: metal layer, $\omega_P = 1.6 \cdot 10^{16}\,\text{s}^{-1}$ [Yeh, 1988] ($f_p = 2.55 \cdot 10^3\,\text{THz}$), for calculations we normally used $\nu = 1.6 \cdot 10^{11}\,\text{s}^{-1}$, but sometimes we chose the value ν small enough to separate the influence the energy leakage to a radiating Q factor of oscillations. Thickness of the metal layers is 35 nm, N is SiO_2, the thickness 232 nm, $n_N = 1.46$ [Kumar *et al.*, 1999], V (outer medium) is *air*, $n_V = 1$. The spherical quantum number is $m = 1$.

Figure 13.11 shows the spectrum of eigenfrequencies for TM waves for such microsphere, received from the numerical solution of the complex eigenfrequency equation (4.2). Found eigenfrequencies f_n have complex values, where real part (Fig. 13.11(a)) defines the frequency oscillations while the imaginary part determines the velocity of damping of eigenoscillations or Q-factor in the form $Q = \text{Re}(f_n)/2\,\text{Im}(f_n)$, where n is the number of roots.

In Fig. 13.11(b) the Q factor is shown in a logarithmic scale. One can see, that already for three-layered metallo-dielectric stack the order of magnitude of Q factor is about 10^5 in the zone of small transmission $|T|^2 << 1$. Our calculations have shown an increase in the number of layers in stack

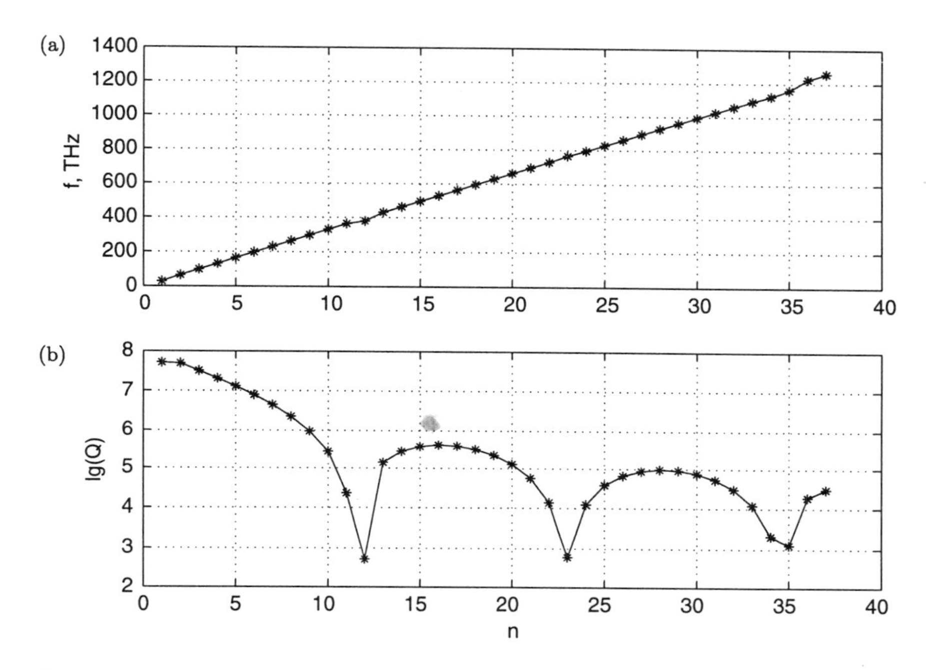

Fig. 13.11 Lowest 37 eigenfrequencies and its Q-factors ($Q = \mathrm{Re}(f_n)/2\mathrm{Im}(f_n)$) for coated microsphere with 3-layered metallo-dielectric stack (see details in text) and spherical quantum number $m = 1$.

the Q factor have been increased up to value 10^9 that is typical to Q factor for WGM oscillations in a bare dielectric microsphere.

The imaginary part of oscillations in microspheres is not zero even for lossless material due to leakage energy (radiation) to a surrounding space. One can see from Fig. 13.11(a) the eigenfrequencies f_n increase as their number n increases. However the behavior of Q factor is more complicated. At small frequencies the dielectric permittivity of metal has a large negative value. In this a case the depth of penetration of field in metal layers (skin depth) is small. Eigenoscillations have very high Q factor in such area, similar to a case of metallized microsphere in which optical fields practically do not leave the dielectric substrate. However at higher frequencies the wavelength of fields becomes comparable with skin-depth and thickness of metal layer. Such fields can already leak through thin metal layers and interfere with oscillations in other layers. Therefore the interference picture arises in the spectrum.

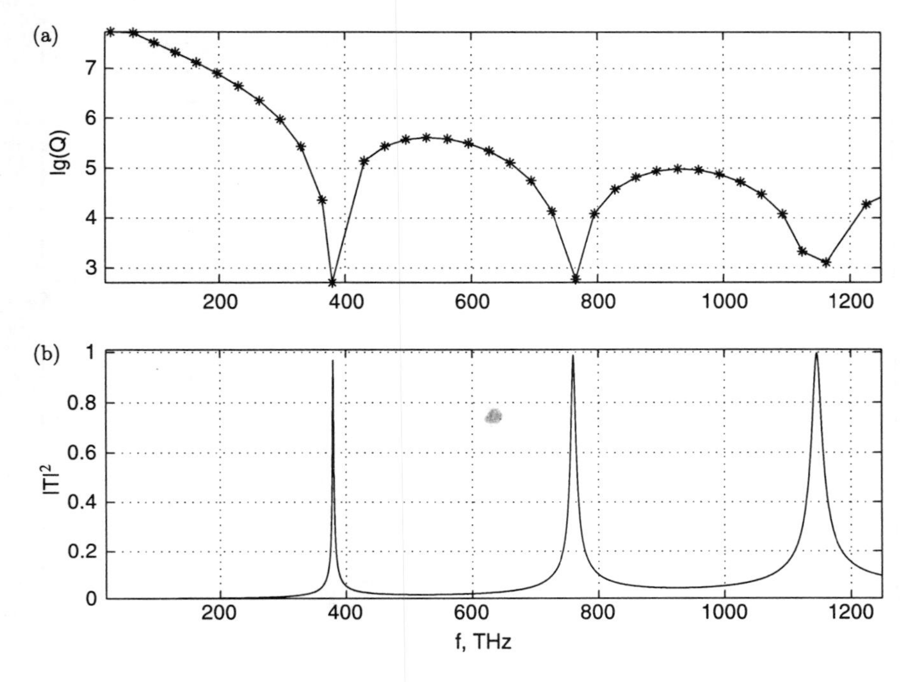

Fig. 13.12 Frequency dependence of Q-factors and the transmittance coefficient $|T|^2$ for coated microsphere with 3-layered metallo-dielectric stack.

To explain the non-monotonic behavior of the Q factor we compare it with the frequency dependence of the transmittance coefficient of the stack T in the same frequency range (see Fig. 13.12). Figure 13.12(a) shows Q factor of eigenoscillations of coated microsphere versus corresponding real part of eigenoscillations (frequency of oscillations), constructed with the use of data from Fig. 13.11, while in Fig. 13.12(b) shows the frequency dependence of transmittance coefficient $|T|$ of the metallo-dielectric spherical stack. Both Figs. 13.12(a,b) use the same frequency range on the argument axis. One can see that in a range $f < \omega_p/2\pi$, $\varepsilon(f) < 0$ in metal layers. Nevertheless the narrow peaks of almost complete transparency arise from this frequency range. One can see the position of the minima of Q factor coinciding well with the position of the transmittance peaks. It means that the reduction of Q factor is caused by the leakage photons through the spherical stack in a range of peaks. Beyond such resonances the reflection from the stack appears nearly completed $R \approx 1$ and $T = 0$. Calculations show an increase in the number of layers when the Q factor increase up to

10^9 while within the transmittance peaks a fine structure is formed due to the intrinsic re-refractions of optical waves inside the stack.

We refer to experiments [Bouwmeester *et al.*, 1997; Kwiat *et al.*, 1995; Pan *et al.*, 2003]. Now we apply above mentioned electromagnetic peculiarities to dynamics of the photons in coated microspheres. The first high-frequency peak of a transparency in Fig. 13.12(b) one can use for the ultraviolet pulsed pump laser with a frequency of $f = 784\,\text{THz}$ ($\lambda = 382\,\text{nm}$). Such peak thus opens access to the interior of microsphere while through the second peak of transparency for photons having frequency $f = 392\,\text{THz}$ ($\lambda = 762\,\text{nm}$) can be radiated from microsphere in surrounding space. The narrowness of the transparency band $\Delta\lambda \sim 5\,\text{nm}$ allows us to separate such photons with spherical number $m = 1$ from other photons in microsphere. In Fig. 13.13(a,b) the frequency dependence of transmittance coefficient $|T|^2$ for various spherical numbers m is shown for slightly

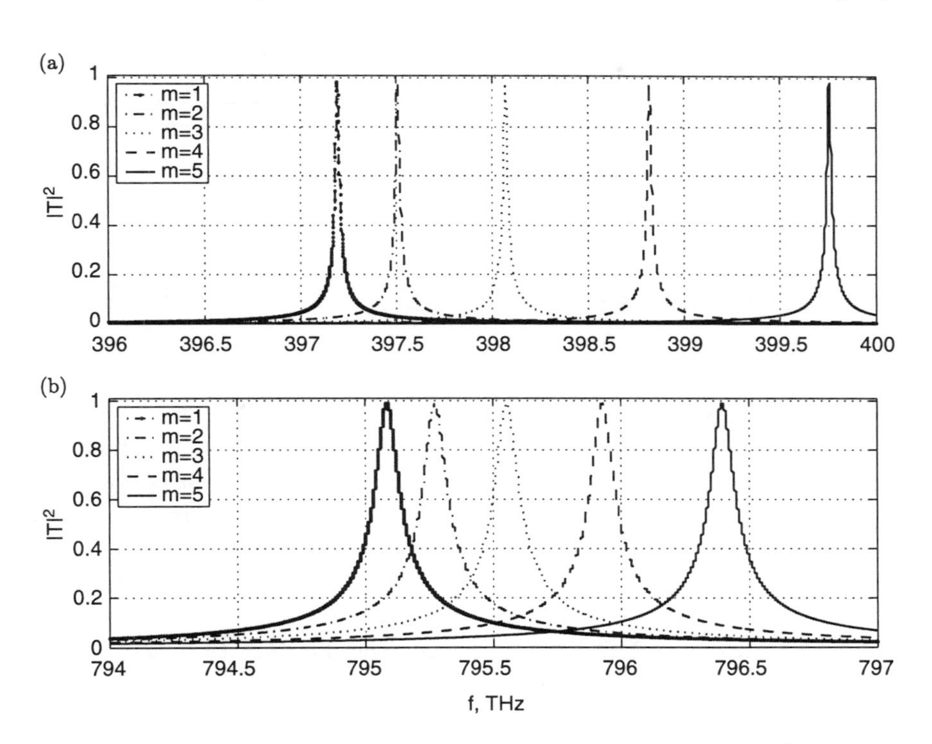

Fig. 13.13 Frequency dependence of the transmittance coefficient $|T|^2$ of coated microsphere with 3-layered metallo-dielectric stack for various spherical quantum numbers in different frequency ranges.

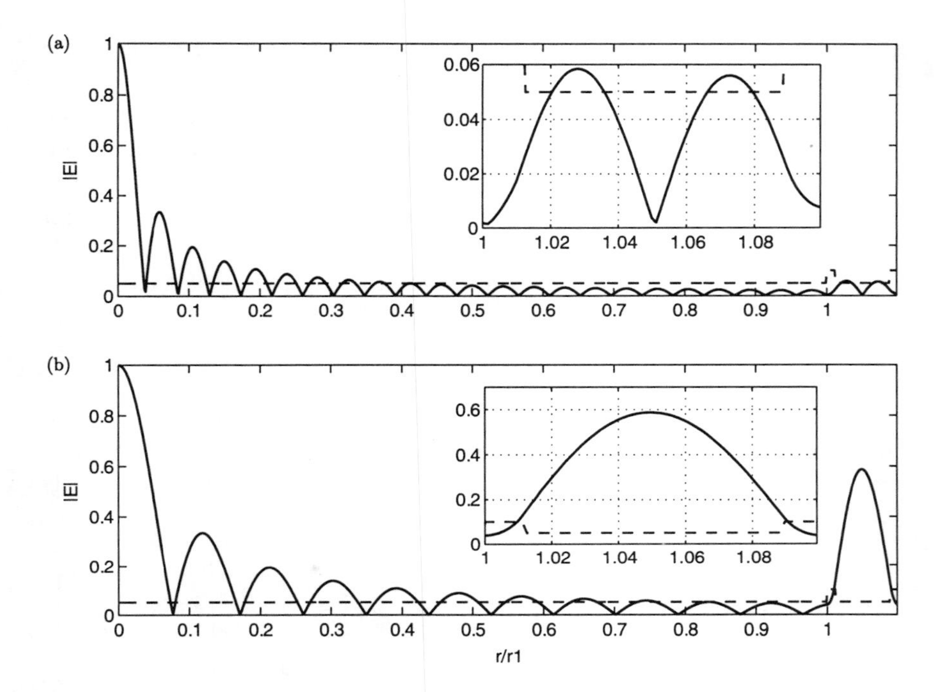

Fig. 13.14 The radial structure of eigenfield $|E_\theta|$ in coated microsphere with 3-layered metallo-dielectric stack for (a) eigenfrequency $f = 784\,\text{THz}$ ($\lambda = 382\,\text{nm}$); and (b) eigenfrequency $f = 392\,\text{THz}$ ($\lambda = 764\,\text{nm}$). Spherical number $m = 1$. Inset shows the fields in the stack. Dash-dot line is the structure of the spherical stack.

distinct parameters. One can see, that the overlapping is not significant and is discernible only for $m = 1$ and $m = 2$ modes. Our calculation the spectrum of eigenmodes shows that for the above-mentioned parameters only one eigenfrequency for $m = 1$ belongs the area of a peak transmittance. Other eigenfrequencies become rather far from the corresponding peaks and do not appreciably influence a field inside the microsphere. Such feature can be exploited to separate necessary $\omega_{\nu 1}$ photons mode and thus to open access to the interior of microsphere.

Waves with other frequencies which are placed beyond the peaks become confined inside the microsphere or will be reflected from its boundaries.

Figures 13.14(a,b) shows the structure of the eigenfields in the zone of both mentioned transparency peaks of the metallo-dielectric stack on frequencies $f = 784\,\text{THz}$ ($\lambda = 382\,\text{nm}$) and $f = 392\,\text{THz}$ ($\lambda = 784\,\text{nm}$). The nonlinear material, used for parametric generation of the photons, has to be placed in maxima of field. From Fig. 13.14, as one can see, such maxima are not only inside the microsphere, but the high peak of field is located in the dielectric layer of the stack. It allows the placing of a nonlinear material also in such layer closely to surface of the microsphere. In the case of large amplitudes it can allow us to produce the pairs of such photons closely of the surface of layered microsphere. Note the threshold of fields generation in coated microsphere with active substrate can also be reduced if the number of layers in stack is large enough [Burlak, 2002].

Let's note an opportunity to slightly control by position of the transparency peaks by applying a weak external magnetic field B. In such field in the permittivity of metal layers (4.1) occurs at $\omega^2 \to \omega^2 - \omega_H^2$ [Kittel, 1976], where $\omega_H = eB/\varepsilon_0 m$ is of cyclotron frequency. So far as the field B is weak enough it can allow us to neglect the anisotropy (diagonal-off entries in tensor of permittivity of metal layers) brought about by such magnetic field. As a result an opportunity to control the radiation of photons pair surfaces. Note that although eigenfrequencies TE waves differ from eigenfrequencies TM waves, nevertheless other properties of TE waves really do not effectively change as compared to TM waves for studied frequency range.

13.4.2 *Wigner function*

Here we study the Wigner function for general triple state (13.15). The Schrödinger equation was deduced from classical mechanics in the Hamilton–Jacobi form. Thus it is no surprise that it yields classical equations of motion when $\hbar$ can be regarded as small. However, establishing the quantum-classical correspondence involves the states as well as the equations of motion. Quantum mechanics is formulated in Hilbert space, which can accommodate localized wavepackets with sensible classical limits as well as the most bizarre superpositions. By contrast, classical dynamics happens in phase space [Wigner, 1932; Zurek, 1991].

To facilitate the study of this aspect of the problem, it is convenient to employ the Wigner transform of a wavefunction $\psi(x,t)$

$$W(x,p,t) = \frac{1}{2\pi\hbar} \int_{-\infty}^{\infty} ds \cdot e^{-ips/\hbar} \psi\left(x - \frac{s}{2}, t\right) \psi^*\left(x + \frac{s}{2}, t\right). \qquad (13.17)$$

which expresses quantum states as functions of position x and momentum p.

The Wigner distribution or Wigner function $W(x,p,t)$ is real, but it can be negative. Hence it cannot be regarded as a probability distribution. Nevertheless, when integrated over either of the two variables, it yields the probability distribution for the other (for example, $\int W(x,p,t)dp = |\psi(x)|^2$). For a minimum-uncertainty wave-packet $\psi(x) \sim \exp\{-(x-x_0)^2/2\delta^2 + ip_0x/\hbar\}$, the Wigner distribution is a Gaussian in x and p:

$$W(x,p,t) = \frac{1}{\pi\hbar}\exp\left\{-\frac{(x-x_0)^2}{\delta^2} - \frac{(p-p_0)^2\delta^2}{\hbar^2}\right\}. \tag{13.18}$$

One can see that a system described by this type of Wigner distribution (13.18) is localized in both x and p. Furthermore we set $\hbar = 1$. We suppose that (13.15) is prepared by an external classical source from a vacuum state, and therefore it is in a coherent state and it is a superposition of three displaced oscillators [Scully & Zubairy, 1996]. For simplicity, we only consider here the case $\Phi_1 = 0$, $\Phi_2 = \pi$, and rewrite (13.15) as

$$|\psi\rangle = r(|-1,\alpha\rangle - |0,-\alpha\rangle + |1,\alpha\rangle), \tag{13.19}$$

where $|\alpha| = \sqrt{n}$, n is average photons number in mode $\omega_{\nu 1}$. The decoherence can destroy such a state [Zurek, 2003], but as we have evaluated above, its lifetime may be large enough if the Q-factor of the corresponding photon mode in microsphere has a great value. The quantization of field in coated microsphere is made in Chapter 12, so we are in a position to perform the study the behavior of triple state (13.19) by means of the Wigner function formalism. Assuming that a field (13.19) is in a coherent state and taking into account (10.18),(10.19) we write

$$\langle\alpha| E_s |\alpha\rangle = F_s\left(\langle\alpha| a^+_{\nu m} + a_{\nu m} |\alpha\rangle\right) = \mathcal{E}(\kappa_{\nu 1})F_s(r,\theta,\varphi)(\alpha^* + \alpha), \tag{13.20}$$

$$\langle\alpha| H_s |\alpha\rangle = if_s\left(\langle\alpha| a^+_{\nu m} - a_{\nu m} |\alpha\rangle\right) = i\mathcal{E}(\kappa_{\nu 1})f_s(r,\theta,\varphi)(\alpha^* - \alpha), \tag{13.21}$$

where $\kappa_{\nu 1} = \omega_{\nu 1}a_0/c$. Further we use $x = (\alpha^* + \alpha)$, $p = i(\alpha^* - \alpha)$. We will study the evolution of the Wigner function $W(x,p,t)$ for various times (we renormalize $t \to \omega t$, $\omega = \omega_{\nu 1}$). Since (13.19) is a pure state, we can use the Wigner function in the form (13.17).

Figures 13.15 and 13.16 illustrate the structure of phase space and a spatiotemporal evolution of state (13.19).

Figure 13.15 shows the Wigner function of (13.19) for parameters $\alpha = 5$ and $t = 0.1$. There are the superposition of two Gaussians peaks on periphery, corresponding to a coherent ground state of oscillator, and coherent

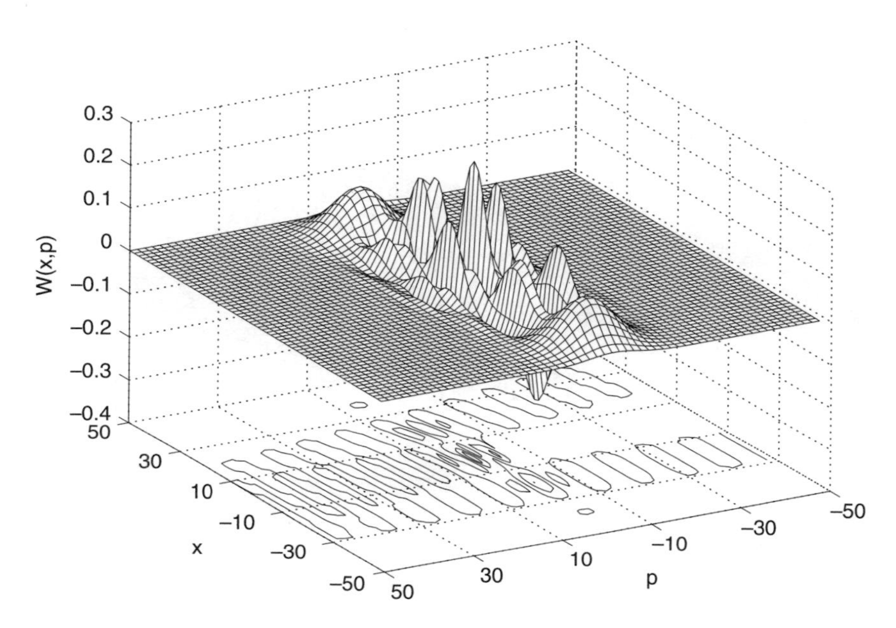

Fig. 13.15 Spatiotemporal evolution of the Wigner function of triple state for $\alpha = 5$ and $\omega t = 0.1$. See details in the text.

oscillations between them. Oscillating ridges and valleys of the interference pattern are parallel to the separation between the two Gaussians. Such interference usually appears for the initial superposition of minimal uncertainty Gaussians in position. Though quantum superposition of Gaussians hills is generated by states, close to classical, the quantum interference between them reflects the quantum mutual correlations and demonstrates deep nonclassical aspects in the behavior of such a system. Figure 13.15 shows that in some parts of the phase space the Wigner function takes on a negative value and a system has nonclassical interferences. From Fig. 13.15 one can see that the form of the Wigner function for triple state (13.19) has a rather complicated structure and essentially differs from typical two-gaussian case states (see [Brune *et al.*, 1996; Zou *et al.*, 2001; Zurek, 2003] and references therein). In Fig. 13.16 the Wigner function is shown at the same parameters, as on Fig. 13.15, but for time $t = 2$. From Figs. 13.15 and 13.16 one can see that the Wigner function undergoes the rotation as the whole (that reflects the classical behavior of the average center oscillators motion [Landau & Lifscitz, 1977]), but also possesses the internal local dynamics, which changes its form. Figure 13.16 shows that at $t = 2$ in the

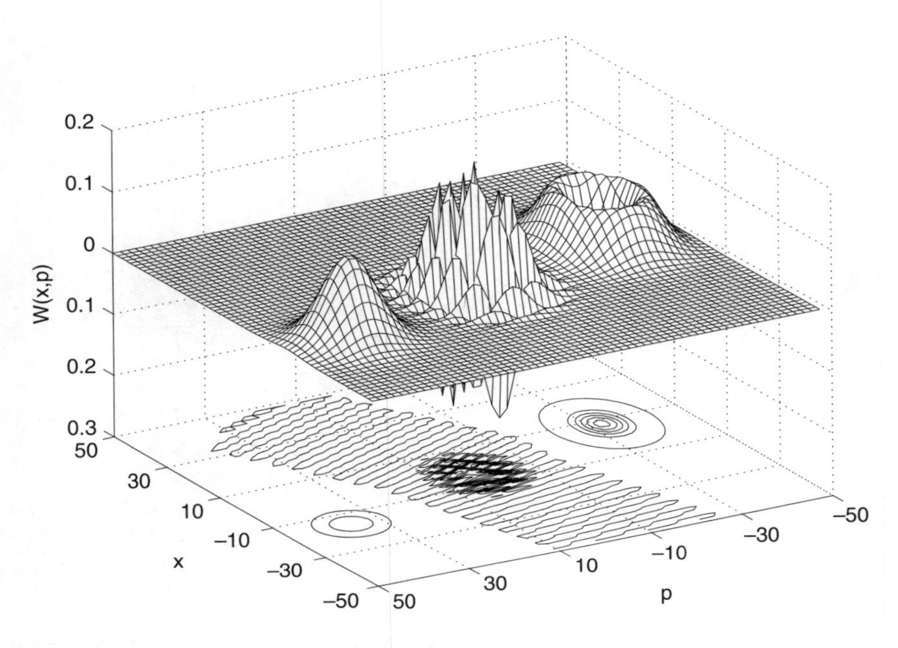

Fig. 13.16 The same as in Fig.13.15 but for $\omega t = 2$. See details in the text.

center of one of Gaussians hills the dip is formed, in which the Wigner function takes negative values. Such a feature is the signature of a nonclassical behavior of such a state. Such behavior of the Wigner function corresponds to formation high nonclassical states and was already observed in experiments on studying of one-photon states of a microwave field [Nogues *et al.*, 2000] and parametric down-conversion [Lvovsky *et al.*, 2001]. Our calculations show that the periodicity of such the dynamics is about $\Delta t \simeq 3$, after that the form of the Wigner function restores.

Note in general the transmission peaks in T spectrum allow the necessary separation of photon states with various spherical number m. Note that the triple photons also can be produced due to spontaneous emission of excited atoms inside of microsphere. A set of such microspheres with photons can allow us to create the optical waveguides, which consists of periodically arrayed dielectric microspheres similarly [Furukawa & Tenjimbayashi, 2001]. In our case such waveguides could provide with facilities to control and communicate the photons inside of arrays of coated microspheres.

13.5 Conclusions

We have studied in this chapter the dynamics of the spontaneous emission of a two-level atom placed in the metallized microsphere in the multimode field case by direct solving of the complete set of equations for probability amplitudes (without continuum modes approximation). It allows us to examine the dynamics of spontaneous emission in the uniform frameworks without a separation into resonant or non-resonant cases.

We found that quite complex temporal dynamics of the probability amplitudes arise at almost tangential orientation of the atomic dipole matrix element. In this case various photonic modes having different principal and spherical quantum numbers take part in the atom-field interaction. The distribution of probability densities and temporal dynamics of the atomic ground states associated with various emitted photon is investigated.

We found that for a true microsphere case the continuum mode approximation appears suitable only at the initial stage. At a longer time the oscillatory behavior of probabilities begins, reflecting a complex dynamics alternating between collapses and revivals due to various Rabi effects (both oscillations and splitting).

In such a system (two-level atom coupled with field) we found it suitable to choose the dynamic electrical polarization of atom as a quantity reflecting the general collective atom-field dynamics. We found that the details of temporal dynamics strongly depend on the dipole orientation. The contribution of the highest eigenfrequencies becomes small at large numbers of radial and spherical quantum modes. The frequency spectrum and Rabi splitting of the atomic dynamic polarization during spontaneous emission is studied at various parameters of the atom.

In this chapter we have concentrated in a spontaneous emission in a metallized microsphere. By virtue of the transparency and simplicity of such a model, it can serve as a reference problem at studying the details of dynamics of spontaneous emission, since only a few factors influence this dynamics. Nevertheless even in such a simple model the large variety of dynamic properties can be found. In the more complicated dielectric microspheres the high-quality electromagnetic whispering gallery mode (WGM) oscillations exist. The contribution of WGM spectrum in such a microsphere requires the consideration of complete spectrum modes with various values of spherical quantum numbers. In this case the radiative losses begin to depend on the frequency of oscillations. In a bare dielectric microsphere the photon modes having a small spherical quantum numbers have a low Q factor [Vainstein, 1988]. They can have a Q factor value only for compound structures similar to the alternative stack deposited onto the dielectric microsphere [Sullivan, 1994; Sullivan & Hall, 1994; Burlak *et al.*, 2000, 2001a,b]. Therefore the dynamics of spontaneous emission will be sensitive to the concrete model of boundary conditions on the surface of microsphere.

We found that if time is small enough the temporal behavior of the probability amplitude of excited state can be approximated by a continuum mode decay rate of the Weisskopf-Wigner approach. At a later time the peak-like sequence of revivals of the excited state can occur.

PART III

Numerical Methods and Object-Oriented Approach to the Problems of Multilayered Microsystems

Preface

The solution of nonlinear (and sometimes and linear) problems of modern physics requires the investigation of various kinds of models described by a complex set of equations. Often it is impossible to obtain the solution of such a problem in a useful form, involving only the analytical methods. However even if such a solution is found, it may be written in a complicated form, involving infinite series or integrals. Sometimes the problem of evaluating such series or integrals requires efforts commensurable with the numerical solution of the initial task. In such cases it is impossible to produce to some useful result without exploiting the numerical methods or simulation. Such problems deal with the ordinary use of numerical methods and programming. Now one has to reformulate the question by the next way, from which a stage of an investigation it is expedient to exploit the numerical methods more intensively.

Now in view of more active application of various inhomogeneous structures in technology the investigation of the complex models generates the question: in which the way one can take into account the various equally important factors simultaneously? Besides the use of the regular numerical methods here already poses a problem about the optimal organization of the program technology. Now such an aspect of a problem often acquires the main importance.

In this chapter some approaches to such a problem are discussed from point of view of object-oriented technology already widely used in modern programming. We try to demonstrate the benefits of the use of object-oriented approach in computer simulation for the development of the formal theory of complex systems and understanding the philosophy of the problem. One can say that sometimes we know how to solve the problem, but we do not have ideas on how to incorporate it to an errorless program code.

Overall there are numerous examples in the chapter, which do not stand independently from the content of the book. We use them as an organic part of the physical and mathematical methods used for the investigation of problems, and we supply the reader with some examples of ready-to-use code.

Neither the features of numerical methods nor the details of the programming languages are discussed in depth in this part. All these can be found in textbooks. In [Calvert *et al.*, 1998; Yang, 2000; Deitel & Deitel, 2002b; Arciniegas, 2002; Hollingworth *et al.*, 2000; Nicolai, 1999; Ortega & Grimshaw, 1999; Young, 1998] one can find more information about *C++* programming. In [Besset, 2001; Deitel & Deitel, 2002a; Eckel, 2002; Lau, 2003; Morgan, 1999] Java programming is presented. About *C#* new programming language in [Tom, 2001; Deitel *et al.*, 2001; Lippman, 2002]. In [Besset, 2001; Hanselman & Littlefield, 2001; Lau, 2003; Ortega & Grimshaw, 1999; Potter, 1973; Press *et al.*, 2002; Thijssen, 2001] one can find more information about numerical methods and numerical simulation.

Our main purpose is to show how one can try to realize the complex and dynamic variability of such new direction as the spherical microstructures by means of the program technology. Finally this has to provide the possibilities in the use of a power of numerical simulations, when one realizes the motto "*The Purpose of Computing is Insight, not Numbers*" [Hamming, 1987].

CHAPTER 14

Use of Numerical Experiment

In this book the C++ is used as base language. Generally speaking there is a wide range of programming languages, from the old FORTRAN to the modern object-oriented C++, Java and a new language C#. Many books have written about the suitability of each language on each application. Besides often the choice of language is determined by the taste of the programmer. If you do not have a very large task, which one has to calculate with the maximal speed, there is nothing better than the fast FORTRAN [Press *et al.*, 1992]. However we often meet the situation when one plans to develop the code and modernize it further, when the code occupies many thousand lines, and when both the logic of calculation and the expected result are not clear even for the developer. In these cases it is recommended to choose one of the object-oriented languages. In such languages an author has the opportunity to construct and develop the logic of the program in a close conformity with the physics of the problem. The code can be naturally shared into the separate parts (classes) which are independent of one another, reflecting the independence of the objects of the real world. Each part (object) can have a complicated internal structure. Furthermore such objects begin to interact by a logical and hereditary way. As a result the depth of the real problem can be achieved. The possibilities of various levels of access are available in the object-oriented languages. It allows one to establish the necessary logical interaction between the objects with a desirable level of completeness.

Let's consider some examples. In the problem about an auto-shop on the sale of bicycles, motorcycles and cars it is necessary to concentrate the base for all moving means in the basic class TVehicular. Here it is possible to place such data which are common for all objects: quantity of wheels, color, mark, the maximal speed and the price. After that in classes-successors one can describe only the data, which are individual for given object, for example, quantity of doors for the object car. If the shop in future decides to sell the military tanks in this case one has to add only the derivative class TTank where it is necessary to write down the availability of the reservation and the type of arms. The base class, and together with it most parts of the code (classes) remain without any changes.

For a second example, let's assume that we consider the dynamics of the quantized electromagnetic field in a cavity. In a simplest case we should create a separate class of electromagnetic field, say TField, containing the next data: number of quanta, the frequency, and also the creation and the annihilation quantum operators. The second following object is class TCavity which is responsible for the form of a cavity, i.e. for eigenfrequencies of the electromagnetic field in a concrete form cavity (spherical, rectangular, etc.). Obviously in such a class both eigenfrequencies and eigenvectors pertaining to the internal geometry can be calculated separately. Establishing the first class as the base, and the second as a derivative class, we obtain the opportunity of the description of the quantized electromagnetic field in a spherical (or other) geometry. It is clear that such approach can be improved further. However the details concerning a field can be left confined to the first class TField. The second derived class TCavity will have access only to the necessary fields and methods (for example to the frequency, the number of quanta in a mode) and not knowing about other details. Thus we are free to change the latter details. Such an approach opens up opportunities for the researcher on the further development of such a scheme with any desirable level of realization. So the details of the structure of quantized fields can reflect a present level of knowledge. However once a new knowledge arrives one has to change only the contents of the class TField, and then leave without changing the other independent parts of a code. Clearly, such an approach allows for the creation of a code stabled to errors at further developments of a program. Therefore it is very important for the preliminary analysis of the intrinsic logic and interaction of the whole project with parts. Then an opportunity to operate objects as the independent parts of a code at any modifications in the project can be done rather freely in the program.

It is impossible to provide the recommendations suitable for all occasions. Therefore the author will further describe some factors, which he regards to be important in a practical work in a given scientific area. In the process of such a work the author has come to the conclusion about the use of *C++*. But this is only the personal conclusion.

(i) One of the important possibilities of the present is an opportunity to create and develop the graphical user interface (GUI). At the presence of such an interface supplies an opportunity to operate in many parameters simultaneously and to see the results in a desirable working area of the screen. Such an approach provides a lot of profits since a well-organized numerical experiment is one of the modern concepts to obtain the information from modeling.

(ii) Portability of a program code. If I want to work at some other universities, the problems of whether is there available corresponding software, should not worry me. I should have an opportunity to take the important software with me and to install it easily on a computer which is available. Now I can easily install Borland *C++* Builder (bcb) or Java (Java in general can be downloaded from the Internet freely) from one CD. However I cannot say the same about languages from Net7 if for my problem it is necessary to install all of packages Net7. Nevertheless now I can download from the Microsoft site (www.microsoft.com) the necessary toolkit for Net7 and find the environment for *C#* in the Borland site www.borland.com. Therefore, now the new language *C#* looks attractively.

(iii) Support of the object-oriented technology. Due to the opinions of specialists and the experience of the author only such technology allows one to create easily a large scaled program code. The most important opportunity is to adequately reflect the physical structure of the problem in the structure and the hierarchy of classes. This cuts off the simple Fortran, *C* and leaves *C++*, Java and *C#*.

(iv) The opportunity to work with vectors, matrices and complex numbers in the natural mathematical notation (as it is presented in books or in articles) is very important for the scientific sphere of programming. In the present Java version there is no complex type (as for instance in FORTRAN), also there is no opportunity in overloading the mathematical operators as in *C++* or in *C#*. It makes Java suitable for integer or float calculations, but it is of little use for problems where calculations with complex numbers or vectors and matrixes are essential. We hope this problem can be solved by the authors of Java in future. For instance, they have to create a specialized package, similar to the package with strings. However at the author's best knowledge such a package is not yet available. Sometimes similar complaints can be heard in various news on the Internet. So it leaves us with two languages $C{+}{+}$ and $C\#$.

(v) People predict a great future for $C\#$ language, but now $C{+}{+}$ remains the "working horse" due to the fact that a lot of code and numerical libraries are written both in $C{+}{+}$ and C.

It turns out that at present only $C{+}{+}$ satisfies to the mentioned requirements. In this book we have chosen Borland $C{+}{+}$ Builder. There are no loss of time for the creation or reorganization of GUI due to the convenience of the environment and since then a certain practical experience is acquired. The compiler generates the huge part of the code automatically. There are inbuilt opportunities to construct the graphical plots in run-time. All these enable one to concentrate more completely on the physical sense of the problem, for which the work has been launched.

Furthermore, we shall make the brief review of the elements of the $C{+}{+}$ language, which has only a help character. However with this clause even brief enumeration of elements of the language would occupy a lot of pages. We refer the reader to the textbooks, where one has the possibility of finding more details and examples [Calvert *et al.*, 1998; Yang, 2000; Deitel & Deitel, 2002b; Arciniegas, 2002; Hollingworth *et al.*, 2002; Hollingworth *et al.*, 2000; Nicolai, 1999; Ortega & Grimshaw, 1999; Press *et al.*, 2002; Young, 1998]. Moreover, in this book we restrict ourselves only to the review of frequently used program operators. Besides the elementary operators, there are the control operators, cycles, operators of choice, functions, classes and the overloading of mathematical operators. The reader can find much more information in many books [Yang, 2000; Deitel & Deitel, 2002b; Young, 1998].

Also we do not touch on the theory of numerical methods. The statement of such theory together with examples of the code can be found in [Press *et al.*, 2002] and references therein.

Some providers offer the various graphic libraries, see for instance the Internet site *www.steema.com*. The author of this book is more closely acquainted with the package TeeChart which is a built-in in *bcb*. On the latter site one can find more information about the professional version of this package.

14.1 Introduction

Here we briefly describe only a number of language operators, which are used in the book. Such description serves only the help purposes and can capture only a small part of the language which a lot of thick books are devoted to. The author also tried to avoid the language operators and constructions which are specific only for *C++* and which, as a rule, are not used in Java and *C#* languages. (Life is too complicate and who knows the language we have to work tomorrow.) Here you will not see the friend functions, the template, the virtual destructors and a lot of other features. The experts of *C++* language can state many reproaches. Answering them, the author can object only that he tried to concentrate on those means of language which allow one to solve a physical problem without a special deepening in a formal structure of such multilayered language as *C++*. The use of too complicate features is fraught with difficult perceptible errors. Often much more time is necessary to debug such programs than what the researcher — physicist or the engineer usually has.

It is necessary to note one important and instructive feature of language *C++*. Experience of the author shows that many computing physical problems can be more precisely formulated at the use of the object-oriented approach — which is the base of the modern languages of programming. Moreover, such well-formulated and structured basic model unexpectedly receives an opportunity to be developed further to a model of higher order. As a result on a basis of such well-debugged basic model there is the opportunity of transition to a new problem at which it was not supposed to find a suitable approach at an initial stage.

There are a lot of operators in *C++*. However there are some based operators. What features of a language are more important and what are less important will only be determined through your experience. In comparison with experimental physics all is much easier: nothing can be broken. Nevertheless it is a good idea to save backup copies of your favorite files (the author does it every time upon termination of work and before crucial modifications). Then programming becomes the pleasant chess game when it is always possible to make a move back as one desires.

14.2 The Brief Review of $C{+}{+}$ Operators

A program consists of a collection of statements. Each statement must end with a semicolon. Any comments up to the end of the line start with symbols //. Long comments in multilines start with /* and finish by */. You can name your creations (types, variables and functions) anything you wish. Here there are some unostentatious rules. For example, avoid names that put the number on the first place, and also avoid including a blank or signs like +, [] and ? inside the name. Agree, that name Luis8 sounds much more pleasant, than 8Luis or Lui[]s. As well as all words in $C{+}{+}$, names are sensitive to the case so John and john are different names. We strongly recommend to provide all variables with some sensible names. For example, if n means the number of layers in a system it needs to be named as nLayers. Our experience shows that in large programs it allows one to reduce both the number of errors and the time for debugging.

To avoid accidental misuse of a variable, it is usually a good idea to introduce the variable into the smallest scope possible. In particular, it is usually best to delay the declaration of a local variable until one can give it an initial value. Both if-else statements and for statements provide such a mechanism to avoid accidental misuse of variables and to increase readability. Also for this purpose one can use the compound operator.

14.2.1 *Data*

OK, let's start. The declaration

TType n;

declares the variable n to be an TType type. In a simple case TType can be int, float, double. In general case TType can be the name of the class or structure. Also TType can arise from the operator typedef (see below). The program must know what is the TType before using it. Any variables you have to declare before the use have to be initialized in one statement such as

int n = 10;

One declaration statement can declare and initialize several variables of the same type, separated by commas, such as

double x, y=0, z; // declare x, y, z in one declaration.

There are two numbers associated with some variable x: its value and its location or address of the variable in memory, written as &x. The symbol & is called the address-of operator. It gives the address a variable.

Arrays

For type TType A[n] is a one-dimensional array of n elements of type TType, where n is a positive integer and the quantity of the entries of array. The elements are indexed from 0 to n-1. For example,

```
float A[3]; // array of 3 floats: A[0],A[1],A[2]
int s[30]; // array of 30 ints: s[0],..., s[29]
```

The first two statements declare A and s to be one-dimensional arrays with 3 and 30 elements of type float and int, respectively. A one-dimensional array can be used to store elements of a (mathematical) vector. To work with elements of array one often uses the for loop. For example,

```
for (int i = 0; i < 30; i++) s [i] = i*i + 7;
```

Two-dimensional arrays having n rows and m columns (looking like a matrix) can be declared as TType b[n][m]. The row index changes from 0 to n-1 and the column index from 0 to m-1. For example,

```
double mt1[4] [7];        // 2D array of 4 rows and 7 columns
mt1[0] [0] = 5;           // access entry at row 0, column 0
mt1[2] [6] = 6;           // access entry at row 2, column 6
double a = mt [0] [0] ; // access entry [0] [0] of mt
for (int i = 0; i < 2; i++)
for (int j = 0; j < 5; j++) mt[i][j] = i + j;
```

The first statement above declares mt1 to be a two-dimensional array with 2 rows and 5 columns, whose element at row i and column j can be accessed by using mt1[i][j]. A nested for loop is often used to access all elements of a 2D array. A two-dimensional array is a natural data structure for storing elements of a matrix.

Three and more dimensional arrays can be declared similarly. Arrays can be initialized at the time of declaration. For example,

```
int v[] = {1, 2, 4, 5};                     // initialization of 1D array
int a[3] = {2, 4, 5};                       // initialization of 1D array
int u[ ][3] = { {1, 2, 3}, {4, 5, 8} };     // initialization of 2D array
```

In the declaration of v above, the compiler will count the number of elements in the initialization and allocate the correct amount of space for v so that the number of elements can be omitted. In an initialization of a multi-dimensional array such as u[ ][3], only the first dimension (the number of rows for u) can be omitted.

Here is an example of an initialization of a three-dimensional array:

```
double ax [4] [2] [3] = {
{ {11, 22, 0}, {55, 0,     0} },  // ax[0]
{ {−1, −2, 0}, {−5, −6,    0} },  // ax[l]
{ {23, −7, 8}, {0, 13,     8} },  // ax[2]
{ {−3, 19, 0}, {9, −5,     3} }   // ax[3]
};
```

In particular, az[0][0][0] = 11, az[2][0][2] = 8, ax[3][l][l] = −5, and az[0] is a two-dimensional array with entries {{11, 22, 0}, {55, 0, 0}}. Its entries may be accessed by a triple loop:

```
for (int i = 0; i < 4; i++)
for (int j = 0; j < 2; j++)
for (int k = 0; k < 3; k++) ax[i] [j][k] = i + 2*j + 3*k;
```

The dimensions of an array must be known at the time it is declared so that the compiler can allocate space for it.

Structures

Unlike an array that takes values of the same type for all elements, a struct can contain values of the same or different types. It can be declared as

```
struct TPolynomial2{float a, b, c;};
struct TPerson {String Name; int age; float height;};
```

The first example defines a new type called TPolynomial2 and answers to the coefficient of a quadratic polynomial $a^*x^2 + b^*x + c$. The second one defines the personal information as type TPerson. Note the semicolon after the right brace. The structure TPerson has three fields or members, one String, int and floats. Its members are accessed by the . (dot) operator. For example,

```
TPerson Smith;
Smith.Name = "John"; Smith.age = 23; Smith.height = 180;
```

A variable of a struct represents a single object and can be initialized by and assigned to another variable. For example,

```
TPerson Smith2 = Smith;
```

It can also be initialized in a way similar to arrays: TPerson Bolonsky = {"Jay", 33, 165.5}; . If a struct is a dynamical variable we have to use it as in the next example

```
struct TPolynomial2{float a, b, c;} *pol;
pol = new TPolynomial2;
pol->a = 1;
pol->b = 2;
pol->c = pol->a + pol->b;
delete pol;
```

In this case the members are accessed by the -> (arrow) operator. Note that the record pol->b = 2 is equivalent to (*pol).b = 2;. If any variable is created by the new operator, it must later be destroyed by the delete operator. Otherwise the leakage of memory occurs.

14.2.2 *Operators*

Increments and Decrements

When increasing or decreasing a variable by unity, more compact notation can be used. If n is a variable, n++ and ++n both increase the value of n by 1. In the postfix form n++, the original value of n is used in the expression in which it occurs before its value is incremented. In the prefix form ++n, the value of n is first incremented by 1 and then the new value of n is used. A similar explanation holds for the decrement operator –.

Conditional Statements

if-else Statements

The simplest conditional statement is the if-statement, whose syntax is:

if (condition) statement

where condition is a logical expression having the value true or false, and the statement can be a single or compound statement. The condition must be enclosed in parentheses. The operators are: logical OR, written ||; logical AND, written &&; and logical NOT, written !.

The if-statement can have an optional else part. Its syntax is

if (condition) statement1 else statement2

where condition is a logical expression and statement1 and statement2 are single or compound statements. statement1 or statement2 is executed depending the value of condition is true or false. The if-else statement can be nested.

The ternary conditional operator ? : is defined as follows. The expression z?x : y has a value x when z is true and it has a value y when z is false. As an example, the following ternary conditional statement and if-else statement are equivalent b = (a > 0)? a : −a;

if (a > 0) b = a; else b = −a; (i.e. b is abs(a)).

Other useful example is the Kronneker symbol in form

int k = (i − j)? 0: 1;
for any variables int i and j.

Switch Statements

Computer programs, might present you with a choice of more than two selections. You can extend the *C*++ if else statement to meet that need. A switch statement is used for multiple-choice situations. It tests the value of its condition, supplied in parentheses after the keyword switch. Then this value will be used to switch control to a case label matching the value. If the value does not match any case label, control will go to the default case. The break statement is often used to exit a switch statement. For example we have the piecewise function

$$\mathrm{f} = \begin{cases} 0, & x < x_0 \\ x, & x_0 \leq x < x_1 \\ 2x, & x_1 \leq x < x_2 \\ 3x, & x_2 \leq x < x_3 \\ 4, & x_3 \leq x \\ -1, & \text{in other cases} \end{cases}$$

It is necessary to calculate the value of this function y(x) for some x, and also to determine the number of the interval n.

```
float PiecewiseLinearFunc (float x, int &n)
{ float x0 = 0, x1 = 1, x2 = 2, x3 = 3, x4 = 4;
if (x < x0) n = 0;
if ((x0 <= x) && (x < x1)) n = 1;
if ((x1 <= x) && (x < x2)) n = 2;
if ((x2 <= x) && (x < x3)) n = 3;
if (x3 <= x) n = 4;
float result;
```

```
switch (n) {
case 0: result = 0; break;
case 1: result = 1*x; break;
case 2: result = 2*x; break;
case 3: result = 3*x; break;
case 4: result = 4; break;
default: n = -1;
}
return result;
}
```

One can call refer to the function in such a way

```
int k;
float x = 1.5;
float y = PiecewiseLinearFunc(x,k);
```

In result it is received

k = 2 y = 3.

In general, the switch expression may be of any type for which integral conversion is provided and the case expression must be of the same type. A switch statement can always be replaced by an if-else statement, but a switch statement can sometimes make the code more readable.

Iteration Statements for Loops

C++ offers three varieties of loops: the for loop, the while loop, and the do-while loop. A loop cycles through the same set of instructions repetitively as long as the loop test condition evaluates to true or non-zero, and the loop terminates execution when the test condition evaluates to false or zero. The for loop and the while loop are entry-condition loops, meaning they examine the test condition before executing the statements in the body of the loop. The do-while loop is an exit-condition loop, meaning it examines the test condition after executing the statements in the body of the loop.

The syntax for each loop calls for the loop body to consist of a single statement. However, that statement can be a compound statement, or block, formed by enclosing several statements within paired curly braces.

The for loop general form is for (initialize; condition; expression) the statement where initialize represents statements initializing the control variable or variables, condition provides loop control to determine if the statement is to be executed, and expression indicates how the variables initialized, which initialize are to be modified and evaluated (each time) after the statement is executed. The loop proceeds until the condition evaluates to false. If the first evaluation of condition is false, neither expression nor statement is evaluated at all. For example, in this switch statement, the value of i is tested. When i is 0, the statement corresponding to case 0 is executed. When i is 5, the statement corresponding to case 5 is executed. When i is neither 0 nor 5, the default case is executed. Notice the appearance of the break statement after each case. Without it, every following statement in the switch statement would be executed, until a break statement is encountered. There are occasions when this is desired. A default choice is optional but is sometimes useful to catch errors. For example, to count up the sum of numbers from 3 up to 50

```
double y = 0;
for (int i = 3; i < 51; i++) y+=i;
```

It gives y = 1272. The initial value of i is 3. Since the condition i < 51 is true, y will accumulate i. This process repeats until i = 51, at which the condition i < 51 is no longer true.

The break statement can be used to stop a loop, and the continue statement stops the current iteration and jumps to the next iteration. The break and continue statements enable a program to skip over parts of the code. You can use the break statement in a switch statement and in any of the loops. It causes program execution to pass to the next statement following the switch or the loop. The continue statement is used in loops and causes a program to skip the rest of the body of the loop and then start a new loop cycle. For example,

```
for (int i = 3; i < 50; i*=2) {
if (i == 6) continue; // jump to next iteration when i == 6
cout << i << '\n';
if (i == 24) break; // break the loop when i is 24
}
```

It will print out integers 3, 12, 24. When i is 6, the value of i is not printed out since the continue statement causes control to go to the next iteration with i = 6 * 2 = 12. When i is 24, the loop is exited due to the break statement.

While Loops

The while loop is a for loop stripped of the initialization and update parts; it has just a test condition and a body:

```
while (test-condition)
body;
```

Relational expressions, which compare two values, are often used as loop test conditions. Relational expressions are formed by using one of the six relational operators: $<, <=, ==, >=, >$, or !=. Relational expressions evaluate to the type bool values true and false.

First, a program evaluates the test-condition expression. If the expression evaluates to true, the program executes the statement(s) in the body. As with a for loop, the body consists of a single statement or a block defined by paired braces. After it finishes with the body, the program returns to the test-condition and re-evaluates it. If the condition is non-zero, the program executes the body again. This cycle of testing and execution continues until the test-condition evaluates to false. Clearly, if you want the loop to terminate eventually, something within the loop body must do something to affect the test-condition expression. For example, the loop can increase a variable used in the test condition or read a new value from the keyboard input. Like the for loop, the while loop is an entry-condition loop. Thus, if the test-condition evaluates to false at the beginning, the program never executes the body of the loop. We shall calculate exp(x) by the summation of term with the accuracy eps. The loop will come to the end, if current composed x begins less than eps.

```
float y, s = 1, x = 1;
int j = 1;
float eps = 1e-4;
y = x;
while (y > eps) {
y = y/j;
s+ = y;
j++;
}
Result is s = 2.7182
```

Do-while Loops

The third *C++* loop is the do-while. It's different from the other two because it is an exit-condition loop. That means this devil-may-care loop first executes the body of the loop and only then evaluates the test expression to see if it should continue looping. If the condition evaluates to false, the loop terminates; otherwise, a new cycle of execution and testing begins. Such a loop always executes at least once because its program flow has to pass through the body of the loop before reaching the test.

The do-while statement resembles the while loop, but has the form

do statement while (expression);

The execution of statement continues until the expression evaluates to false. It is always executed at least once. For example, the same, as in the previous example one can realize so

```
float y, s = 1, x = 1;
int j = 1;
float eps = 1e-4;
y = x;
do{
y = y/j;
s+=y;
j++;
}
while (y > eps);
```

It will stop after y > eps.

The continue statement can also be used in a while or do-while loop to cause the current iteration to stop and the next iteration to begin immediately, and the break statement to exit the loop. Note that continue can only be used in a loop while break can also be used in switch statements.

14.2.3 *Functions*

Function Declarations and Definitions

The theme of functions is too diverse in *C++* language to capture all subtleties. Therefore we shall stop on some simple, but important rules.

C++ functions come in two varieties: those with return values and those with none. Functions with return values require that you use a return statement so that the value is returned to the calling function. The value itself can be a constant, a variable, an object, or a more general expression. The definition of a function cannot appear inside another function similar to Pascal.

The functions are the *C++* programming modules. To use a function, you need to provide a definition and a prototype, and you have to use a function call. The function definition is the code that implements what the function does. The function prototype describes the function interface: how many and what kinds of values to pass to the function and what sort of return type, if any, to get from it. The function call causes the program to pass the function arguments to the function and to transfer program execution to the function code. Let's write two examples for the calculation of a four power of number

```
float pow4(float x){
float y = x*x*x*x; return y;
}
void pow4(float x, float &result){
float y = x*x*x*x;
result = y;
}
```

In the first case the program receives argument x and the result returns as value of function. In the second case function would seem does not give result (the word void is written). Nevertheless the result comes back by means of the second parameter in which the address of a variable where it is necessary to return the result is specified. The fragment of the program looks like

```
float p = pow4(2);    // p = 16
float z = 0;
pow4(5,z);            // z = 625
```

By default, *C++* functions pass arguments by value. This means that the formal parameters in the function definition are new variables that are initialized to the values provided by the function call. Thus, *C++* functions protect the integrity of the original data by working with copies.

All these declarations float pow4(float x) and void pow4(float x, float&result) are called prototypes. It is only the types that matter in a prototype. Every function that is called in a program must be defined somewhere and only once. The instructions that a function performs must be put inside a pair of braces. The return statement is used for a function to produce an output. After the execution of a function, the program control reverts to the statement immediately following the function call.

Note this important point for practical programming. Usually one saves the functions in separate files. The prototype (or declaration) of function is usually placed in a head file with extension h, for example file1.h. One can write the definition (or body) of this function in the same file. However in the case of long functions it is preferable to write it down in the program file with expansion .cpp, for example file1.cpp. If you want such a function to be accessible in the other file, say file2.cpp, it is necessary to write down the operator #include "file1.h" in the heading of a file2.cpp, otherwise the compiler will report on error ("unknown function"). Then that or other functions can be called by any other functions in files, where the declaration including "file1.h" is written. In this way you can create the libraries of functions and such approach is normally used in this book.

For instance:

```
//file1.h, in this file the function pow4 is declared
double pow4(x); // the prototype of this function
...                 // other prototypes

// file1.cpp
#include " file1.h"
double pow4(x){double y=x*x; return y*y;} // the definition (body)
    of function pow4

//file2.h, in this file the function pow4 will be used
#include "file1.h" // we write down a name of a file, where the function
    pow4 is declared
double pow8(x); // the prototype of other function

// file2.cpp // in this file the function pow4 will be called
#include " file2.h"
double pow8(x){return pow4(x)*pow4(x);} // function pow4 is used in body
    of pow8
```

Note the declaration #include <BigFile.h> normally enables one to make some necessary functions from the system library BigFile.h accessible.

Return Values

A function that is not declared void must return a value. Each time a function is called, new copies of its arguments and automatic variables are created. Since automatic variables will be destroyed upon the return of the function call, it is an error to return a pointer or reference to an automatic variable. The compiler normally gives a warning message so that such errors can be easily avoided.

Observe that objects created by operator new exist on the dynamic memory (heap, or free store). They still exist after the function in which they are created is returned. These objects can be explicitly destroyed by calling delete.

For a void function, it is an error to return a value. However, an empty return statement can be used to terminate a function call, or a call to a void function can be made in a return statement.

As a more meaningful example of a function that returns a pointer, an n by n matrix matr multiplication with an n-column vector vecr1 plus vecr2 can be written as follows:

```
double* MatrVect (double** const matr, const double* const vecr1
   , const double* const vecr2, int n)
   {double* tvecr = new double [n] ; // tvecr will store result
   for (int i = 0; i < n; i++)
   {// find the product and sum
    tvecr [i] = 0;
    for (int j = 0; j < n; j++)
    tvecr [i] += matr [i] [j] * vecr1[j] + vecr2[i];
   return tvecr; // return the address of the result
   }}

main()
{
   int n = 100;
   double** a = new double* [n] ; // create space for matrix
   for (int i = 0; i < n; i++) a[i] = new double [n] ;
   double* b1 = new double [n] ; // create space for vector1
   double* b2 = new double [n] ; // create space for vector2
   double* rslt = new double [n] ; // create space for result

   for (int i = 0; i < n; i++) // assign values to matrix
    for (int j = 0; j < n; j++)
     a[i] [j] = i*i + j;
```

```
  for (int j = 0; j < n; j++){
    b1[j] = 3*j + 5;
    b2[j] = 5*j + 3;
}
rslt = MatrVect(a,b1,b2,n); //calculation
}
```

Recursive Functions

A function that calls itself in its definition is said to be a recursive function. For example, a function that calculates the factorial n! of a non-negative integer n can be defined recursively:

```
long factorial(int n) {
if (n == 0) return 1;
return n*factorial(n - l); }
```

Here factorial(n) is recursively defined as n * factorial(n − 1) and the recursion stops when it reaches factorial(0). Thus a call like factorial(2) can be interpreted as

factorial(2) = 2*factorial(l) = 2*1*factorial(Q) = 2*1*1 = 2.

Note that without the stopping statement: if (n == 0) return 1, the program would cause an infinite recursion.

Let's consider a more interesting example. Calculate the rational part of the spherical Hankel function of first type order m. Here we use a teamwork of operators switch and recursion. Formulas are written in Sec. 1.1.6 we calculate the complex spherical Hankel function. The program code is the next

```
#include <complex.h>
typedef Extended RealType;
typedef complex<RealType> TcompD;
TcompD P1(int m, TcompD y)
// Calculation the polynom part of H(l,1,x)
// where H is Hankel index l number k = 1, argument x
{ TcompD Res;
const TcompD C1(1,0),C2(2,0),C3(3,0);
switch (m)
{
```

```
case −1: Res = C1/y; break; /* for m <= 4 the calculations are carried
out implicitly .*/
case 0: Res = −I/y; break;
case 1: Res = (−I − y)/pow(y,2); break;
case 2: Res = (−I*C3 + (−C3 + I*y)*y)/pow(y,3); break;
case 3: Res = (−(TcompD)15*I + (−(TcompD)15 + ((TcompD)6*I +
y)*y)*y)/pow(y,4); break;
case 4: Res = −((TcompD)105*I + ((TcompD)105 + (−(TcompD)45*I
+(−(TcompD)10 + I*y)*y)*y)*y)/pow(y,5); break;
// However, if m > 4, the recursion starts up
default: Res = (C2*(TcompD)m-C1)/y*P1(m − 1, y) − P1(m − 2, y);
}
return(Res);
} // P1
```

Overloading

Apparently from the previous example two functions with identical names pow4, but different prototypes coexist.

These versions of the pow4 function can coexist in a program. The compiler can decide which version to use by comparing the type of the argument with the type of parameter. Using the same name for similar operations on different types is called overloading.

However, when a function is called before its definition, the function declaration must appear before the statement that calls the function. Such a declaration is called a forward declaration. For example,

```
main {
float pow4(float x); // forward declaration
void pow4(float x, float &result);
float p, t = 2, u = 5;
p = pow4(t); // p = 16
float z = 0;
pow4(u,z);
}
float pow4(float x){
float y = x*x*x*x; return typ;
}
void pow4(float x, float &result){
```

```
float y = x*x*x*x;
result = y;
}
```

In the definition of the function pow4, x is called a parameter (or formal argument) to the function. It does not take on any value until pow4() is called, when it is replaced by the argument (or actual argument) from the calling environment, as in p = pow4(x). In this call, the parameter x is replaced by the argument t. The type of t is also checked against the type of x and some conversion may be performed when their types do not match.

Function Types

It is sometimes useful to define a type synonym using the keyword typedef.

By such a way *C++* can establish a new name as an alias for a type. The method is to use the keyword typedef to create an alias. For instance, to make byte an alias for char, do this:

```
typedef char byte; // makes byte an alias for char
```

Here's the general form:

```
typedef typeName aliasName;
```

In the previous example it was used two operators:

```
typedef Extended RealType;
typedef complex<RealType> TcompD;
```

The sense of these operators is next. I want to work with numbers of large accuracy such as Extended. For this purpose I write the alias RealType and I use it everywhere in the program. However, if for any reasons (for example because of shortage of memory or at debugging) I need to use the float type I have to rewrite the operator typedef Extended RealType; only as typedef float RealType; This can save my time essentially.

In other words, if you want aliasName to be an alias for a particular type, declare aliasName as if it were a variable of that type and then prefix the declaration with the typedef keyword. For example, to make byte_pointer an alias for pointer-to-char, declare byte_pointer as a pointer-to-char and then stick typedef in front:

```
typedef char * byte_pointer; // pointer to char type
```

Notice that typedef does not create a new type. It just creates a new name for an old type. If you make word an alias for int, cout treats a type word value as the int it really is.

A declaration prefixed by typedef introduces a new name for a type rather than declaring a new variable. Sometimes it can be useful for debug purposes. For example I plan to calculate with double type variables but now I temporally have to calculate with float variables. I write typedef float RealType; RealType a, b, c, d; do something with a, b. After debug is finished I have to change only the operator typedef double RealType to perform calculations with double type.

Passing Functions as Argument

The following is very useful. A pointer to a function can be passed as an argument to another function like any other pointer. For instance I have to evaluate a root z for the complex equation f(z, p) = 0, where p is some parameter. For this purpose I shall use Muller's method as a solver. How can I pass the name of this function in a solver? I write the corresponding typedef operator

typedef TcompD (*TMyFuncEq1)(TcompD z, TcompD &p);

After that I can describe the prototype of a solver as

TcompD Muller(TMyFuncEq1 MyFuncEq, TcompD yParam, TcompD & newRoot);

Now the reference one can write as following. For example, for function z*z*p + 1 we write

TcompD f(TcompD z, TcompD p){return z*z*p + 1;} /* Evaluate z*z*p + 1 */
TcompD p(0,0);
TcompD newRoot;
int k = Muller(f, p, newRoot); // evaluate the root of function newRoot

The program gives out whether it has found a root (k is zero or not). In the latter case the value of newRoot is the root of the given equation. Such technique is frequently used, for example at numerical integration of some function or at the solution of set of the differential equations.

14.2.4 *Interacting the data of class with member functions*

Now we shall pay attention to functions — members of a class in a context of their interaction with the data of a class. There is a reasonable recommendation to hide the data of a class behind the maximum high level of access, private level. Why and how one can get the access to the data, for example, to learn its value? The answer is that the operative system can usually supervise a correctness of performance of a code of the program and in any doubtful cases under abnormal condition it can terminate the work or even to warn about probable infringements. However the system cannot estimate a correctness of the change of data. Therefore the care of integrity of the data lays entirely on the shoulders of the programmer. For example, if the data is a date of some event. An error will be to assign the number of day in a month the number 32, and to assign the number of month in an year the number 15. However from the point of view of the system the assignment of the integer value to other integer is a completely regular operation. But from the semantic point of view such infringement of a correctness of the data can have critical consequences and should be rejected.

The problem can be solved by such a way. One has to give the data a private level, but establish for methods of access to the data lower level protected or even public. In a body of such methods it is necessary to write down a code checking the correctness of new values, for example check of the range. If conditions of correctness will be broken, the data will not change, and the user has to obtain a signal about such incorrectness. Our example is as follows.

```
class TMyDate
{
private:
int _nDayInWeek, _DayInMonth, _nMonthInYear; // <- data
public:
TMyDate() // constructor of class
{_nDayInWeek = 1; _DayInMonth = 1; _nMonthInYear = 1;}
int get_nDayInWeek(){return _nDayInWeek;} // obtain data
void set_nDayInWeek(int n)              // modificate data
{
if ((0 < n) && (n < 8))                 // check the validity
_nDayInWeek = n;                        // OK
else                                    // wrong
throw "The attempt of incorrect operation in" // raising the exception
"set_nDayInWeek.\r\nOperation is rejected.";
}
};
```

There are three fields: int _nDayInWeek, _DayInMonth and _nMonthInYear in the class TMyDate. Of course the values of all dates can't be arbitrary and should be in a certain range. We shall discuss the field _nDayInWeek, the value of which must be in the range from 1 to 7. There are two public methods, having access to this field: int get_nDayInWeek() gives us the value of _nDayInWeek, and void set_nDayInWeek(int n) can change its value. The latter method has to prevent eventual errors and it verifies the range of new value before updating the field _nDayInWeek. If a new value is outside the interval [1,7], such a modification is rejected and the exception with the message "The attempt of incorrect operation in set_nDayInWeek. Operation is rejected." is raised (see Fig. 14.4). You can see the fragment of the program, in which the attempt of such modification is undertaken and the error's message emerges.

```
TMyDate myDate;
try
{
myDate.set_nDayInWeek(15);
}
catch(char *e)
{
MessageDlg((String)e, mtError, TMsgDlgButtons() << mbYes, 0);
}
```

As a result the following message on the error will be shown.

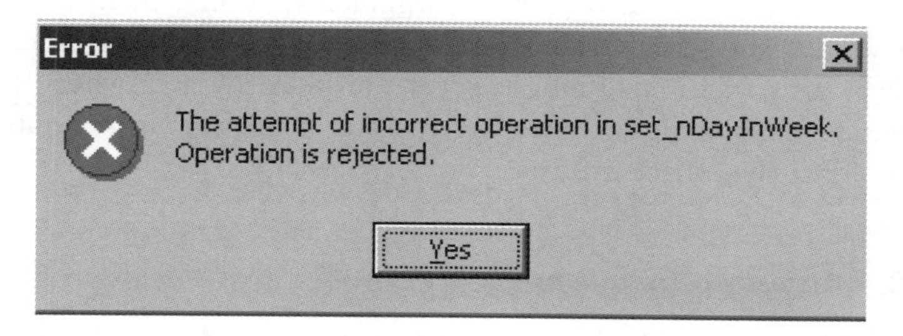

Fig. 14.1 The error message in a case of wrong operation (see TMydate class).

14.2.5 *Classes*

A class is a user-defined specification for a data type. This specification details how information is to be represented and also the operations that can be performed with the data. An object is an entity created according to a class prescription, just as a simple variable is an entity created according to a data type description. The *C++* construct class provides an encapsulation mechanism so that a user can encapsulate data and functions manipulating the data together to define new suitable types. This can also increase the readability and maintainability of a program, and separate implementation details from its user interface.

Usually, you separate the class declaration into two parts, typically kept in separate files *.h and *.cpp. The class declaration goes into a header file, with the methods represented by function prototypes. The source code that defines the member functions goes into a methods file. This approach separates the description of the interface from the details of the implementation. In principle, you only need to know the public class interface to use the class. Of course, you can look at the implementation (unless it has been supplied to you in compiled form only), but your program should not rely on details of the implementation, such as knowing that a particular value is stored as an int. As long as a program and a class communicate only through methods defining the interface, you are free to improve either part separately without worrying about unforeseen interactions.

A class is a user-defined type, and an object is an instance of a class. That means an object is a variable of that type or the equivalent of a variable. *C++* tries to make user-defined types as similar as possible to standard types, so you can declare objects, pointers to objects, and arrays of objects. You can pass objects as arguments, return them as function return values, and assign one object to another of the same type. If you provide a constructor method, you can initialize objects when they are created. If you provide a destructor method, the program executes that method when the object expires.

14.2.6 *Access to members*

With derived classes, the keyword protected is introduced. A member (it can be a function, type, constant, etc., as well as a data member) of a class can be declared to be private, protected, or public.

A private member can be used only by members of the class in which it is declared. A protected member can be used by members of the class in which it is declared. Furthermore, it becomes a private, protected, or public member of classes derived from this class; as a member of a derived class, it can be used by members of the derived class and maybe by other members in the class hierarchy depending on the form of inheritance. Note the keyword private in class declarations is the default access control for class objects.

A public member can be used freely in the program. It also becomes a private, protected, or public member of classes derived from this class, depending on the form of inheritance.

- When you define a member function out the class's body, you have to use the scope operator (::) to identify the class to which the function belongs, for example void TBigStore::toString( ).
- Class methods can access all components of the class independently whether it is private or public.

In our example, we have the base class TBigStore and two derived classes TShop and TSmallShop. The base class TBigStore have three types of waress: computers, printers and books. It can supply with the printers TShop and with the books TSmallShop to sell separately. These classes only have to know how many printers or books are there in bigStore, but they have no access to information about other wares available in bigStore. To separate such information we use the key words private, protected and public.

```
class TBigStore {
private:
int _computers;
protected:
int _printers;
int _books;
private:
int get_computers( ){return _computers;}
void set_computers(int s){ _computers = s;}
protected:
int get_printers( ){return _printers;}
void set_printers(int s){ _printers = s;}
int get_books( ){return _books;}
void set_books(int s){ _books = s;}
public:
TBigStore( ){ // initialization
_computers = 55;
_printers = 66;
_books = 77;
```

```
}
newWaress(int newcomputers, int newprinters, int newbooks){
_computers = newcomputers;
_printers = newprinters;
_books = newbooks;
}
virtual int haveWares(){return get_computers();}
virtual String toString(){
String s = "";
s = "In bigStore we have:\r\nthe computers: " + IntToStr(_computers)
+"\r\nthe printers: " + IntToStr(_printers)
+"\r\nthe books: " + IntToStr(_books);
return s;
}
};
// - - - - - class TShop - - - - - - - -
class TShop : public TBigStore{
public:
int get_printers(){return _printers;}
void set_printers(int s){ _printers = s;}
virtual int haveWares(){return get_printers();}
virtual String toString(){return "In the shop we have the printers :"
+IntToStr(get_printers());}
};
// - - - - - class TSmallShop - - - - - - - -
class TSmallShop : public TShop{
public:
int get_books(){return _books;}
void set_books(int s){ _books = s;}
virtual int haveWares(){return get_books();}
virtual String toString(){return "In the smallshop we have the books :"
+IntToStr(get_books());}
};
```

Protected members of a class are designed for use by derived classes and are not intended for public use. They provide another layer of information hiding, similar to private members. The protected part of a class usually provides operations for use in the derived classes. Public derivation makes the derived class a subtype of its base class and is the most common form of derivation. Protected and private derivations are used to represent implementation details. No matter what form the derivation is, a derived class contains all the members of a base class (some of which may have been overridden), although members inherited from the private part of the base class are not directly accessible even by members of the derived class. The form of derivation controls access to the derived class's members inherited from the base class and controls the conversion of references from the derived class to the base class.

14.2.7 *Virtual functions*

Virtual functions can be declared in a base class and may be redefined (also called overridden) in each derived class when necessary. They have the same name and same set of argument types in both base class and derived class, but perform different operations. When they are called, the system can guarantee that the correct function is invoked according to the type of object at run-time. For example, the base class TBigStore and derived classes TShop and TSmallShop contain two virtual functions toString( ) and haveWaress( ) (see Fig. 14.2). The function toString( ) provides information of the availability of waress, therefore we will study it with more details.

When toString( ) is called, the compiler will generate code that chooses the right toString( ) for an actual object's type TBigStore, TShop or TSmallShop. The function toString( ) will be called automatically by the system for the actual object. The following example illustrates the use of private, protected and public derivations. Figure 14.1 shows the output of this program.

```
#include "BigShop1.h"
#include "clases.h"
void __fastcall TBigshopForm::RunButtonClick(TObject *Sender)
{
TBigStore *bigStore = new TBigStore( ); // Starting initialization
TSmallShop *smallShop = new TSmallShop( );
TShop *shop = new TShop( );
ReportMemo->Lines->Add("First report:"); // print the first
information
ReportMemo->Lines->Add(bigStore->toString( )+"\r\n");
ReportMemo->Lines->Add(shop->toString( ));
ReportMemo->Lines->Add(smallShop->toString( ));
bigStore->newWaress(100, 1000, 5000); // new waress are come
```

What is my type? **TBigStore**
What I have? **_computers, _printers, _books;**
How to change a quantity?
set_computers(int n); set_printers(int n); set_books(int n);
How to know a quantity?
get_computers(); get_printers(); get_books();
New wares cameto me:
NewWares(int nc, int np, int nb);
What I have? **haveWares();**
How I can write my actual state?
ToString();

What is my type? **TShop**
What I have? **Nothing, but I have access to the base class;**
How to change a quantity?
set_printers(int n);
How to know a quantity?
get_printers();
New wares came
NewWares(int nc, int np, int nb);
What I have? **haveWares();**
How I can write my actual state?
ToString();

What is my type? **TSmallShop**
What I have? **Nothing, but I have access to the base class;**
How to change a quantity?
set_books(int n);
How to know a quantity?
get_books();
New wares came
NewWares(int nc, int np, int nb);
What I have? **haveWares();**
How I can write my actual state?
ToString();

Fig. 14.2 The structure of classes in BigStore.

```
ReportMemo->Lines->Add("- - - - - - - - - - -\r\nSecond report: new
wares come");
// print the second information
ReportMemo->Lines->Add(bigStore->toString( )+"\r\n");
ReportMemo->Lines->Add(shop->toString( ));
ReportMemo->Lines->Add(smallShop->toString( ));
shop->set_printers(222); // new printers are come
ReportMemo->Lines->Add("\r\n- - - - - - - - - -\r\nThird report: new
wares come");
ReportMemo->Lines->Add(bigStore->toString( )+"\r\n");
//shop->set_computers(1000); // !! is not accessible
ReportMemo->Lines->Add(shop->toString( ));
ReportMemo->Lines->Add(smallShop->toString( ));
delete smallShop;
delete shop;
delete bigStore;
}
```

Since the Run button is clicked first the program passes the initialization stage. Three dynamical objects are created: bigStore, shop and smallShop. While initialization there are in the base class TBigStore _computers = 55; _printers = 66; _books = 77. From output in Fig. 14.3 one can see namely such information they give.

After new waress come to bigStore:
bigStore ->newWaress(100, 1000, 5000)

We have the second message, but from output in Fig. 14.1 one can see that the quantity of waress is changed for bigStore, but nothing is changed for smallShop and shop. It is not strange that the objects smallShop, shop or bigStore live as independent creatures! Finally 222 new printers come to the shop. From the output in Fig. 14.3 one can see that wares are changed for the shop only, but nothing is changed for smallShop and bigStore. Moreover if shop wants to obtain 1000 new computers it should compute

```
shop->set_computers(1000); // error
```

the compiler gives the message that the object shop has no access to the computers as such wares "is not accessible". One can see from the example that it is true since the data_computers and member functions get_computers and set_computers are private.

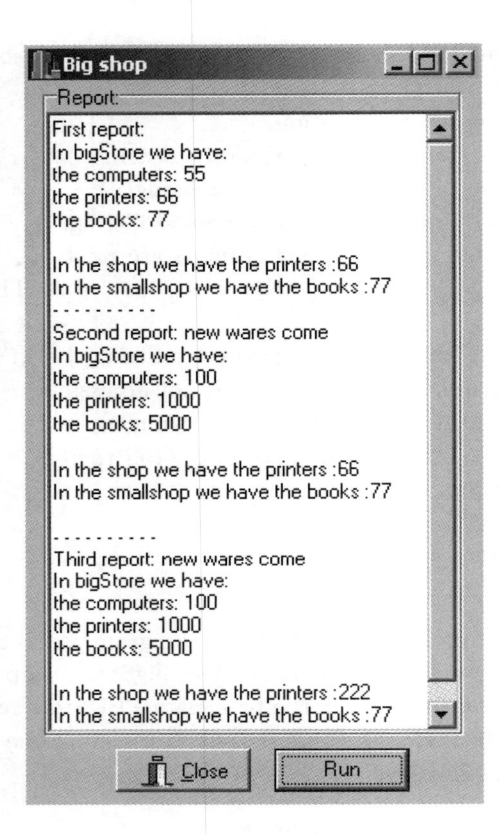

Fig. 14.3 The output from BigShop program.

We can conclude that inheritance enables one to adapt the programming code to particular needs by defining a new class (a derived class) from an existing class (the base class). Public inheritance models is an is-a relationship, meaning that a derived-class object should also be a kind of base-class object. As part of the is-a model, a derived class inherits the data members and most methods of the base class. However, a derived class does not inherit the base class constructors, destructor, and assignment operator. A derived class can access the public and protected members of the base class directly and the private base class members via the public and protected base-class methods. You can then add new data members and methods to the class, and you can use the derived class as a base class for further development. Each derived class requires its own constructors.

When a program creates a derived class object, it first calls a base class constructor and then the derived class constructor. A key benefit of using of a constructor is that it is guaranteed that the object will go through proper initialization before being used. When a program deletes an object, it first calls the derived class destructor and then the base class destructor.

If a class is meant to be a base class, you may choose to use protected members instead of private members so that derived classes can access those members directly. However, using private members will, in general, reduce the scope for programming bugs. If you have the intention that a derived class can redefine a base class method, make it a virtual function by declaring it with the keyword virtual. This enables objects accessed by pointers or references to be handled on the basis of the object type rather than on the basis of the reference type or pointer type. In particular, the destructor for a base class should normally be virtual.

14.2.8 *Overloading the mathematical operator*

We shall now see why language *C++* so is well adapted to a record of mathematical expressions in the natural form.

When the schoolboy sees the mathematical expression $c = a * b$, he represents that there are some numbers a and b at which values can be substituted in this expression and to receive c. However the student of a university will already reflect on what the symbols a, b, c mean. These are the usual numbers (scalars), the complex numbers, vectors or the matrices. Depending on their type a rule of calculation will be different. The large advantage of language *C++* is that it gives an opportunity to define for each type (class) the different interpretation of the sense of mathematical operators $+, -, *, /, []$, etc. It can be seen easily for a complex type. For example, we shall define a class

```
class TCmpl
{
private:
float _re, _im; /* I prefer to use the prefix symbol _ for fields of a class.
```

If I see such a variable outside a class that I know, 'aha...' I must not see the data of this class here. I have to see a method which gives access to this field. */

```
public:
TCmpl( ){_re = 0; _im = 0;} // the default initialization
TCmpl(float x, float y){_re = x; _im = y ;} // initialization by the new
values x and y
TCmpl operator + (TCmpl x){
// we determine rules of action of a sign + as usually for complex
numbers
TCmpl result; // some intermediate variable
result._re = _re + x._re;
result._im = _im + x._im;
return result;
}
TCmpl operator * (TCmpl x) {
// we determine rules of action of a sign * as usually for complex
numbers
TCmpl result; ; // some intermediate variable
result._re = _re * x._re − _im * x._im;
result._im = _re * x._im + _im * x._re;
return result;
}
};
```

Now at performance of the fragment of the program we obtain

```
TCmpl x(1,2),y(3,4),z, I(0,1); //I is imaginary unit
z = x + y; //z = (4, 6)
z = y*I; //z = (−4, 3)
```

In the same way one can define the other mathematical operators for this class. Operating with a similar way it is possible to write the classes of a vector, a complex matrix, and also its mathematical operators accordingly. Furthermore the example of the program utilizing it is shown. The general remarks are as follows.

C++ extends overloading to operators by letting you define special operator functions that describe how particular operators relate to a particular class. An operator function can be a class member function. *C++* lets you invoke an operator function either by calling the function or by using the overloaded operator with its usual syntax. An operator function for the operator op has this form operator op(argument-list), where the argument-list represents operands for the operator.

Figure 20.2 shows the program for the checking of various vectorial and matrices operations with complex vectors and matrices. Bases of the program are separated into two classes: class TBVector and class TBMatrix. The graphic user interface of this program is shown in Fig. 20.1. The result of work for calculation of an inverse matrix is shown in Fig. 20.2.

CHAPTER 15

Exception Handling

Everybody knows that errors can occur even in simple programs. But what really matters is what happens after the error occurs. How is the error handled? Can we control its? Can our program recover, or should it just die?

The aim of exception handling is to separate run-time error reporting and error-handling. Errors that cannot be resolved locally may be reported somewhere in a program while they are handled somewhere else. The error-handling mechanism can be seen as a run-time analogue to the compile-time type checking. The code developed using it should have a better chance of running as expected and producing more reliable results.

Exception handling requires the use of keywords: try, catch and throw. Programs prepare to catch exceptions by trying statements that might generate a special condition. When a *C++* program throws an exception, you can transfer or throw control to another part of the program called the exception handler that handles that type of exception. The handler is said to catch the exception. If no exception is encountered during the execution of the code in the try-block, the catch-block will be ignored. Otherwise, the flow of control will go to the catch-block and it can return the message. A catch-block is called an exception handler. It can be placed only immediately after a try-block or after another exception handler.

For example, let us write the program Calculator to calculate z = x@y, where @ is one of $+, -, *$, or /. Only for division operation one expects something to go wrong. In the case of x/0 the exception occurs. Therefore only in handler DivButtonClick the try/catch feature is used. The work of program is shown in Fig. 15.1.

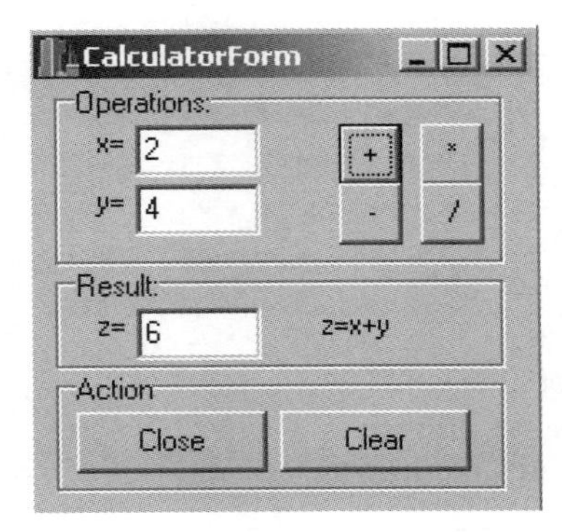

Fig. 15.1 Graphical user interface (GUI) for the program calculator.

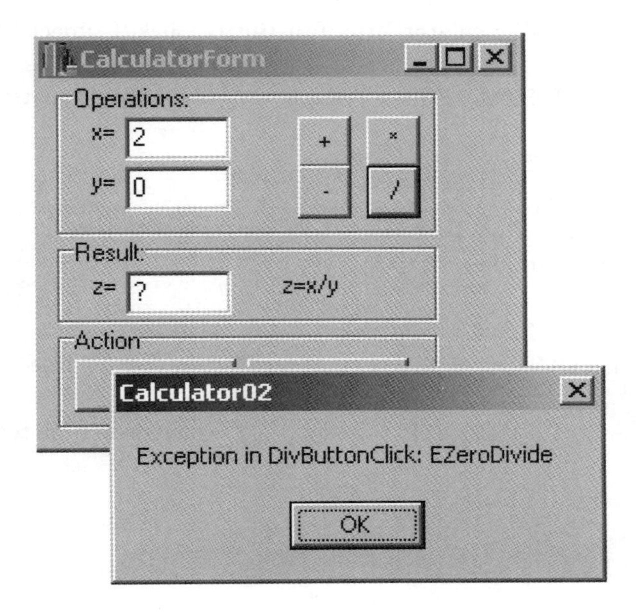

Fig. 15.2 The error message for wrong operation in the program calculator.

15.1 Code

```
//- - - - - - - - - - - - - - - - - - - - - -
    #include <vcl.h>
    #pragma hdrstop
    #include "Calculator2.h"
    //- - - - - - - - - - - - - - - - - - - - - -
    #pragma package(smart_init)
    #pragma resource "*.dfm"
    TCalculatorForm *CalculatorForm;
    //- - - - - - - - - - - - - - - - - - - - - -
    __fastcall TCalculatorForm::TCalculatorForm(TComponent* Owner)
    : TForm(Owner)
    {}      // constructor of form
    //- - - - - - - PlusButtonClick- - - - - - - - - - - - - - - -
    void __fastcall TCalculatorForm::PlusButtonClick(TObject *Sender)
    {
    float x = StrToFloat(xEdit->Text);      // conversion from text to num-
ber
    float y = StrToFloat(yEdit->Text);
    float z = x + y;
    zEdit->Text = FloatToStr(z);      // conversion to text
    InfoLabel->Caption = "z = x + y";
    }
    //- - - - - - - MinusButtonClick- - - - - - - - - - - - - - -
    void __fastcall TCalculatorForm::MinusButtonClick(TObject *Sender)
    {
    float x = StrToFloat(xEdit->Text);      // conversion from text to num-
ber
    float y = StrToFloat(yEdit->Text);
    float z = x - y;
    zEdit->Text = FloatToStr(z);
    InfoLabel->Caption = "z = x - y";
    }
    //- - - - - - - - MultButtonClick- - - - - - - - - - - - - - -
    void __fastcall TCalculatorForm::MultButtonClick(TObject *Sender)
    {
    float x = StrToFloat(xEdit->Text);
    float y = StrToFloat(yEdit->Text);
```

```
float z = x*y;
zEdit->Text = FloatToStr(z);
InfoLabel->Caption = "z = x*y";
}
// - - - - - - - - DivButtonClick- - - - - - - - - - - - - -
void __fastcall TCalculatorForm::DivButtonClick(TObject *Sender)
{
float x = StrToFloat(xEdit->Text);
float y = StrToFloat(yEdit->Text);
float z;
try {
InfoLabel->Caption = "z = x/y";
z = x/y;
zEdit->Text = FloatToStr(z);
}
catch (Exception &e) {      // something was wrong
zEdit->Text = "?";
ShowMessage("Exception in DivButtonClick: " + e.Message);   // our
message
}
}
// - - - - - - - ClearButtonClick- - - - - - - - - - - - - - - -
void __fastcall TCalculatorForm::ClearButtonClick(TObject *Sender)
{
xEdit->Text = "0";
yEdit->Text = "0";
zEdit->Text = "?";
InfoLabel->Caption = "?";
}
// - - - - - - - CloseButtonClick- - - - - - - - - - - - - - - -
void __fastcall TCalculatorForm::CloseButtonClick(TObject *Sender)
{
Application->Terminate();
}
```

CHAPTER 16

Visual Programming: Controls, Events and Handlers

16.1 DOS and Visual Programming

Imagine somebody living in a two-dimensional black-and-white world with Spartan conditions in a rather crowded surroundings. There are only two interesting events: to start the program and to see that it has come to the finishing point. It is rather boring to live in such a world, isn't it so?

And now imagine, the wizard has transferred this somebody to a three-dimensional color world. There is a lot of free space, he can breathe easily and it is possible to place many useful things here which can appear at hand when he needs them. All things around you are ready to serve you at your commands (or mouse clicks). In short, who wants to come back to the Spartan conditions of a two-dimensional world after tasting such a three-dimensional environment?

The programmer will feel similarly after transition from a two-dimensional world with black-and-white-screen programming to a color world of Visual programming. In the latter you can open and write the information into any desirable window and store it there. At the necessary moment such a window appears above your other windows (in fact the world is quasi-three-dimensional!). The buttons await your click to produce a desired event. The head of a long column with numbers in a window is not lost; you can bring it back and do much more. I can say to some adherents of a command-line style, "OK, I see you can produce excellent C++ object-oriented code in a command-line regime. Nevertheless I believe that your programs will be richer if you start to develop them in the Visual style."

What does one have to do to switch from DOS black-and-white-screen style to Visual style programming? First it is necessary to understand which are the controls, events and the handlers.

16.2 Controls, Events and Handlers

First a little piece of history. At once after the invention of the mouse the speed of access to each point of a computer screen, and accordingly, the desire to place necessary information there has increased sharply. Moreover programmers wanted to place on screen not only the information, but also the control elements similarly to a button on the computer keyboard. Then a click on such a control could engage a necessary specific event. Programmers have grasped this idea and a lot of controls elements, having various types and sizes have been produced. As a result the screens became slightly mad. There was chaos. And so, the user, already well-known with the computer, became the beginner on a neighbor computer, not knowing what he has to do to achieve a desirable result.

In a result of team-work of leading computer firms the standards of such control elements were developed. The words: graphical user interface (GUI), together with a set of standard controls have appeared. Such a set consists of buttons, labels, menu, an input line, dialog panels, etc. Now the employee, having come to the neighbor computer already knows what is the function and what is the event produced by the mouse click in some area of screen, occupied with a control element (e.g. button).

Now we shall explain some terms connected to such an approach. The simplest manner to work is described as follows. The mouse click into the screen area occupied with any control element produces the specific event. The operating system listens, whether there was an event and whether it can transfer the control to the necessary program code (procedure). Such a code is referred to as a handler. The code and result produced by a handler is written by the programmer according to the requirements of the problem.

Now the most advanced packages show the big attention for the programmers. For example in Borland *C++* Builder (BCB) it is enough to drag-and-drop the necessary items for you control element from the panel where they are stored. After that it becomes already integrated into your program. This technology is called Rapid Application Development (RAD). If you double click at the control into your form the BCB-environment will create a skeleton of the necessary programs code (the handler) and it will build it into your program. The programmer should only insert the desirable code into this skeleton, and the operating system will breathe a new life into this handler. BCB will add a corresponding piece of code to your program automatically. Thus RAD tools are designed for programmers who have a job and want to get it done quickly. After some efforts to study the RAD you will find that you will start to produce scientific programs in an incredibly short period of time. For instance, some steps of RAD technology in BCB environment you can see Figs. 16.1 and 16.2.

When you start C++Builder, you are immediately placed within the integrated development environment, also called the IDE. This environment provides all the tools you need to design, develop, test and debug applications (see Fig. 16.1). C++Builder's development environment (IDE)

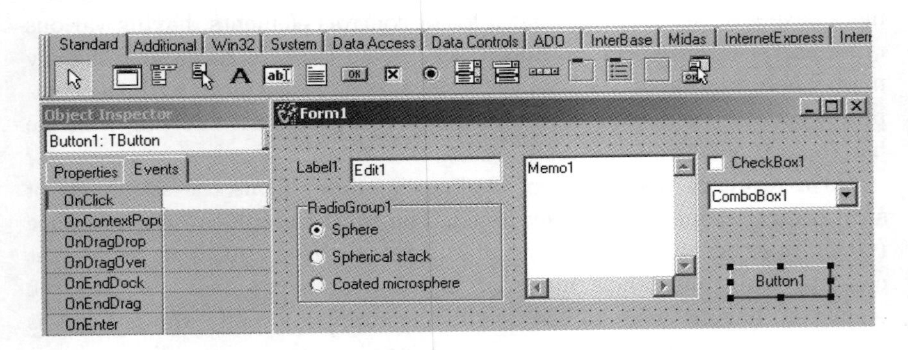

Fig. 16.1

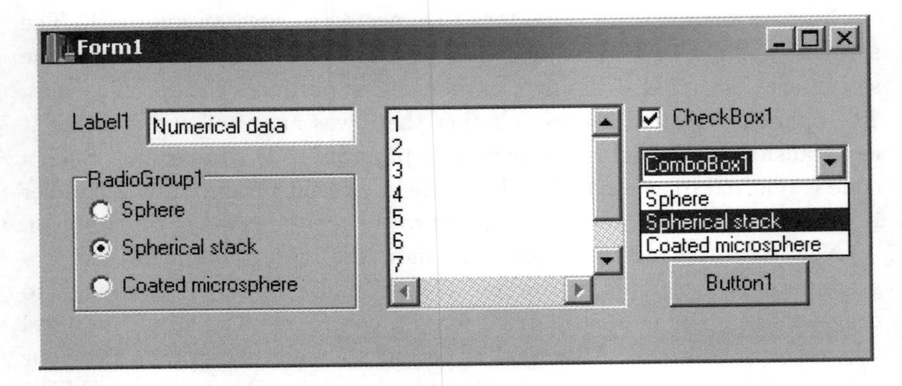

Fig. 16.2

includes a visual form designer, Object Inspector, Component palette, Project Manager, source code editor, debugger, and installation tool. You can move freely from the visual representation of an object (in the form designer), to the Object Inspector to edit the initial runtime state of the object, to the source code editor to edit the execution logic of the object (Button F12). Changing code-related properties, such as the name of an event handler, in the Object Inspector automatically changes the corresponding source code. In addition, changes to the source code, such as renaming an event handler method in a form class declaration, is immediately reflected in the Object Inspector.

A blank window, known as a form, is set up on which to design the GUI for your application. With C++Builder, you create a user interface (UI) by selecting components from the Component palette and dropping them onto the form (see Fig. 16.1).

There are in bcb an extensive class libraries with many reusable objects: an Object Inspector for examining and changing object traits, a Code editor that provides direct access to the underlying program logic, a Project Manager for managing the files that make up one or more projects. Many other tools such as an image editor on the toolbar and an integrated debugger on menus support application development in the IDE. When you compile a project, an executable (.EXE) file is created (see Fig. 16.2). The executable file usually provides the basic functionality of your program, and simple programs often consist of only an EXE.

Admirers of Console programming have the possibility to develop the programs in a command-line mode. Console applications are 32-bit Windows programs that run without a graphical interface, usually in a console window. These applications typically don't require much user input and perform a limited set of functions.

The kinds of events that can occur can be divided into two main categories: user events and system events. Regardless of how the event was called, C++Builder looks to see if you have assigned any code to handle that event. If you have, then that code is executed; otherwise, nothing is done.

User events. User events are actions that are initiated by the user. Examples of user events are OnClick (the user clicked the mouse), OnKeyPress (the user pressed a key on the keyboard), and OnDblClick (the user double-clicked a mouse button). These events are always tied to a user's actions.

System events. System events are events that the operating system fires for you. For example, the OnTimer event (the Timer component issues one of these events whenever a predefined interval has elapsed), the OnCreate event (the component is being created), the OnPaint event (a component or window needs to be redrawn), etc. Usually, system events are not directly initiated by a user action.

To make the following programming code more clear we should touch on the Standard Template Library (STL) topic. Since its development by Dr. Bjarne Stroustrup in the 80s, the C++ language has been widely used by professional programmers to develop large, complex applications in various fields. The final standardization of the C++ library now makes it easier to learn C++ and to use it across a wide variety of platforms.

The Standard *C++* Library is a large and comprehensive collection of classes and functions for low-level programming. Within this library, you will find the following components: The large set of data structures and algorithms formerly known as the Standard Template Library (STL): An iostream facility, a locale facility, a templatized string class, a templatized complex class for representing complex numbers, etc. In this book we often use complex and vector classes.

Like an array, a vector is an indexed data structure, with index values that range from 0 to one less than the number of elements contained in the structure. Also like an array, values are most commonly assigned to and extracted from the vector using the subscript operator. However, the vector differs from an array in the following important respects. In particular, a vector can be queried about its size, about the number of elements it can potentially hold, which can differ from its current size, and so on. The size of the vector can change dynamically. New elements can be inserted on to the end of a vector, or into the middle. Whenever you use a vector type, you must include the vector header file: # include <vector>.

In this book, it is not possible to teach the programming process or to explain the process of creating the GUI. In case of problems you may consult the corresponding parts of the help system in BCB products. A lot of good books explain it. Also, I provide the readers with codes accordingly to the topics of this book assuming that they can understand in more details just by looking at the figures which accompanies the description of problems and by reading the program code provided.

My point here is that you should be able to glean information of this kind either from manuals that come with the product or from online help (see www.boRLand.com). You could also turn to *C++*Builder books, such as [Calvert *et al.*, 1998; Hollingworth *et al.*, 2002; Hollingworth *et al.*, 2000].

16.3 Graphical User Interface

Figure 16.1 shows an example of GUI creations which serves the demonstration purposes only. In this case GUI consists of Label (usually this is some caption, TLabel is a non-windowed control that displays text on a form), the input-line TEdit which is a Windows single-line edit control. The switch TRadioGroup represents a group of radio buttons that function together, a little editor TMemo is a windows multiline edit control, TCheckBox represents a Windows check box, TComboBox combines an edit box with a scrollable list, and TButton is a push button control (may be most popular control). All these controls were taken from the Component palette and by drap-and-drop where put on the form (see Fig. 16.1), which represents a typical window. The object inspector (left part in Fig. 16.1) shows the possible values of the properties and also the events which the control creates or responds for the selected control (in this case Button).

Recall that an event is a mechanism that links an occurrence while calculations to some code. More specifically, an event is a closure that points to a method in a specific class instance.

Figure 16.2 shows the same form as on Fig. 16.1, but after the program is run. During work I wrote to Edit the text "Numerical data". I checked the second position in Radiogroup, and wrote down the column of 9 numbers in the Memo window (numbers 8 and 9 are not visible; the Scrollbars therefore has appeared automatically). I have also checked the Checkbox. I have unwrapped Combobox for the choice of one of its items. Certainly such program has no special sense; I only show it to demonstrate how you can create a GUI. It is important to note that while doing so your GUI builder automatically produces a skeleton of codes or handlers and inserts it in your program. You need to only write some code reflecting the details of your problem in the corresponding places of this skeleton. As a result of such technology you can write the programs with a high speed.

CHAPTER 17

Quantum Electromagnetic Field

17.1 Introduction

In this chapter the dynamics of scalar multi-mode quantum fields is considered. It examines a quantized field which can have photons in various states (see). Some theoretical basics are stated in Part II.

Each field of class is described by the next states parameters:

- _fockState (vector) is a multi-mode state of the field with numbers $|n_1\rangle \ldots |n_k\rangle = \{|n\rangle\}$
- _eigValFockState (vector) are eigenvalues of states
- _nmbCurrentStatethe (int) is a number of the current state
- _sTitleOfState (String) is the name of the current state (String)
- _sTitleOfQuantField (String) is the name of the given field (String, this parameter is convenient while working with several fields)

Functions:

- the operators of construction of a vacuum field with a given number of states and, possible, the name, (various constructors TQuantField)
- the operators of construction of a copy of given vacuum field (overloading operator '='),
- the operators of creations (aX is a^+) and annihilations (a) of photons,
- set/get are various functions to access and modify the state parameters of the field,
- the operators of summation or subtraction of the fields states (overloading '+' and '−' operators)
- toString is function to obtain the information on the current state of the field.

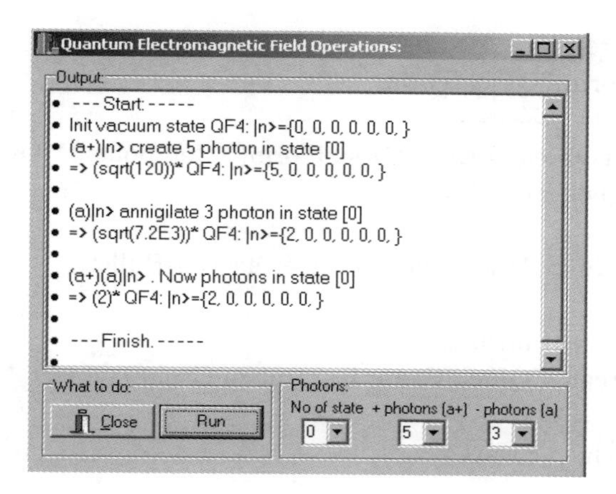

Fig. 17.1 Graphical user interface (GUI) and output for the program 'Quantum Electromagnetic Field'.

In Fig. 17.1 the result of output of this program is shown. At first the vacuum field with six empty states is constructed. (We remind that in C++ the arrays indexes are counted from 0.) The name of this quantum field is QF4. Then five photons in the state with number 0 are created. Furthermore 3 photons in the state 0 are destroyed. Then the number of particles in the state 0 is calculated. One can see that this number is 2. The number of the current state, the number of created or destroyed photons can be changed from the text fields in the graphical user interface (GUI) of the this problem, see Fig. 17.1.

17.2 Code

The code of this program in the following. Both code and structure of classes are shown later. We provide the code with brief comments, since the sense of operators is intuitively clear from its names.

```
//----------------------
#include <vcl.h>
#pragma hdrstop
#include "QuantEl2.h"
#include "QField.h"
//----------------------
#pragma package(smart_init)
```

```
#pragma resource "*.dfm"
TQFieldForm *QFieldForm;
//- - - - - - - - - - - - - - - - - - - - - -
__fastcall TQFieldForm::TQFieldForm(TComponent* Owner)
: TForm(Owner)
{
InfoRichEdit->Paragraph->Numbering = nsBullet;
}
//- - - - - - - RunButtonClick - - - - - - - - - - -
void __fastcall TQFieldForm::RunButtonClick(TObject *Sender)
{
InfoRichEdit->Clear( );
String s;
String ends = "\r\n";
double mult;
InfoRichEdit->Lines->Add(" - - - Start: - - - - -");
TQuantField QF4(6,"QF4");
//construct the field with 6 states and name 'QF4'
InfoRichEdit->Lines->Add("Init vacuum state" + QF4.toString( ));
// provide with information on the field state
int nState = StrToInt(NoStateComboBox->Text);
try
{
QF4.set_nmbCurrentState(nState);
//change the number of states
}
catch (char *e)
//catch the control if something is wrong
{
if (MessageDlg("Error: "+(String)e+". Terminate ?",
mtError, TMsgDlgButtons( ) << mbYes << mbNo, 0) == mrYes)
Application->Terminate( );
}
int nPhotonsPlus = StrToInt(NmbPhotonsPlusComboBox->Text);
// how much the photons to create?
for(int ii = 0; ii<nPhotonsPlus; ii++)
{
+QF4;//QF4 = // (a+)*a                    // creating the
photons in this states; other possibilities are comment off
}
```

```
mult = QF4.get_eigenValue( );
s = "(a+)|n > create "+ NmbPhotonsPlusComboBox->Text +" photon in state ["
+IntToStr(QF4.get_nmbCurrentState( ))+"]\r\n => (sqrt("
+ FloatToStrF(mult*mult,ffGeneral,3,3)+"))* " + QF4.toString( )+ ends; // create the information's string
InfoRichEdit->Lines->Add(s);
// show this info in memo
int nPhotonsMinus = StrToInt(NmbPhotonsMinusComboBox->Text);
// how much photons to destroy?
for(int ii = 0; ii<nPhotonsMinus; ii++)
  QF4 = -QF4;
// destroying the photons in this state
mult = QF4.get_eigenValue( );
s = "(a)|n > annigilate"+ NmbPhotonsMinusComboBox->Text +" photon in state ["
+IntToStr(QF4.get_nmbCurrentState( ))+"]\r\n => (sqrt("
+FloatToStrF(mult*mult,ffGeneral,3,3)+"))* " + QF4.toString( )+ ends; // create the information's string
InfoRichEdit->Lines->Add(s);
// show this info in memo
QF4.clear_eigValFockState( );
QF4 = +(-QF4);
// calculation the number of photons, a+ a
mult = QF4.get_eigenValue( );
s = "(a+)(a)|n > . Now photons in state ["
+IntToStr(QF4.get_nmbCurrentState( ))+"]\r\n => ("
+FloatToStrF(mult,ffGeneral,3,3)+")* " + QF4.toString( )+ends;
// create the information's string
InfoRichEdit->Lines->Add(s);
// show this info in memo
InfoRichEdit->Lines->Add(" - - - Finish. - - - - -");
}
// - - - - - - - - - - - - - - - - - - - -
```

Below is the written code of classes for the ‘Quantum electromagnetic field’ problem.

17.3 Classes

```
#include<vector>
   typedef std::vector<int> TFockState;      // we use STL, see above
   typedef std::vector<float> TEigValFockState;
   // - - - - - class TQuantField - - - - - - -
   class TQuantField
   {
   private:
   TFockState _fockState;          //_fockState (vector) is a multi-mode state of the field {|n⟩}
   TEigValFockState _eigValFockState; //_eigValFockState (vector) are eigenvalues of states
   int _nmbCurrentState;           //_nmbCurrentStatethe (int) is number of the current state
   String _sTitleOfState;          //_sTitleOfState (String) is the name of the current state (String)
   String _sTitleOfQuantField;     //_sTitleOfQuantField (String) is the name of the given field (String, this parameter is convenient while working with several fields)
   private:
   // - - - aXServe - -
   float aXServe(int nState)       // to serve the creation operators a+
   {
   int sz = _fockState.size( );
   if (nState >= sz) throw "nState >= Fock size";
   float eigVal = _fockState[nState];
   _fockState[nState]++;
   _sTitleOfState = "Creation";
   float eigOld = (_eigValFockState[nState] == 0)? 1:
_eigValFockState[nState];
   float eis = (eigOld == 0)? 1: eigOld*sqrt(eigVal+1);
   _eigValFockState[nState] = eis;
   return eis;
   }
   // - - - aServe - -
   float aServe(int nState)        // to serve the annihilation operators a
   {
   int sz = _fockState.size( );
```

```
    if (nState >= sz) throw "nState >= size";
    float eigVal = _fockState[nState];
    _fockState[nState]--;
    _sTitleOfState = "Annigilation";
    float eigOld = (_eigValFockState[nState] == 0)? 1:
_eigValFockState[nState];
    float eis = (eigVal == 0)? 1: eigOld*sqrt(eigVal+0);
    _eigValFockState[nState] = eis;
    return eis;
    }
    public:
    // - - - TQuantField - -
    TQuantField( )   // default constructor0 of vacuum field with 5 states
    { int m = 5;
    _fockState.resize(m);
    _eigValFockState.resize(m);
    int n = _fockState.size( );
    _nmbCurrentState = 0;
    _sTitleOfState = "Initialized";
    for(int ii = 0; ii<n; ii++)
    _eigValFockState[ii] = _fockState[ii] = 0;      //create vacuum
    }
    // - - - TQuantField - -
    TQuantField(int n)      // constructor1 of vacuum field with n states
    {
    _fockState.resize(n);
    _eigValFockState.resize(n);
    _nmbCurrentState = 0;
    _sTitleOfState = "Initialized";
    for(int ii = 0; ii<n; ii++)
    _eigValFockState[ii] = _fockState[ii] = 0;
    }
    // - - - TQuantField - -
    TQuantField(int n, String title)          // constructor2 of vacuum field
with n states and name title
    {
    _fockState.resize(n);
    _eigValFockState.resize(n);
    _nmbCurrentState = 0;
```

```
_sTitleOfState = "Initialized";
_sTitleOfQuantField = title;
for(int ii=0; ii<n; ii++)
_eigValFockState[ii] = _fockState[ii] = 0;
}
// - - - clear_eigValFockState - -
void clear_eigValFockState( )  //clear the eigValFockState
{
_sTitleOfState = "Initialized";
int n = _fockState.size( );
for(int ii = 0; ii<n; ii++)
_eigValFockState[ii] = 0;
}
// further are written various operators to access and modifications of
the fields parameters
int get_nmbCurrentState( ){return _nmbCurrentState;}
void set_nmbCurrentState(int i)
{
int sz = _fockState.size( );
if (i >= sz) throw "In set_nmbCurrentState: nState >= Fock size";
_nmbCurrentState = i;
}
//
String get_sTitleOfQuantField( ){return _sTitleOfQuantField;}
void set_sTitleOfQuantField(String i){ _sTitleOfQuantField = i;}
float get_eigenValue( ){return _eigValFockState[_nmbCurrentState];}
float get_eigenValue(int n){return _eigValFockState[n];}
// - - - set_vacuumState - -
void set_vacuumState( )  //set vacuum
{
_sTitleOfState = "Vacuum";
int n = _fockState.size( );
for(int ii = 0; ii<n; ii++)
_eigValFockState[ii] = _fockState[ii] = 0;
}
// Further are written various modifications of the creation and
annihilation operators
// - - - aX - -
```

```
float aX(int nState) //operator of the creation of the fields particles (the creation operator)
{ try {return aXServe(nState);}
catch(char *e) {throw e;}
}
// - - - a - -
float a(int nState) //operator of the destruction of the fields particles (the annihilation operator)
{ try {return aServe(nState);}
catch(char *e) {throw e;}
}
// - - - aX - -
TQuantField aX(TQuantField qf) // the creation operator (a+)
{ int ns = get_nmbCurrentState( );
aXServe(ns);
for(int jj = 0; jj<ns; jj++)
_fockState[jj] = qf._fockState[jj];
return *this;
}
// - - - a - -
TQuantField a(TQuantField qf) // the annihilation operator (a)
{ int ns = get_nmbCurrentState( );
aServe(ns);
for(int jj = 0; jj<ns; jj++)
_fockState[jj] = qf._fockState[jj];
return *this;
}
// - - - aX - -
TQuantField aX( ) // the creation operator (a+)
{ int ns = get_nmbCurrentState( );
aXServe(ns);
return *this;
}
// - - - a - -
TQuantField a( ) // the annihilation operator (a)
{ int ns = get_nmbCurrentState( );
aServe(ns);
return *this;
}
```

```
//
TFockState get_fockState( ){return _fockState; }
TEigValFockState get_eigValFockState( ){return _eigValFockState; }
int get_sizeFockState( ){_fockState.size( );}
// - - - set_sizeFockState - -
void set_sizeFockState(int n)
{
int oldN = _fockState.size( );
_fockState.resize(n);
_eigValFockState.resize(n);
int newN = _fockState.size( );
for(int ii = oldN-1; ii<newN; ii++)
_eigValFockState[ii] = _fockState[ii] = 0;
}
// - - - '+' unarn - -
TQuantField operator+ ( )        //overloaded '+' is summation of
the fields states
{ int ns = get_nmbCurrentState( );
int itmp = _fockState[ns];
aXServe(ns);
itmp = _fockState[ns];
return *this;
}
// - - - - '–' unarn -
TQuantField operator- ( ) //overloaded '–' is subtraction of the fields
states
{ int ns = get_nmbCurrentState( );
aServe(ns);
return *this;
}
// - - - '*' - -
String operator* (const TQuantField &u) //overloaded *
{
int n = u.get_sizeFockState( );
float r = 0.;
String su = u.get_sTitleOfQuantField( );
String s = ": [";
for(int ii0; ii < n; ii++)
```

```
{
float rRhgt = _fockState[ii], ru = u._fockState[ii];
float eig = u._eigValFockState[ii] + _eigValFockState[ii];
float nrm = (_fockState[ii] == u._fockState[ii])? 1: 0;
r = nrm ;
s+ = FloatToStr(eig*r) + ", ";
}
return s+"]";
}
// - - - '=' - -
TQuantField & operator = (const TQuantField &u) //overloaded=. Is construction of a copy of given vacuum field
{
if (this == &u) return *this;
int n = u.get_sizeFockState( );
for(int ii = 0; ii < n; ii++)
{
_fockState[ii] = u._fockState[ii];
_eigValFockState[ii] = u._eigValFockState[ii];
}
_nmbCurrentState = u._nmbCurrentState;
return *this;
}
// - - - toString - -
String toString( ) // receive the information on the current state of the field.
{
String s = _sTitleOfQuantField+": |n >= {";
for (int ii = 0; ii<get_sizeFockState( ); ii++)
{ int itmp = _fockState[ii];
s+ = IntToStr(get_fockState( )[ii]) + ", ";
}
return s+"}";
}
};//end of class TQuantField
```

CHAPTER 18

Root Finding for Nonlinear and Complex Equations

18.1 Introduction

In this chapter the problem of calculation of complex roots of a function $F(x, p) = 0$, where p is a parameter is considered. Function F is not necessarily a polynomial on x, but it can be general form function. Such task often appears in many fundamental and applied problems. In this book it is applied to a finding of the eigenfrequencies in both bare and multilayered microsphere. The real part of a found root x defines the eigenfrequency of oscillation, while with the help of the imaginary part one can calculate the Q (quality) factor in the form $Q = \mathrm{Re}(x)/2\mathrm{Im}(x)$, as it is written in Part I, Sec. 3.3, Eq. (3.5).

The program is a translation into *C++* from *Turbo Pascal Numerical Methods Toolbox* (*1986, 87 Borland International, Inc.*) and uses the Muller's method. A procedure is rewritten in *C++* and is contained in class TMullerComplEq. Since in *C++* the complex type is available (see STL [Nicolai, 1999]) it allows us to simplify the program code. Section 18.2 shows the program code (with GUI), which allows one to evaluate the complex roots of function $y(p) = 0$, where both y and p are complex numbers. The example of the function $y = \cosh(zp), z = 1$ is considered. The reader can construct their own function (see EqToSolve) to test and search for the roots. Figure 18.1 shows the form where it is allowed to set: the number of required roots, the initial guesses, the maximal number of iterations, function y, both real and imaginary parts of parameter p. The figure also shows the output. One can see that it has correctly found the first three roots $x_n = i(2n + 1)\pi/2$, $n = 0, 1, 2$.

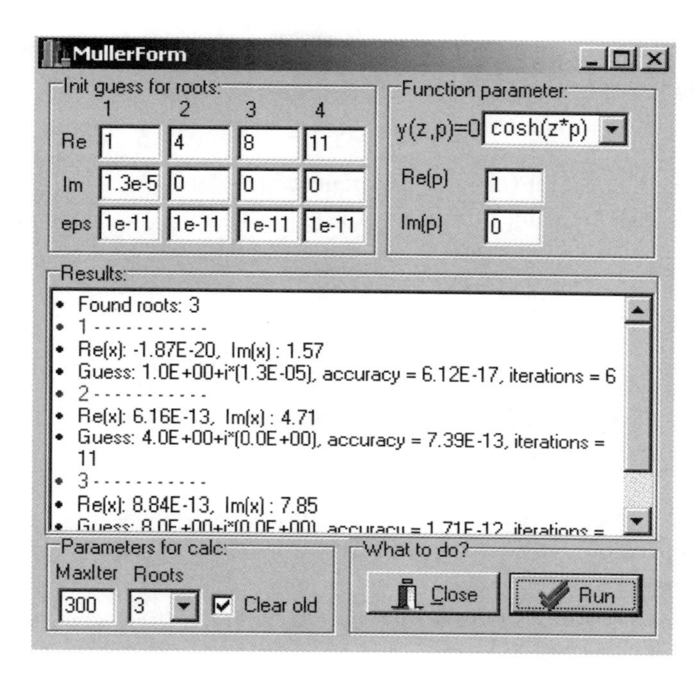

Fig. 18.1 Graphical user interface (GUI) and output for the program 'Complex Roots Finding of the Nonlinear Complex Equation'.

18.2 Code

```
#include "MullerAlgor1.h"
    #include "BGNutil1.h"
    #include "CMullerEq1.h"
    // - - - - - - - - - - - - - - - - - - - - - -
    #pragma package(smart_init)
    #pragma resource "*.dfm"
    TMullerForm *MullerForm;
    TcompD EqToSolve(TcompD x,TcompD &a);
    // - - - - - - - - - - - - - - - - - - - - - -
    __fastcall TMullerForm::TMullerForm(TComponent* Owner)
    : TForm(Owner)
    {
    RootsInfoRichEdit->Paragraph->Numbering = nsBullet;
    }
```

```
// - - - - - - -CompexToStr - - - - -
String CompexToStr(TcompD c)
{
int i = 2;
return FloatToStrF(c.real( ),ffExponent,i,i)+ "+i*("+
FloatToStrF(c.imag( ),ffExponent,i,i)+ ")";
}
// - - - - - - -CompexToStr2 - - - - -
String CompexToStr2(TcompD c)
{
int i = 3;
return FloatToStrF(c.real( ),ffGeneral,i,i)+ "+i*("+
FloatToStrF(c.imag( ),ffGeneral,i,i)+ ")";
}
// - - - - - - - - - EqToSolve - - - - - - - - - - - - -
TcompD EqToSolve(TcompD x,TcompD &a)
{ TcompD y;
y = MullerForm->ComplEqn(x,a);
return y;
} //EqToSolve
//- - - - - - ComplEqn - - - - -
TcompD TMullerForm::ComplEqn(TcompD f,TcompD yParam)
/* Search for the complex roots of function
Input parameter f -complex frequency
{
TcompD y(0,0);
y = cosh(f*yParam);
return y;
}//ComplEqn
// - - - - - - - CalcComplRoots - - - - - - - - - -
void TMullerForm::CalcComplRoots(TMyFuncEq1 f)
{
int NMaxRoots = StrToInt(NumberRootsCmbBx->Text);
//
const MAXROOTSinMULLER = 15;
TcompD *AnswersMuller, *GuessesMuller, *yAnswersMuller;
AnswersMuller = cvector(1, MAXROOTSinMULLER);
GuessesMuller = cvector(1, MAXROOTSinMULLER);
yAnswersMuller = cvector(1, MAXROOTSinMULLER);
```

```
RealType *epsilonsMuller;
epsilonsMuller = dvector(1, MAXROOTSinMULLER);
int *IterMuller, *erroresMuller;
IterMuller = ivector(1, MAXROOTSinMULLER);
erroresMuller = ivector(1, MAXROOTSinMULLER);
RealType eps0 = StrToFloat(epsEd1->Text);
for (int ii1 = 1; ii1<=MAXROOTSinMULLER; ii1++)
{
IterMuller[ii1] = 0; // to clean
erroresMuller[ii1] = 0;
epsilonsMuller[ii1] = eps0;
}
{
RealType ReGuess1 = StrToFloat(ReInitEd1->Text); //Get init guess
from form
RealType ImGuess1 = StrToFloat(ImInitEd1->Text);
RealType ReGuess2 = StrToFloat(ReInitEd2->Text);
RealType ImGuess2 = StrToFloat(ImInitEd2->Text);
RealType ReGuess3 = StrToFloat(ReInitEd3->Text);
RealType ImGuess3 = StrToFloat(ImInitEd3->Text);
RealType ReGuess4 = StrToFloat(ReInitEd4->Text);
RealType ImGuess4 = StrToFloat(ImInitEd4->Text);
AnswersMuller[1] = GuessesMuller[1] = ReGuess1+I*ImGuess1;
AnswersMuller[2] = GuessesMuller[2] = ReGuess2+I*ImGuess2;
AnswersMuller[3] = GuessesMuller[3] = ReGuess3+I*ImGuess3;
AnswersMuller[4] = GuessesMuller[4] = ReGuess4+I*ImGuess4;
}
TMullerComplEq *muller = new TMullerComplEq( ); //create the
solver
int NFoundRoots = 0, ErrorTotal = 0;
try
{
TcompD SomeParameter = 0; //evaluating the complex roots
NFoundRoots = muller->Muller6(GuessesMuller, epsilonsMuller,
get_MaxIter( ),
AnswersMuller, yAnswersMuller, IterMuller, erroresMuller,
f, get_param( ), NMaxRoots);
ErrorTotal = 0;
```

```
for (int ii1 = 1; ii1 <=MAXROOTSinMULLER; ii1++)
{
ErrorTotal + = erroresMuller[ii1];
int nmbErr = erroresMuller[ii1];
if (nmbErr> 0)
ShowMessage("Found roots: " + IntToStr(NFoundRoots)
+". \r\nWhile searching "+ IntToStr(NFoundRoots+1)
+ "-th the error is detected:
\r\n"+muller->MullerMessage(nmbErr, _MaxIter));
}
}// try
catch (const exception& e)
{ String s = muller->MullerMessage(ErrorTotal, _MaxIter);
String s2 = e.what( );
MessageDlg(s+". s2 = " + s2, mtError,
TMsgDlgButtons( ) << mbYes , 0);
}
delete muller;
RootsInfoRichEdit->Lines->Add("Found roots: " +
IntToStr(NFoundRoots));
for (int mm = 1; mm<=NFoundRoots; mm++)
{
RootsInfoRichEdit->SelAttributes->Color = clBlue;
RootsInfoRichEdit->Lines->Add(IntToStr(mm)+" - - - - - - - - - - -");
TcompD xx = AnswersMuller[mm];
RealType xr = xx.real( );
RealType xi = xx.imag( );
RealType r = abs(yAnswersMuller[mm]);
if (erroresMuller[mm])
{ RootsInfoRichEdit->Lines->Add(" - - - - - - - - - - -");
RootsInfoRichEdit->SelAttributes->Color = clRed;
RootsInfoRichEdit->Lines->Add("BAD accuracy = "+
FloatToStrF(r,ffGeneral,3,2));
RootsInfoRichEdit->Lines->Add("Try again.");
}
else
{
String s = " ";
s + ="Re(x): " + FloatToStrF(xr,ffGeneral,3,3)
```

```
+ ", Im(x): " + FloatToStrF(xi,ffGeneral,3,2)+"\r\n";
s + ="Guess: " + CompexToStr(GuessesMuller[mm])
+ ", accuracy = " + FloatToStrF(r,ffGeneral,3,2)
+ ", iterations = "+
FloatToStrF(IterMuller[mm]*1.,ffGeneral,5,0);
RootsInfoRichEdit->Lines->Add(s);
}
}// for mm
free_cvector(AnswersMuller,1,MAXROOTSinMULLER);
free_cvector(GuessesMuller,1,MAXROOTSinMULLER);
free_cvector(GuessesMuller,1,MAXROOTSinMULLER);
free_dvector(epsilonsMuller,1,MAXROOTSinMULLER);
free_ivector(erroresMuller,1,MAXROOTSinMULLER);
free_ivector(IterMuller,1,MAXROOTSinMULLER);
} //CalcComplRoots
// - - - - - - - - RunBitBtnClick - - - - - - - - - - -
void __fastcall TMullerForm::RunBitBtnClick(TObject *Sender)
{
if(ClearOldResCheckBox) RootsInfoRichEdit->Clear( );
set_MaxIter(StrToInt(MaxIterEd->Text));
RealType xr = StrToFloat(RePrmEd->Text);
RealType xi = StrToFloat(ImPrmEd->Text);
set_param(xr+I*xi);
CalcComplRoots(EqToSolve); //Calculation Eigen frequency system;
}
```

18.3 Classes

```
//——————————————————————————————
#ifndef CMullerEq1H
#define CMullerEq1H
#include "BGNDef9.h"
//——————————————————————————————
class TMullerComplEq
{
private:
struct TNquadratic {TcompD A, B, C;};
int _NFoundRoots;
```

```
public:
TMullerComplEq(){}
int Muller6(TcompD Guess2[], RealType Tol2[] , int MaxIter,
TcompD Answer[], TcompD yAnswer[], int Iter2[],
int Error2[],
TMyFuncEq1 MyFuncEq1, TcompD yParam, int NMaxRoots);
String MullerMessage(int Err, int MaxIter);
private:
void Initial(TcompD, RealType, int, int&, int&, bool&,
TcompD&, TcompD&, TcompD&,TcompD&, TcompD&,
TNquadratic&,
TMyFuncEq1 MyFuncEq1, TcompD yParam);
// - - - - -
void QuadraticFormula(TNquadratic, TcompD, TcompD &,
TMyFuncEq1 MyFuncEq1, TcompD yParam);
// - - - - - - - - - - - - - - - - - - - - - - - - -//
bool TestForRoot(TcompD, TcompD, TcompD, RealType);
// - - - - - - - - - - - - - - - - - - - - - - - - -//
void MakeParabola(TcompD,TcompD,TcompD,TNquadratic&,
int& Error,
TMyFuncEq1 MyFuncEq1, TcompD yParam);
};
#endif

//—————————————————————————————————————
#include <vcl.h>
#pragma hdrstop
#include "CMullerEq1.h"
//- - - - - - - - -Muller6 - - - - - - - - - -
int TMullerComplEq::Muller6(TcompD Guess2[], RealType Tol2[],
int MaxIter,
TcompD Answer[], TcompD yAnswer[], int Iter2[],
int Error2[],
TMyFuncEq1 MyFuncEq1, TcompD yParam, int NMaxRoots)
{
TcompD X0, X1, OldApprox;
TcompD NewApprox, yNewApprox; // Iteration variables //
TNquadratic Factor; //Factor of polynomial }
bool Found; // Flags that a factor }
```

```
int nFound = 0, Iter;
_NFoundRoots = 0;
for (int ii = 1; ii<=NMaxRoots; ii++)
{ //for all possible roots
TcompD Guess = Guess2[ii];
RealType Tol = Tol2[ii];
int Error;// = Error2[ii];
Initial(Guess, Tol, MaxIter, Error, Iter, Found, NewApprox,
yNewApprox,
X0, X1, OldApprox, Factor, MyFuncEq1, yParam);
while ((! Found)&&(Error == 0)&&(Iter < MaxIter))
{
Iter++;
QuadraticFormula(Factor, OldApprox, NewApprox, MyFuncEq1,
yParam);
yNewApprox = MyFuncEq1(NewApprox, yParam);
// Calculate a new yNewApprox
Found = TestForRoot(NewApprox, OldApprox, yNewApprox, Tol);
X0 = X1;
X1 = OldApprox;
OldApprox = NewApprox;
Factor.C = yNewApprox;
MakeParabola(X0, X1, OldApprox, Factor, Error, MyFuncEq1,
yParam);
} //while
if (Found)
{Error = 0;}
else
{
if ((Error == 0)&&(Iter >= MaxIter))
{Error = 1;};
Error2[ii] = Error;
}
if (Error == 0)
{
nFound++;
bool Found1 = abs(yNewApprox)<Tol;
Answer[ii] = NewApprox;
yAnswer[ii] = yNewApprox;
```

```
Iter2[ii] = Iter;
Tol2[ii] = Tol;
Error2[ii] = Error;
}
}//for (ii...
return nFound;
} // int Muller6
// - - - - - - MakeParabola - - - - - - - - - - -
void TMullerComplEq::MakeParabola(TcompD X0, TcompD X1,
TcompD Approx,
TNquadratic &Factor, int &Error,
TMyFuncEq1 MyFuncEq1, TcompD yParam)
//————————————————————————————————//
//- Input: X0, X1, Approx -//
//- Output: Factor, Error -//
//- -//
//- This void constructs a parabola to fit the -//
//- three points X0, X1, Approx. The intersection of -//
//- this parabola with the x-axis will yield the next -//
//- approximation. If the parabola is a horizontal line -//
//- then Error = 2 since a horizontal line will not -//
//- intersect the x-axis. -//
//————————————————————————————————//
{
TcompD H1, H2, H3, H;
TcompD Delta1, Delta2;
TcompD Dum1, Dum2, Dum3, Dum4; // Dummy variables //
H1 = X0 – Approx;
H2 = X1 – Approx;
H3 = X0 – X1;
Dum1 = H2* H3;
H = H1*Dum1;
if (abs(H) < TNNearlyZero)
{Error = 2;}; // Can't fit a quadratic to these points //
Dum1 = MyFuncEq1(X1, yParam);
Delta1 = Dum1– Factor.C; // C was passed in //
Dum1 = MyFuncEq1(X0, yParam);
Delta2 = Dum1– Factor.C;
if (Error == 0) // Calculate coefficients of quadratic //
```

```
{
Dum1 = H1* H1; // Calculate B //
Dum2 = Dum1* Delta1;
Dum1 = H2* H2;
Dum3 = Dum1* Delta2;
Dum4 = Dum2−Dum3;
Factor.B = Dum4/ H;
Dum1 = H2* Delta2; // Calculate A //
Dum2 = H1* Delta1;
Dum3 = Dum1 − Dum2;
Factor.A = Dum3/ H;
}
if ((abs(Factor.A) <=TNNearlyZero) && (abs(Factor.B)
<=TNNearlyZero))
{Error = 2;}; // Test if parabola is actually a constant //
} // void MakeParabola //
// - - - - - - - - Initial - - - - - - - - -
void TMullerComplEq::Initial(
TcompD Guess,
RealType Tol,
int MaxIter,
int &Error,
int &Iter,
bool &Found ,
TcompD &NewApprox ,
TcompD &yNewApprox,
TcompD &X0,
TcompD &X1,
TcompD &OldApprox,
TNquadratic &Factor,
TMyFuncEq1 MyFuncEq1, TcompD yParam
)
//————————————————————————
//- Input: Guess, Tol, MaxIter
//- Output: Error, Iter, Found, NewApprox, yNewApprox,
//- X0, X1, OldApprox, Factor
//-
//- This void initializes all the above variables. It
//- sets OldApprox equal to Guess, X0 and X1 are set close to
```

```
//- Guess. The void also checks the tolerance (Tol) and
//- maximum number of iterations (MaxIter) for errors.
//————————————————————————
{
RealType rds = 1.;
TcompD cZero(0.,0.), yOldApprox,zI(0.,1.);
Error = 0;
Found = false;
Iter = 0;
NewApprox = cZero;
yNewApprox = cZero;
// X0 and X1 are points which are close to Guess //
//X0 = (-0.75 * Guess.real( ) -1)+zI*Guess.imag( );
//X1 = (0.75 * Guess.real( ))+zI*(1.2 * Guess.imag( )+1);
X0 = ((-0.75 * Guess.real( ) -1)+zI*Guess.imag( ))*rds;
X1 = ((0.75 * Guess.real( ))+zI*(1.2 * Guess.imag( )+1))*rds;
OldApprox = Guess;
yOldApprox = MyFuncEq1(OldApprox, yParam);
Factor.A = cZero;
Factor.B = cZero;
Factor.C = yOldApprox;
MakeParabola(X0, X1, OldApprox, Factor, Error, MyFuncEq1,
yParam);
if (Tol <=0)
{Error = 3;};
if (MaxIter < 0)
{Error = 4;};
} // void Initial //
// - - - - - - QuadraticFormula - - - - - - - - - - -
void TMullerComplEq::QuadraticFormula(TNquadratic Factor,
TcompD OldApprox,
TcompD &NewApprox,
TMyFuncEq1 MyFuncEq1, TcompD yParam)
//————————————————————————-//
//- Input: Factor, OldApprox -//
//- Output: NewApprox -//
//- -//
//- This void applies the TcompD quadratic formula -//
//- to the quadratic Factor to determine where the parabola -//
```

```
//- represented by Factor intersects the x-axis. The solution -//
//- of the quadratic formula is subtracted from OldApprox to -//
//- yield NewApprox. -//
//-----------------------------------------------------------//
{
TcompD Discrim, Difference, Dum1, Dum2, Dum3, zI(0.,1.);
Dum1 = Factor.B* Factor.B; // B^2 //
Dum2 = Factor.A* Factor.C;
Discrim = Dum1 - (TcompD)4. * Dum2; // B^2 - 4AC
Discrim = sqrt(Discrim );
Dum1 = Factor.B-Discrim; // B +/- sqrt(B^2 - 4AC) //
Dum2 = Factor.B+Discrim;
// Choose the root with B +/- Discrim greatest //
if (abs(Dum1) < abs(Dum2))
{Dum1 = Dum2;};
Dum3 = Factor.C+Factor.C;
if (abs(Dum1) < TNNearlyZero) // if B +/- sqrt(B^2 - 4AC) = 0 //
{
NewApprox = (0,0);}
else
{ // 2C/[B +/- sqrt(B^2 - 4AC)] //
Difference = Dum3/ Dum1;
// Calculate NewApprox //
NewApprox = OldApprox - Difference;
}
} // void QuadraticFormula //
// - - - - - - - - TestForRoot - - - - - - - - -
bool TMullerComplEq::TestForRoot (TcompD X, TcompD OldX,
TcompD Y, RealType Tol )
{
//-----------------------------------------------------------
//- These are the stopping criteria. Four different ones are
//- provided. If you wish to change the active criteria, simply
//- comment off the current criteria (including the preceding OR)
//- and remove the comment brackets from the criteria (including
//- the following OR) you wish to be active.
//-----------------------------------------------------------
TcompD Dif = X - OldX;
TcompD FracDif = X* Tol;
```

```
RealType r1 = abs(Y);
bool cr1 = r1 <=TNNearlyZero;
bool cr2 = abs(X – OldX) < abs(OldX * Tol);
bool cr3 = abs(X – OldX) < Tol;
bool cr4 = r1 <=Tol;
/* bool Res = // = TestForRoot =
(
(r1 <=TNNearlyZero) // – Y = 0
||
(r1 <=Tol) // – Absolute change in y
); */
bool Res =
cr1 // Y = 0
//|| cr2 // Relative change in X
//|| cr3 // Absolute change in X
|| cr4 // Absolute change in Y
;
return(Res);
//————————————————————
/* TestForRoot : = {——————————}
(ABS(Y) <=TNNearlyZero) {– Y = 0 –}
{– –}
or {– –}
{– –}
(ABS(X – OldX) < ABS(OldX * Tol)) {– Relative change in X –}
{– –}
{– –}
or {– –}
{– –}
(ABS(X – OldX) < Tol) {– Absolute change in X –}
{– –}
or {– –}
{– –}
(ABS(Y) <=Tol) {– Absolute change in Y –}
*/
//- The first criteria simply checks to see if the value of the
//- funct is zero. You should probably always keep this criteria
//- active.
//- The second criteria checks the relative error in x. This criteria
```

```
//- evaluates the fractional change in x between interations. Note
//- that x has been multiplied through the inequality to avoid Divide
//- by zero errors.
//- The third criteria checks the absolute difference in x between
//- iterations.
//- The fourth criteria checks the absolute difference between
//- the value of the funct and zero.
//-----------------------------------------------------------
} // funct TestForRoot //
// - - - - -MullerMessage - -
String TMullerComplEq::MullerMessage(int Err, int MaxIter)
{ String s;
switch (Err)
{
case 1 : s = "This will take more than"+ IntToStr(MaxIter)+
" iterations."; break;
case 2 : s = "A parabola which intersects the x-axis can not be
constructed";
//"through these three points.";
break;
case 3 : s = "The tolerance must be >0."; break;
case 4 : s = "The maximum number of iterations must be >0.";
break;
}; // switch }
return s;
} //MullerMessage
//-----------------------------------------------------------
#pragma package(smart_init)
```

CHAPTER 19

Evaluation of Complex ODE

19.1 Introduction

In this section we evaluate the complex system of ordinary differential equations (ODE) by the Runge-Kutta method. This method can be applied to various linear and nonlinear problems [see e.g. Jian Zou, 2001a; Salasnich, 2003a; Burlak *et al.*, 2002; Kartashov *et al.*, 2003a,b]. The following program is a translation into $C{+}{+}$ from [Press *et al.*, 1993] and is contained in class CRunge2. Since in $C{+}{+}$ the complex type is available (see STL [Nicolai, 1999]) it is allowed to rewrite the mentioned code for complex type ODE. Section 19.2 shows the program code (with GUI), which allows one to evaluate the solution of ODE $dy_i/dx = f_i(x, y_1, \ldots, y_N)$, where y_i and f_i are complex numbers. Readers can construct their own function (see TDiffEqsSystem2, void MySystem2 ets) to their own calculation. The next test program for the calculation for ODE is considered

$$\frac{dy_1}{dx} = y_2 - y_3, \quad \frac{dy_2}{dx} = y_1^2 + y_2, \quad \frac{dy_3}{dx} = y_1^2 + y_3 \tag{19.1}$$

The exact solution is given by [Kamke, 1983]

$$\begin{aligned} &y_1 = C_1 + C_2 e^x, \quad y_2 = -C_1^2 + \left(2C_1^2 x + C_3\right) e^x + C_2^2 e^{2x}, \\ &y_3 = y_2 - y_1 + C_1. \end{aligned} \tag{19.2}$$

Figure 19.1 shows the initial data and output. In this GUI one can write down the initial conditions x_0, y_{01}, y_{02}, y_{03}, the order of ODE, accuracy of calculations ε and the value of the independent variable x to find the

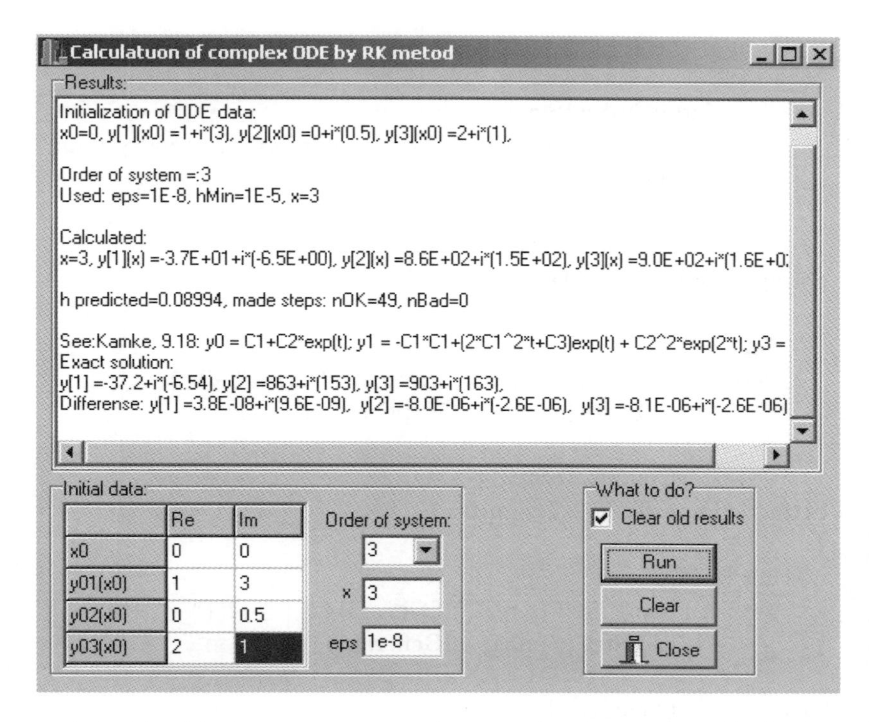

Fig. 19.1 Graphical user interface and output for the program 'Evaluation of Complex ODE'.

solution. In this case $\varepsilon = 10^{-8}$ and $x = 3$. From Fig. 19.1 one can see that the calculation accuracy is sufficient.

19.2 Code

```
//- - - - - - - - - - - - - - - - - - -
    #include <vcl.h>
    #pragma hdrstop
    #include "ClassRunge7.h"
    //- - - - - - - - - - - - - - - - - - -
    #pragma package(smart_init)
    #pragma resource "*.dfm"
    TClassRungeForm *ClassRungeForm;
    //- - - - - - - - - - - - - - - - - - -
    __fastcall TClassRungeForm::TClassRungeForm(TComponent* Owner)
```

```
: TForm(Owner)
{
    _nOrderOfCurrentSystem = 3;
    OrderOfSystemComboBox->Text =_nOrderOfCurrentSystem;
    CreateGrid(_nOrderOfCurrentSystem);
}
// - - - - - - -CompexToStr - - - - -
String CompexToStr(TcompD c)
{
    int i = 2;
    return FloatToStrF(c.real( ),ffExponent,i,i)+ "+i*("+
        FloatToStrF(c.imag( ),ffExponent,i,i)+ ")";
}
// - - - - - - -CompexToStr2 - - - - -
String CompexToStr2(TcompD c)
{
    int i = 3;
    return FloatToStrF(c.real( ),ffGeneral,i,i)+ "+i*("+
        FloatToStrF(c.imag( ),ffGeneral,i,i)+ ")";
}
// - - - - - - -CreateGrid - - - - - - - -
void TClassRungeForm::CreateGrid(int aRows)
{
    int cRows = aRows+2, c10 = 3;//globalSt::NPoints;
// - - create table - - - }
    NumericStringGrid->ColCount = 3; //Number of columns
    NumericStringGrid->RowCount = cRows+0; //Number of rows
    for (int ii = 0; ii<NumericStringGrid-> RowCount; ii++)
    NumericStringGrid->RowHeights[ii] = 19;
    NumericStringGrid->ColWidths[0] = 60;
    for (int ii = 1; ii<NumericStringGrid-> ColCount; ii++)
    NumericStringGrid->ColWidths[ii] = 40;
    NumericStringGrid->Cells[0][1] = "x0";
    NumericStringGrid->Cells[1][0] = "Re";
    NumericStringGrid->Cells[2][0] = "Im";
    for (int ii = 0; ii<=cRows; ii++)
        NumericStringGrid->
Cells[0][ii+2] = "y0"+IntToStr(ii+1)+"(x0)";
    for (int j = 1; j<=c10; j++)
```

```
        for (int j2 = 1; j2<=cRows; j2++)
            NumericStringGrid->Cells[j][j2] = "0";
        NumericStringGrid->Cells[1][2] = "1"; //Re y10
        NumericStringGrid->Cells[2][2] = "3"; //Im y10
    }
    // - - - - - - - - RunButtonClick - - - - - - - - - -
    void __fastcall TClassRungeForm::RunButtonClick(TObject *Sender)
    {
        int nvar = StrToInt(OrderOfSystemComboBox-> Text);
//get_OrderOfCurrentSystem;
        TcompD *ystart0, *yFin, *consts2;//,*consts3;
        TcompD i(0,1);
        RealType
            x0 = StrToFloat(NumericStringGrid->Cells[1][1]),
            x = StrToFloat(tEd->Text),
            eps = StrToFloat(epsEd->Text),
            hmin = StrToFloat(hMinEd->Text);
        ystart0 = cvector(1, nvar);
        yFin = cvector(1, nvar);
        consts2 = cvector(1, nvar);
        for (int ii = 1; ii<=nvar; ii++)
        ystart0[ii] = StrToFloat(NumericStringGrid-> Cells[1][ii+1])+
            i* StrToFloat(NumericStringGrid->Cells[2][ii+1]);
        switch (nvar)
        {
        case 2: consts2[1] = ystart0[1], consts2[2] = ystart0[2]; break;
        case 3: consts2[1] = ystart0[1]-ystart0[2]+ystart0[3];
                consts2[2] = ystart0[1] - consts2[1];
                consts2[3] = ystart0[2] - pow(ystart0[1],2) +
        (TcompD)2*ystart0[1]*consts2[1];
        break;
        }
        if(ClearCheckBox->Checked) ReflMemo-> Clear( );
        ReflMemo->Lines->Add(" - - - - - - - - - - - - - - - - -");
        ReflMemo->Lines->Add("Initialization of ODE data:");
        String sTmp = "x0 = "+ NumericStringGrid->Cells[1][1] + ", ";
        for(int ii = 1; ii<=nvar; ii++)
```

```
        sTmp + = "y["+IntToStr(ii)+"](x0) =
"+CompexToStr2(ystart0[ii])+",
";
    ReflMemo->Lines->Add(sTmp);
    bool Error = true;
    TcompD *par = NULL;
    par = cvector(1, nvar);
    TDiffEqsSystem2  *ff2 = NULL;
    TDiffEqsSystem3  *ff3 = NULL;
    CRunge2          *runge = NULL;
    switch (nvar)
    {
    case 2: ff2 = new TDiffEqsSystem2( );
        runge = new CRunge2(ystart0, ff2-> get_nOfSystem( ), x0,
eps, hmin,
        ff2->MySystem2, par); //initial.
        break;
    case 3: ff3 = new TDiffEqsSystem3( );
        runge = new CRunge2(ystart0, ff3-> get_nOfSystem( ), x0,
eps, hmin, ff3->MySystem3, par); //initial.
        break;
    }
    yFin = runge->get_DiffEqSolution(x, Error); // calc.of solution
    ReflMemo->Lines->Add("\r\nOrder of
system = :"+IntToStr(runge-> get_nOfSystem( )));
    ReflMemo->Lines->Add("Used:
eps = "+FloatToStrF(eps,ffGeneral,3,2)+
        ", hMin = "+FloatToStrF(hmin,ffGeneral,3,2)+", x = "+
tEd- >Text);
    ReflMemo->Lines->Add("\r\nCalculated:");
    if (Error)
     ReflMemo->Lines->Add("ERROR of calc!");
    else
    {
    String sTmp = "x = "+ tEd->Text+", ";;
     for(int ii = 1; ii<=nvar; ii++)
     sTmp + = "y["+IntToStr(ii)+"](x) = "
+CompexToStr(yFin[ii])+", ";
     ReflMemo->Lines->Add(sTmp + ",\r\n\r\nh predicted = "+
```

```
        FloatToStrF(runge->get_stepPredicted( ),ffGeneral,4,4)+
         ", made steps: nOK = ",+FloatToStrF(runge->
get_NmbGoodSteps( ),ffGeneral,4,4) +
         ", nBad = "+FloatToStrF(runge->
get_NmbBadSteps( ),ffGeneral,4,4)
      );
      ReflMemo->Lines->Add(" ");
      TcompD *execSol;
      execSol = cvector(1, nvar);
      switch (nvar)
      {
      case 2: ff2->exectSolution(x, consts2, execSol); // calculate a
solution in point x
            ReflMemo->Lines->Add("See:" +ff2- > strExectSolution( ));
            break;
      case 3: ff3->exectSolution(x, consts2, execSol);// calculate a
solution in point x
            ReflMemo->Lines->Add("See:" +ff3- > strExectSolution( ));
            break;
      }
      sTmp = " ";
      for(int ii = 1; ii<=nvar; ii++)
            sTmp + =
"y["+IntToStr(ii)+"] = "+CompexToStr2(execSol[ii])+", ";
      ReflMemo->Lines->Add("Exact solution: \r\n" + sTmp);
      sTmp = "Difference:";
      for(int ii = 1; ii<=nvar; ii++)
            sTmp + = " y["+IntToStr(ii)+"]  = "
+CompexToStr(yFin[ii]-execSol[ii])+", ";
      ReflMemo->Lines->Add(sTmp);
      free_cvector(execSol,1,nvar);
      }//if
      if(ff2! = NULL) delete ff2;
      if(ff3! = NULL) delete ff3;
      delete runge;
      free_cvector(ystart0,1,nvar);
      free_cvector(yFin,1,nvar);
      free_cvector(consts2,1,nvar);
      free_cvector(par,1,nvar);
```

```
    }
    // - - - - - - - - ClearButtonClick- - - - - - - - - -
    void __fastcall TClassRungeForm::ClearButtonClick(TObject *Sender)
    {
        ReflMemo->Clear( );
    }
    // - - - - - - - OrderOfSystemComboBoxChange- - - - - - - - - - -
    void __fastcall TClassRungeForm::OrderOfSystemComboBoxChange(
    TObject *Sender)
    {
        int nOrderOfSystem = StrToInt(OrderOfSystemComboBox->
Text);
        set_OrderOfCurrentSystem(nOrderOfSystem);
        CreateGrid(nOrderOfSystem);
    }
```

19.3 Classes

```
// - - - - - - - - - - - - - - - - - -
    Class TAbstrDiffEqsSystem in File ODESystem.h has form
    #include "BgnDef9.h"
    /// - - - - - - TAbstrDiffEqsSystem - - - - - - -
    class TAbstrDiffEqsSystem
    {
    private:
        int _nOfSystem;
    public:
        TAbstrDiffEqsSystem( ){}
        virtual String strExectSolution( ){};// = 0;
    };
    // - - - - - - TDiffEqsSystem2 - - - - - - -
    class TDiffEqsSystem2: TAbstrDiffEqsSystem
    {
    private:
        int _nOfSystem;
    public:
        TDiffEqsSystem2( ){_nOfSystem = 2;}
```

```
      int get_nOfSystem(){return _nOfSystem;}
   //---
   static void MySystem2(RealType t, TcompD XValue[],
TcompD XDeriv[], TcompD param[])
   {     TcompD aa,bb;
      aa = XDeriv[1] = (-t*XValue[1] + XValue[2])/(t*t+1);
      bb = XDeriv[2] = -(XValue[1] + t*XValue[2])/(t*t+1);
   }
   //---
   void exectSolution(RealType t, TcompD consts[], TcompD exectSol[])
   {
      TcompD aa,bb;
      TcompD C1 = consts[1], C2 = consts[2];
      aa = exectSol[1] = (C1+C2*t)/(t*t+1);
      bb = exectSol[2] = (-C1*t+C2)/(t*t+1);
   }
   //---
   virtual String strExectSolution()
   {
      return "Kamke, 9.11: y0 = (C1+C2*t)/(t*t+1); y1 =
(-C1*t+C2)/(t*t+1);";
   }
   };
   //------- TDiffEqsSystem3 ----------
   class TDiffEqsSystem3 : TAbstrDiffEqsSystem
   {
   private:
      int _nOfSystem;
   public:
      TDiffEqsSystem3(){_nOfSystem = 3;}
      int get_nOfSystem(){return _nOfSystem;}
   //--
      static void MySystem3(RealType t, TcompD XValue[], TcompD
XDeriv[], TcompD param[])
        {
           TcompD x = XValue[1], y = XValue[2], z = XValue[3];
           XDeriv[1] = y - z;
           XDeriv[2] = x*x+y;
           XDeriv[3] = x*x+z;
```

```
        }
    //--------
      void exectSolution(RealType t, TcompD *consts, TcompD exectSol[])
        {TcompD x,y,z;
            TcompD C1 = consts[1], C2 = consts[2], C3 = consts[3];
            x = exectSol[1] = C1+C2*(TcompD)exp(t);
            y = exectSol[2] =
-C1*C1+((TcompD)2*C1*C2*t+C3)*(TcompD)exp(t)
                + C2*C2*(TcompD)exp(2*t);
            z = exectSol[3] = y - x+C1;
        }
    //-----------
      virtual String strExectSolution()
        {
            return "Kamke, 9.18: y0 = C1+C2*exp(t);"
                " y1 = -C1*C1+(2*C1^2*t+C3)exp(t)"
                " + C2^2*exp(2*t); y3 = y - x+C1";
        }
    };
```

CHAPTER 20

The Complex Vectorial and Matrix Operations

20.1 Introduction

One of the most important problems is the realization of matrix and vectorial operations, when both matrixes and vectors may have complex structure. In this book the technique of work with a transfer matrix, and also various vectorial representations for field's component is widely used. As a rule we exploit one of the important features of *C++* which is its adaptability to write the mathematical expressions in a natural form by means of the technique of the operators overloading. For instance, we can write the expression for three matrices A, B, C in the form $A * B + C$, if the arithmetical operators for matrices and vectors are defined by the correct way. Further, the program and classes for vectorial and matrix complex operations

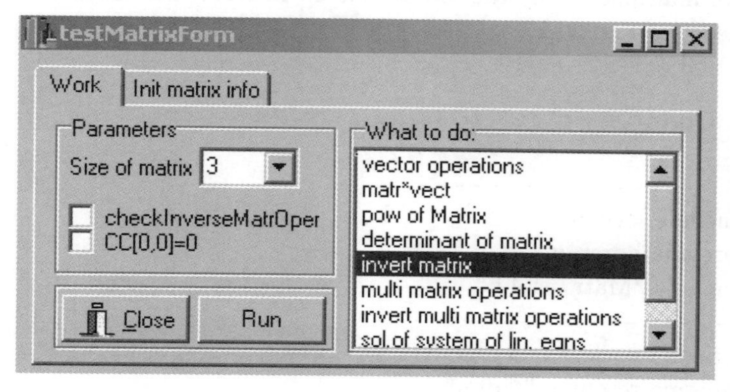

Fig. 20.1 Various matrix and vector operations, where both matrices and vectors may have a complex structure.

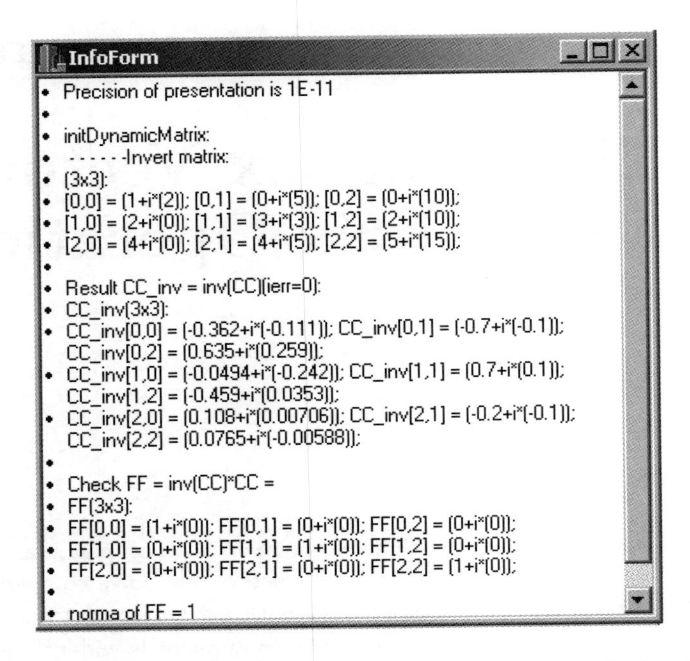

Fig. 20.2 The matrix inversion for a complex matrix order 3×3.

are presented (see Fig. 20.1). Various operations of summation, subtraction or multiplication of vectors or matrices do not represent any difficulties. More difficult task is the matrix inversion. Figure 20.2 shows the output for such a program. We have inverted a complex matrix 3×3 order and then we have multiplied it by the initial matrix. In result the unitary matrix is obtained (see Fig. 20.2).

20.2 Code

```
//-----------------------
   #include <vcl.h>
   #pragma hdrstop
   #include "MatrTest4.h"
   //-----------------------
   #pragma package(smart_init)
   #pragma resource "*.dfm"
   TtestMatrixForm *testMatrixForm;
   //-------- TtestMatrixForm--------------
```

```
__fastcall TtestMatrixForm::TtestMatrixForm(TComponent* Owner)
: TForm(Owner)
{
TasksListBox->Selected[0] = true;
initInfoRichEdit->Paragraph->Numbering = nsBullet;
}
//- - - - - - - - - - -
void __fastcall TtestMatrixForm::TasksListBoxDblClick
(TObject *Sender)
{
    Run();
}
//------------------------------------------------------------
void __fastcall TtestMatrixForm::RunButtonClick(TObject *Sender)
{
    Run();
}
//- - - - - - - - - - -Run - - - - - - - - - - -
void TtestMatrixForm::Run()
{
    InfoForm->Show();
    InfoForm->InfoRichEdit->Clear();
    int nSize, N;
    N = StrToInt(SizeComboBox1->Text);
    nSize = N;
    set_N(N);
    _AA1 = new TBMatrix(N, "_AA1");
    _BB1 = new TBMatrix(N, "_BB1");
    _CC1 = new TBMatrix(N, "_CC1");
    _DD1 = new TBMatrix(N, "_DD1");
    printToMemo("Precision of presentation is "
+ FloatToStr(ACCUR)+" \r\n");
//
    int itms = TasksListBox->Items->Count;
    initDynamicMatrix();
    for(int ii = 0; ii<itms; ii++) // for all items from the TasksListBox
    {
        String s = TasksListBox->Items-> Strings[ii];
        // if (TasksListBox->Selected[ii])
```

```
            // Memoprintf(" -> Task #"+IntToStr(jj) +
" - \r\nCalculation: " + s);
            switch(ii)
            {
                case 0:
                if (TasksListBox->Selected[ii])
                {
                    printToMemo("- - - - - Vector operations: - - - -" );
                    testVectorOperations( );
                }
                break;
                case 1:
                if (TasksListBox->Selected[ii])
                {
                    printToMemo("\ - - - - - -nMatr*vect: ");
                    testMatrVectOper( );
                }
                break;
                case 2:
                if (TasksListBox->Selected[ii])
                {
                    printToMemo("\ - - - - - -testMatrPow( ): ");
                    testMatrPow( );
                }
                break;
                case 3:
                if (TasksListBox->Selected[ii])
                {
                    printToMemo("- - - - - - - Determinant of matrix:" );
                    testDeterminant( );
                }
                break;
                case 4:
                if (TasksListBox->Selected[ii])
                {
                    printToMemo(" - - - - - -Invert matrix:") ;
                    testImvertMatrix( );
                }
                break;
```

```
            case 5:
            if (TasksListBox->Selected[ii])
            {
                printToMemo(" - - - - - - multi matrix operatios:");
                testMultiMatrOper( );
            }
            break;
            case 6:
            if (TasksListBox->Selected[ii])
            {
                printToMemo(" - - - - invert multi matrix
operatios:");
                testInvertMultiOperation( );
            }
            break;
            case 7:
            if (TasksListBox->Selected[ii])
            {
                printToMemo(" - - - - - - solutions of linear system:");
                linearSystenSolution( );
            }
            break;
        }//switch
    }
    delete _AA1; _AA1 = NULL;
    delete _BB1;
    delete _CC1;
    delete _DD1;
}
//- - - - initDynamicMatrix- - - - - - - - - - - - -
void TtestMatrixForm::initDynamicMatrix( )
{
    initInfoRichEdit->Clear( );
    printToMemo( "initDynamicMatrix:") ;
    int N = get_N( );
    for(int ii = 0; ii < N; ii++)
    for(int jj = 0; jj < N; jj++)
    (*_CC1)[ii][jj] = (TcompD)(ii*2) + (TcompD)jj*(TcompD)5.*I;
    // Enter value A TBMatrix:
```

```
        for(int ii = 0; ii < N; ii++)
        for(int jj = 0; jj < N; jj++)
            (*_AA1)[ii][jj] = (TcompD)(ii*3.14)+(TcompD)(jj*2.)*I;

        (*_CC1)[0][0] += (TcompD)1.+(TcompD)2.*I; //to avoid
singuliarity
        (*_CC1)[1][1] += (TcompD)1. - (TcompD)2.*I; //to avoid
singuliarity
        if (N>2) (*_CC1)[2][2] += (TcompD)1.+(TcompD)5.*I; //to avoid
singuliarity
        if (N>3) (*_CC1)[3][3] += (TcompD)1.-(TcompD)5.*I; //to avoid
singuliarity
        (*_AA1)[0][0] += (TcompD)0.5+(TcompD)2.*I; //to avoid
singuliarity
        (*_AA1)[1][1] += (TcompD)0.5 - (TcompD)2.*I; //to avoid
singuliarity
        if (N>2) (*_AA1)[2][2] += (TcompD)0.1+(TcompD)3.*I; //to
avoid singuliarity
        if (N>3) (*_AA1)[3][3] += (TcompD)0.1-(TcompD)3.*I; //to
avoid singuliarity
    /* */
        printToInitMemo(_AA1->toString( ));
        printToInitMemo(_CC1->toString( ));
        try
        {
            printToInitMemo("_AA1.det = " + CompToStr(_AA1->
determinant( )));
        }
        catch(char * e)
        {
            ShowMessage("initDynamicMatrix: _AA1.det" + (String)e);
        }
        try
        {
            printToInitMemo("_CC1.det = " + CompToStr(_CC1->
determinant( )));
        }
```

```
        catch(char * e)
        {
            ShowMessage("initDynamicMatrix: _CC1.det" + (String)e);
        }
        //printToInitMemo("_CC1.det = " + CompToStr(_CC1->
determinant( )));
        printToInitMemo("Final of initDynamicMatrix.\r \n") ;
    }
    //---------- testInvertMultiOperation ------------
    void TtestMatrixForm::testInvertMultiOperation( );
    {// WORKING WITH DYNAMICAL MATRIXES
        int N = get_N( );
        int ierr = 0;
        *_BB1 = (*_AA1) + (*_CC1);
        *_DD1 = (*_AA1) * (*_CC1);



        TBMatrix *BB_new = new TBMatrix(N,"BB_new");
        TBMatrix *CC_inv = new TBMatrix(N,"CC_inv");
        ierr = 0;
        _CC1->set_isCheckedInvOfMatr
(checkInverseMatrOperCheckBox->Checked);
        try
        {
            *CC_inv = _CC1->inverse(ierr);
        }
        catch (char* E)
        {   String s = E;
            ShowMessage("testInvertMultiOperation, _CC1: " + s);
            return;
        }
        String sierr = IntToStr(ierr);
        if (ierr == 2)
        {
            printToMemo( "!! IT SINGULAR matrix (ierr = " + sierr +")
\r\n");
            ShowMessage( "!!CC: IT SINGULAR matrix (ierr = " +
sierr +") \r\n");
        }
```

```
        else
        {
            TBMatrix *EE3 = new TBMatrix(N,"EE3");
            printToMemo("*EE3 = *_AA1* *_CC1 + *_CC1 * *_BB1 -
*_DD1");
            (*EE3) = (*_AA1) * (*_CC1) + (*_CC1) * (*_BB1) -
(*_DD1);
            printToMemo(EE3->toString( ));
            printToMemo("BB_new = (1/CC)*(EE3+DD-AA*CC): ") ;
            *BB_new = (*CC_inv) * ((*EE3)+ *_DD1 - *_AA1 * *_CC1);
            printToMemo(BB_new->toString( ));
            printToMemo("check the difference:") ;
            TBMatrix *EE4 = new TBMatrix(N,"EE4");
            *EE4 = *_BB1 - *BB_new;
            printToMemo("EE4 = BB-BB_new : ") ;
            printToMemo(EE4->toString( ));
            printToMemo("norma of " + EE4->get_title( )+"="
            +FloatToStrF(EE4->get_norm( ), ffGeneral ,3,3));
            delete EE3;
            delete EE4;
        }
        delete CC_inv;
        delete BB_new;
    }
    // - - testDeterminant - - - - - - - - - -
    void TtestMatrixForm::testDeterminant( )
    {
        TBMatrix AA = *_AA1;
        AA.set_title("AA");
        int N = get_N( );
        printToMemo(AA.toString( ));
        TBMatrix AA_inv(N,"AA_inv");
        int ierr = 0;
        AA.set_isCheckedInvOfMatr(checkInverseMatrOperCheckBox->
Checked);
        try
        {
            AA_inv = AA.inverse(ierr);
        }
```

```
        catch (char* E)
        {     String s = E;
            ShowMessage("testDeterminant, AA: " + s);
            return;
        }
        String sierr = IntToStr(ierr);
        if (ierr == 2)
        {     printToMemo( "!! "+ AA.get_title()+" <- IS a SINGULAR
matrix (ierr =" + sierr +")\r\n");
            ShowMessage( "Determinant of matrix:\r \n!! "+
AA.get_title()
                +" IS A SINGULAR matrix (ierr =" + sierr +")\r\n");
        }
            else
        {
            TcompD GG = AA.determinant();
        int in = 4;
        String s = "("+FloatToStrF(GG.real(), ffGeneral ,in,in)
            + "+i*("+FloatToStrF(GG.imag(), ffGeneral ,in,in)+"))";
        printToMemo("Result GG = det("+AA.get_title()+") = "+s);
        if (abs(GG)<1.e - 10)
            ShowMessage( "Determinant of matrix: "+AA.get_title()+".
abs(det) ="
                +FloatToStrF(GG.imag(), ffGeneral,in,in)+"
<1.e - 10 !!\r\n");
        }
    }
    // - - - - - testImvertMatrix - - - - - -
    void TtestMatrixForm::testImvertMatrix()
    {
        TBMatrix CC = *_CC1;
        printToMemo(CC.toString());
        int ierr = 0;
        int N = get_N();
        TBMatrix *CC_inv = new TBMatrix(N,"CC_inv");
        // *CC_inv = InvertBMatrix(CC, ierr);
        CC.set_isCheckedInvOfMatr(checkInverseMatrOperCheckBox->
Checked);
```

```
        try
        {
            *CC_inv = CC.inverse(ierr);
        }
        catch (char* E)
        {    String s = E;
            ShowMessage("testImvertMatrix, CC: " + s);
            return;
        }
        String sierr = IntToStr(ierr);
        if (ierr == 2)
        {
            printToMemo( "!!CC: IT SINGULAR matrix (ierr = " +
sierr +")\r\n");
            ShowMessage( "!!CC: IT SINGULAR matrix (ierr = " +
sierr +")\r\n");
        }
        else
        {
            printToMemo("Result CC_inv =
inv(CC)(ierr = "+sierr+"):");
            printToMemo(CC_inv->toString( ));
            TBMatrix FF = (*CC_inv) * CC;
            FF.set_title("FF");
            printToMemo("Check FF = inv(CC)*CC = " );
            printToMemo(FF.toString( ));
            printToMemo("norma of " + FF.get_title( )+" = "
                +FloatToStrF(FF.get_norm( ), ffGeneral ,3,3));
        }
    }
    // - - - - - - testMatrVectOper - - - - - -
    void TtestMatrixForm::testMatrVectOper( )
    {
        TBMatrix AA = *_AA1;
        AA.set_title("AA");
        int N = get_N( );
        TBVector a(N, "a"), b(N, "b");
        for(int jj = 0; jj < N; jj++)
        a[jj] = (TcompD)(jj*5.)*I+(TcompD)2.;
```

```
        printToMemo(a.toString( ));
        AA.toString( );
        b = AA*a;
        printToMemo("Result b = A*a :");
        printToMemo(b.toString( ));
    }
    // - - - - testMatrPow - -
    void TtestMatrixForm::testMatrPow( )
    {
        TBMatrix FF = MatrPow(*_AA1, 5);
        FF.set_title("FF");
        printToMemo("Result FF = *_AA1^5 :");
        printToMemo(FF.toString( ));
    }
    // - - - - - testMultiMatrOper - - -
    void TtestMatrixForm::testMultiMatrOper( )
    {
        int N = get_N( );
        TBMatrix EE3 = (*_AA1) * (*_CC1) + (*_CC1) * (*_BB1) –
(*_DD1);
        EE3.set_title("EE3");
        printToMemo("EE3 = AA*CC+CC*BB – DD : ") ;
        printToMemo(EE3.toString( ));
    }
    // - - - - testVectorOperations- - -
    void TtestMatrixForm::testVectorOperations( )
    {
        int N = get_N( );
        TBVector a(N, "a");
        for(int jj = 0; jj < N; jj++)
            a[jj] = (TcompD)jj*(TcompD)5.*I+(TcompD)2.;
        printToMemo(a.toString( ));
        // TBVector b = a;
        TBVector b(N, "b");
        b = a;
        printToMemo(b.toString( ));
        TBVector c(N, "c");
        c = a+b;
        printToMemo("c = a+b, "+c.toString( ));
```

```
    printToMemo("Now "+a.toString());
    printToMemo("Now "+b.toString());
    c = a - b;
    printToMemo("c = a - b, "+c.toString());
    TcompD ss = a*b;
    printToMemo("c = a*b = " + CompToStr(ss));
    printToMemo("Vector finish. " );
}
// - - - - - testImvertMatrix - - - - - -
void TtestMatrixForm::linearSystenSolution()
{
    printToMemo("solutions of linear system: CC*x = a") ;
    int N = get_N();
    TBVector a(N, "a");
    for(int jj = 0; jj < N; jj++)
    a[jj] = (TcompD)jj*(TcompD)5.*I+(TcompD)2.;
    printToMemo(a.toString());
    TBMatrix CC = *_CC1;
    if (CC_CheckBox->Checked) CC[0][0] = 0.;
    CC.set_title("CC");
    printToMemo(CC.toString());
    TcompD GG;
    try
    {
        GG = CC.determinant();
    }
    catch (char* E)
    {   String s = E;
        ShowMessage("testImvertMatrix: CC: " + s);
        return;
    }
    int in = 4;
    String s = "("+FloatToStrF(GG.real(), ffGeneral ,in,in)
        + "+i*("+FloatToStrF(GG.imag(), ffGeneral ,in,in)+"))";
    printToMemo("det("+CC.get_title()+") = "+s+"\r\n");
    int ierr = 0;
    TBMatrix *CC_inv = new TBMatrix(N,"CC_inv");
    // *CC_inv = InvertBMatrix(CC, ierr);
```

```
        CC.set_isCheckedInvOfMatr(checkInverseMatrOperCheckBox->
Checked);
        try
        {
            *CC_inv = CC.inverse(ierr);
        }
        catch (char* E)
        {     String s = E;
            ShowMessage("linearSystenSolution, CC: " + s);
            return;
        }
        printToMemo("Solutions: x = CC_inv*a: " );
        TBVector x(N, "x"),a_new(N, "a_new");
        x = (*CC_inv)*a;
        printToMemo(x.toString( ));
        printToMemo("Check a_new = CC*x " );
        a_new = CC*x;
        printToMemo(a_new.toString( ));
        printToMemo(a.toString( ));
        printToMemo("Check difference a_new - a: " );
        TBVector aDiff(N, "aDiff");
        aDiff = a - a_new;
        printToMemo(aDiff.toString( ));
        printToMemo("norm(aDiff) = "+
FloatToStrF(aDiff.get_norm( ),ffGeneral,3,3));
    }
```

20.3 Classes

In this case the structure of classes is rather simple. There are two separated classes: simple class TBVector and more advanced class TBMatrix. To have a possibility in writing both matrix-matrix and matrix-vector operation in a standard mathematical notation all mathematical operators are overloaded. This means that for instance we can write the product of two matrices in the form A*B. It is clear that first we have to write two variables of type TBMatrix A,B; then we have to create and initialize its

```
    #ifndef MatrOper4H
    #define MatrOper4H
```

```
#include "BGNDef9.h"
RealType const ACCUR = 1.e - 11;
// - - - - - TBVector - - - - - - -
class TBVector
{
private:
      int _vecSize;
      TcompD *_vect;
      String _title;
public:
      TBVector(){_vecSize = 0; _vect = 0; _title =" ";}
// default constructor0, only initialization
      TBVector(int n)
// constructor1, provide with size
            {_vecSize = n; _vect = new TcompD[n]; _title =" "; }
      TBVector(int ii, String s)
// constructor2, provide with size and name
            {_vecSize = ii; _vect = new TcompD[ii];_title = s;}
      ~TBVector()
// default destructor0
            { if(_vect ! = 0) delete [ ]_vect; }


      TBVector(const TBVector &);
// copy constructor
      TBVector operator+ (const TBVector&);              // overloading
operator+ (what means + sign for vector?)
      TBVector operator- (const TBVector&);              // overloading
operator-  (what means – sign for vector?)
      TcompD operator* (const TBVector&);                // overloading
operator*  (what means * sign for vector?)
      TcompD& operator[ ] (int);                         // overloading
operator[]  (what means [] sign for vector?)
      TBVector& operator = (const TBVector&);            // overloading
operator =  (what means = sign for vector?)
      String toString( );                                // information
about state of this object
      RealType get_norm( );                              // calc. norm
      int getVector_size( ) const {return _vecSize;}
```

```
    void read( );
    void print( );
    void allot(int);
};//TBVector
// - - - - -class TBMatrix - - - - - - - -
class TBMatrix
{
private:
    int _matrSize;
    TBVector *_matrix;
    String _title;
    bool _isCheckedInvOfMatr;
public:
    String get_title( )const {return _title;}
    void set_title(String s) {_title = s;}
    void set_isCheckedInvOfMatr(bool b) {_isCheckedInvOfMatr = b;}
    bool get_isCheckedInvOfMatr( ) {return _isCheckedInvOfMatr;}
    int get_matrSize( )const {return _matrSize;}
    RealType get_norm( );
    TBVector* get_Matrix( ){return _matrix;}
    TBMatrix* get_Matrix2( ){return (TBMatrix*)_matrix;}
    TBMatrix( ){_matrix = 0;} //default constructor 0
    TBMatrix(int N); //constructor 1
    TBMatrix(int N, String s); //constructor 2
    ~TBMatrix( )
        { if(_matrix ! = 0) delete [ ] _matrix;} //destructor
    TBVector &operator [ ] (int ii) {return _matrix[ii];}
    TBMatrix(const TBMatrix &); //copy constructor 3
    TBMatrix &operator = (const TBMatrix &);
    TBMatrix &operator+= (const TBMatrix &B); // += operator
    TBMatrix &operator-= (const TBMatrix &B); // -= operator
    TBMatrix &operator*= (const TBMatrix &B); // *= operator
    void read( );
    void print( );
    String toString( );
    TcompD determinant( );
    TBMatrix inverse(int& Error);
    TBMatrix getUnitMatrix( )
```

```
    {
        TBMatrix tmpMatr(_matrSize);
        for(int ii = 0; ii< _matrSize; ++ii)
         for(int jj = 0; jj< _matrSize; ++jj)
            if(ii == jj)
                tmpMatr[ii][jj] = (TcompD)1;
            else
                tmpMatr[ii][jj] = (TcompD)0;
        return tmpMatr;
    }
}; //TBMatrix
#endif
```

Further the realization for both TBMatrix and TBVector classes is written.

```
#include <vcl.h>
#pragma hdrstop
#include "MatrOper4.h"
#pragma package(smart_init)
// - - - - forward - - - -
void InvertBMatrixLoc(TBMatrix& Data1, TBMatrix& Inv, int&
Error);
TcompD DetermBMatrixLoc(TBMatrix& Data1);
RealType MyRound3(RealType x, RealType c = ACCUR) {return
fabs(x) > c? x: 0.;}
// - - -CompToStr -
String CompToStr(TcompD z)
{       int in1 = 3,in2 = 2;
     String s = "("+FloatToStrF(MyRound3(z.real( )), ffGeneral,
in1,in2)
          + "+i*("+FloatToStrF(MyRound3(z.imag( )), ffGeneral,
in1,in2)+"))";
     return s;
}//
// - - - - TBMatrix- - - - - - - - -
TBMatrix::TBMatrix(int N) //constructor
{
     _matrSize = N;
     _matrix = new TBVector [N]; //allocate array of vectors
```

```
    for(int ii = 0; ii < N; ii++) //allocate storage
        _matrix[ii].allot(N);
}//
// - - - - TBMatrix- - - - - - - - -
TBMatrix::TBMatrix(int N, String s) //constructor
{
    _matrSize = N;
    _title = s;
    _matrix = new TBVector [N]; //allocate array of vectors
    for(int ii = 0; ii < N; ii++) //allocate storage
        _matrix[ii].allot(N);
}//
// - - - - get_norm- - - - - - - - -
RealType TBMatrix ::get_norm( )
{
    int N = get_matrSize( );
    RealType rm = 0.;
    for(int ii = 0; ii < N; ii++)
     for(int jj = 0; jj < N; jj++)
        rm += norm(_matrix[ii][jj]);
    return rm/(N);
}//
// - - - - TBMatrix '+='- - - - - - - - -
TBMatrix& TBMatrix::operator+=(const TBMatrix &B)
{
    int N = B._matrSize;
    for(int ii = 0; ii < N; ii++)
     for(int jj = 0; jj < N; jj++)
        _matrix[ii][jj] += B._matrix[ii][jj];
    return *this;
}//
// - - - - - TBMatrix '-=' - - - - - - - -
TBMatrix& TBMatrix::operator-=(const TBMatrix &B)
{
    int N = B._matrSize;
    for(int ii = 0; ii < N; ii++)
     for(int jj = 0; jj < N; jj++)
        _matrix[ii][jj] -=B._matrix[ii][jj];
```

```
    return *this;
}// '-='
// - - - - - TBMatrix '* =' - - - - - - - -
TBMatrix& TBMatrix::operator * = (const TBMatrix &B)
{
    int N = B._matrSize;
    TBMatrix tmp(N);
    for(int ii = 0; ii < N; ii++)
     for(int jj = 0; jj < N; jj++)
    {
        TcompD s = 0;
        for(int kk = 0; kk < N; kk++)
          s += _matrix[ii][kk] * B._matrix[kk][jj];
        tmp[ii][jj] = s;
    }
    for(int ii = 0; ii < N; ii++)
     for(int jj = 0; jj < N; jj++)
        _matrix[ii][jj] = tmp[ii][jj];
    return *this;
}//'* ='
// - - - - TBMatrix '+' - - - - - - - - -
TBMatrix operator+(const TBMatrix &lhs,const TBMatrix &rhs)
{
    TBMatrix tmp(lhs);
    tmp = lhs;
    return tmp += rhs;
}//'+'
// - - - - TBMatrix '-' - - - - - - - - -
TBMatrix operator - (const TBMatrix &lhs,const TBMatrix &rhs)
{
    TBMatrix tmp(lhs);
    tmp = lhs;
    return tmp -= rhs;
}//'-'
// - - - - - TBMatrix '*' - - - - - - - -
TBMatrix operator*(const TBMatrix &lhs,const TBMatrix &rhs)
{
    TBMatrix tmp(lhs);
    tmp = lhs;
```

```
    return tmp * = rhs;
}//'*'
// - - - - - TBMatrix - - - - - - - -
TBMatrix::TBMatrix(const TBMatrix &B)//copy constructor
{
    int N = B._matrSize;
    _matrix = new TBVector[N];      //allocate
    _matrSize = N;
    for(int jj = 0; jj < N; jj++)      //new storage
        _matrix[jj].allot(N);
    for(int ii = 0; ii < N; ii++)      //copy the TBMatrix
     for(int jj = 0; jj < N; jj++)
        _matrix[ii][jj] = B._matrix[ii][jj];
}//
// - - - - - - TBMatrix '=' - - - - - - -
TBMatrix& TBMatrix::operator = (const TBMatrix &B)//operator =
{
    if(this ! = &B)
    {
        if(_matrix! = 0) delete [ ]_matrix; //delete storage
        TBMatrix tmp(B);
        int N =       tmp._matrSize;
        _matrix = new TBVector[N];      //allocate
        for(int jj = 0; jj < N; jj++) //new
        _matrix[jj].allot(N);      //storage
        for(int ii = 0; ii < N; ii++)
          for(int jj = 0; jj < N; jj++)      //copy the TBMatrix
            _matrix[ii][jj] = B._matrix[ii][jj];
    }
    return *this;
}//'='
// - - - - - TBVector '*' - - - - - - - -
TBVector operator*(const TBMatrix AA, const TBVector &B)
{
    int N = B.getVector_size( );
    TBVector C(B);
    TBMatrix AA1(AA);
    for (int ii = 0; ii < N; ii++)
        C[ii] = AA1[ii]*B;
```

```
        return C;
    }//
    // - - - - - - MatrPow - - - - - - -
    TBMatrix MatrPow(const TBMatrix &rhs, int N1)
    {
        TBMatrix tmp(rhs);
        tmp = rhs;
        for (int ii = 1; ii < N1; ii++)
            tmp *= rhs;
        return tmp ;
    }// MatrPow
    // - - - - read- - - - - - -
    void TBMatrix::read( )
    {
        cout << "input " << _matrSize << " squared values for TBMatrix";
        for(int ii = 0; ii < _matrSize; ii++)
            for(int jj = 0; jj < _matrSize; jj++)
                cin >> _matrix[ii][jj];
    }//read
    // - - - - - - print - - - - - - -
    void TBMatrix::print( )
    {
        cout << "TBMatrix:" << endl;
        for(int ii = 0; ii < _matrSize; ii++)
        {
            for(int jj = 0; jj < _matrSize; jj++)
                { TcompD x = _matrix[ii][jj]; //cout << _matrix[ii][jj] ;
            cout<< "(" << MyRound3(x.real( )) << "," <<
MyRound3(x.imag( )) << ")";
        }
        cout << " " << endl;
    }
    }//print
    // - - - - - - toString - - - - - - -
    String TBMatrix::toString( )
    {
        String ss = _title+"("+IntToStr(_matrSize)+"x"+
IntToStr(_matrSize)+")"+":\r\n";
```

```
    for(int ii = 0; ii < _matrSize; ii++)
    {
        for(int jj = 0; jj < _matrSize; jj++)
        {     TcompD x = _matrix[ii][jj];
            ss += _title + "["+IntToStr(ii)+","+IntToStr(jj)+"] = "
                +CompToStr(x)+ "; ";
        }
    ss += "\r\n" ;
    }
    return ss;
}//toString
// - - - - - - TBMatrix InvertBMatrix - - - - - - -
TBMatrix InvertBMatrix( TBMatrix& Data1, int& Error)
{
    int isz = Data1.get_matrSize( );
    TBMatrix Inv(isz);
    if (isz == 2)
    {
        TBMatrix A1(Data1);
        TcompD dt = A1[0][0]*A1[1][1] - A1[0][1]*A1[1][0];
        Inv[0][0] = A1[1][1]/dt;
        Inv[0][1] =- A1[0][1]/dt;
        Inv[1][0] =- A1[1][0]/dt;
        Inv[1][1] = A1[0][0]/dt;
        if ( abs(dt) < TNNearlyZero)
            Error = 2; // Singular matrix
    }
    else
        InvertBMatrixLoc( Data1, Inv, Error);
    if (Error > 0)
    throw"inverce bad";
    return Inv;
}//InvertBMatrix
// - - - - - - DeterminantBMatrix - - - - - - -
TcompD DeterminantBMatrix( TBMatrix& Data1)
{
    int isz = Data1.get_matrSize( );
    TcompD dt;
    if (isz == 2)
```

```
        dt = Data1[0][0]*Data1[1][1] - Data1[0][1]*Data1[1][0];
    else
    dt = DetermBMatrixLoc(Data1);
    return dt;
}//DeterminantBMatrix
// - - - ~TBVector- -
TBVector::TBVector(const TBVector &v) //copy constructor
{
    _vecSize = v._vecSize;
    _vect = new TcompD[_vecSize];
    for(int ii = 0; ii < _vecSize; ii++)
        _vect[ii] = v._vect[ii];
}//
// - - - TBVector + - -
TBVector TBVector::operator+ (const TBVector &u) //overloaded +
{
    int n = u._vecSize;
    TBVector v(n);
    for(int ii = 0; ii < n; ii++)
        v._vect[ii] = _vect[ii] + u._vect[ii];
    return v;
}//
// - - - TBVector - - -
TBVector TBVector::operator- (const TBVector &u) //overloaded -
{
    int n = u._vecSize;
    TBVector v(n);
    for(int ii = 0; ii < n; ii++)
        v._vect[ii] = _vect[ii] - u._vect[ii];
    return v;
}//
// - - - TBVector * - -
TcompD TBVector::operator* (const TBVector &u) //overloaded *
{
    int n = u._vecSize;
    TBVector v(n);
    TcompD s = 0.;
    for(int ii = 0; ii < n; ii++)
        s +=_vect[ii] * u._vect[ii];
```

```
return s;
}//
// - - - TBVector [ ] - -
TcompD& TBVector::operator[ ](int ii) //[ ] definition
{
    return _vect[ii];
}//
// - - - read - -
void TBVector::read( )
{
    for(int ii = 0; ii < _vecSize; ii++)
    {
        cout << "array element ?";
        cin >> _vect[ii];
    }
}//
// - - - - get_norm- - - - - - - - -
RealType TBVector::get_norm( )
{
    int N = _vecSize;
    RealType rm = 0.;
    for(int ii = 0; ii < N; ii++)
        rm += norm(_vect[ii]);
    return rm/(N);
}//
// - - - print - -
void TBVector::print( )
{
    cout << "Vector:" << endl;
    for(int ii = 0; ii < _vecSize; ii++)
    cout << "element" << (ii + 1) << "="<< _vect[ii] << endl;
}//
// - - - toString - -
String TBVector::toString( )
{    String ss = "Vector: "+ _title+ "("+IntToStr(_vecSize)+")"+":
\r\n";
    for(int ii = 0; ii < _vecSize; ii++)
        ss += _title + "["+IntToStr(ii + 1) + "] = " + Comp-
ToStr(_vect[ii]) + "\r\n";
```

```
        return ss;
    }//
    // - - - - allot - - - -
    void TBVector::allot(int n)
    {
        if(_vect ! = 0) delete []_vect;
        _vecSize = n;
        _vect = new TcompD[n];
    }//
    // - - - - - TBVector = - - - - - - - -
    TBVector& TBVector::operator = (const TBVector &v)//operator =
    {
        if(this ! = &v)
        {
            if(_vect ! = 0) delete [ ]_vect;//delete storage
            _vecSize = v._vecSize;
            _vect = new TcompD [_vecSize];          //allocate
            for(int ii = 0; ii < _vecSize; ii++)    //new
                _vect[ii] = v._vect[ii];            //storage
        }
        return *this;
    }//
    // - - - - - - - - - Initialization- - - - - - - - - - - -
    void Initial(TBMatrix& Data, TBMatrix& Inv, int & Error )
    //- Input Dimen, Data
    //- Output Inv, Error
    //- This procedure tests for errors in the value of Dimen
    {
        int Dimen = Data.get_matrSize( );
        Error = 0;
        if (Dimen < 1)
            Error = 1;
        else
        {
        // First make the inverse-to-be the identity matrix
        for ( int ii = 0; ii<Dimen; ii++)
            for ( int jj = 0; jj<Dimen; jj++)
                if (ii == jj)
                    Inv[ii][jj] = 1.;
```

```
            else
                Inv[ii][jj] = 0.;
            if (Dimen == 1)
                if ( abs(Data[0][0]) < TNNearlyZero)
                    Error = 2; // Singular matrix
                else
                {   TcompD cr0 = 1.;
                    Inv[0][0] = cr0/Data[0][0];
                }
    }
} // void Initial
// - - - - - - - - - EROdiv1- - - - - - - - - - - - -
void EROdiv1(TcompD Divisor, TBMatrix& Data, int Row )
//- Input Divisor, Dimen, Row
//-
//- elementary row operation - dividing by a constant
{       int Dimen = Data.get_matrSize( );
     for ( int Term = 0; Term<Dimen; Term++)
         Data[Row][Term] = Data[Row][Term] / Divisor;
} // void EROdiv
// - - - - - - - - EROswitch1- - - - - - - - - - - - - -
void EROswitch1(TBMatrix & Data, /*Row1,Row2 )*/
int ii1, int ii2)
//- Input Row1, Row2
//- Output Row1, Row2
//- Elementary row operation - switching two rows
{
     int c4 = Data.get_matrSize( );
     TBVector DummyRow(c4);
     for(int jj = 0; jj<c4; jj++)
          for(int ii = 0; ii<c4; ii++)
          {
               DummyRow[ii] = Data[ii][ii1];
               Data[ii][ii1] = Data[ii][ii2];
               Data[ii][ii2] = DummyRow[ii];
          }
} // void EROswitch
// - - - - - - - - - EROmultAdd1 - - - - - - - - - - - - -
void EROmultAdd1(TcompD Multiplier, TBMatrix & Data,
```

```
    int& ReferenceRow, int& ChangingRow)
    //- Input Multiplier, Dimen, ReferenceRow, ChangingRow
    //- Output ChangingRow
    //-
    //- Row operation - adding a multiple one row to another
    {
        int Dimen = Data.get_matrSize( );
        for ( int Term = 0 ; Term<Dimen; Term++)
            Data[ChangingRow][Term] = Data[ChangingRow][Term] +
        Multiplier*Data[ReferenceRow][Term];
    } // void EROmultAdd
    // - - - - - - - - - -Pivot - - - - - - - - - - - -
    void Pivot(int ReferenceRow, TBMatrix& Data, TBMatrix& Inv, int&
Error )
    //- Input: Dimen, ReferenceRow, Coefficients,
    //- Lower, Upper, Permute -}
    //- Output: Coefficients, Lower, Permute, Error
    //- -
    //- This procedure searches the ReferenceRow column of the
    //- Coefficients matrix for the element in the Row below the
    //- main diagonal which produces the largest value of
    //- -
    //- Coefficients[Row, ReferenceRow] -
    //- -
    //- Sum K = 1 to ReferenceRow - 1 of
    //- Upper[Row, k] - Lower[k, ReferenceRow]
    //- -
    //- If it finds one, then the procedure switches
    //- rows so that this element is on the main diagonal. The
    //- procedure also switches the corresponding elements in the
    //- Permute matrix and the Lower matrix. If the largest value of
    //- the above expression is zero, then the matrix is singular
    //- and no solution exists (Error = 2 is returned).
    {
        int NewRow;
        int Dimen = Data.get_matrSize( );
        Error = 2; // No inverse exists
        NewRow = ReferenceRow;
        while ((Error > 0)&&(NewRow < Dimen-1))
```

```
// Try to find a row with a non-zero diagonal element
      {
            NewRow = NewRow++;
            if (abs(Data[NewRow][ReferenceRow]) > TNNearlyZero)
            {
                  EROswitch1(Data,NewRow,ReferenceRow);
                  // Switch these two rows
                  EROswitch1(Inv,NewRow,ReferenceRow);
                  Error = 0;
            }
      } // while
} // void Pivot
// - - - - - - - - - - InvertBMatrixLoc - - - - - - - - - - - -
void InvertBMatrixLoc( TBMatrix& Data1, TBMatrix& Inv, int&
Error)
//- Input: Dimen, Data
//- Output: Inv, Error
//-
//- This procedure computes the inverse of the matrix Data
//- and stores it in the matrix Inv. If the matrix Data
//- is singular, then Error = 2 is returned.
{
      TcompD Divisor, Multiplier;
      int ReferenceRow;
      int Dimen = Data1.get_matrSize( );
      TcompD GG = DeterminantBMatrix( Data1);
      Error = 0;
      if (abs(GG) == 0) {Error = 2; return; } //singular matrix
      TBMatrix Data(Data1);
      for (int ii = 0; ii<Dimen; ii++ )
            for (int jj = 0; jj<Dimen; jj++ )
      Data[ii][jj] = Data1[ii][jj];
      Initial(Data, Inv, Error);
// Make Data matrix upper triangular
      ReferenceRow = -1/*0*/;
      while ( (Error == 0)&&(ReferenceRow < Dimen-1))
      {
            ReferenceRow = ReferenceRow++;
// Check to see if the diagonal element is zero
            if (abs(Data[ReferenceRow][ReferenceRow]) < TNNearlyZero)
```

```
            Pivot(ReferenceRow, Data, Inv, Error);
            if (Error == 0)
            {
                  Divisor = Data[ReferenceRow][ReferenceRow];
                  EROdiv1(Divisor, Data,ReferenceRow);
                  EROdiv1(Divisor, Inv,ReferenceRow);
                  for ( int Row = 0 ; Row<Dimen; Row++)
    // Make the ReferenceRow element of this row zero
                        if ((Row ! = ReferenceRow) &&
                              (abs(Data[Row][ReferenceRow])) >
TNNearlyZero)
                        {
                              Multiplier = -Data[Row][ReferenceRow] /
                              Data[ReferenceRow][ReferenceRow];
                              EROmultAdd1(Multiplier, Data,ReferenceRow,
Row);
                              EROmultAdd1(Multiplier,   Inv,ReferenceRow,
Row);
                        }
            }
      }
   } // void InvertBMatrixLoc
   // - - - - - - - - - - - Pivot2 - - - - - - - - - - -
   void Pivot2(int ReferenceRow, TBMatrix& Data,
TcompD& PartialDeter, bool& DetEqualsZero)
   //- Input: Dimen, ReferenceRow, Data, PartialDeter
   //- Output: Data, PartialDeter, DetEqualsZero
   //-
   //- This procedure searches the ReferenceRow column of the
   //- matrix Data for the first non-zero element below the
   //- diagonal. If it finds one, then the procedure switches
   //- rows so that the non-zero element is on the diagonal.
   //- Switching rows changes the determinant by a factor of
   //- -1; this change is returned in PartialDeter.
   //- If it doesn't find one, the matrix is singular and the
   //- Determinant is zero (DetEqualsZero = true is returned).
   {
      int NewRow;
      int Dimen = Data.get_matrSize( );
      DetEqualsZero = true;
```

```
        NewRow = ReferenceRow;
        while (DetEqualsZero && (NewRow < Dimen-1))// Try to find a
row
    // with a non-zero
    // element in this column
        {
            NewRow++; //NewRow = Succ(NewRow);
            if (abs(Data[NewRow][ReferenceRow]) > TNNearlyZero)
            {
                EROswitch1(Data,NewRow, ReferenceRow);
    // Switch these two rows
                DetEqualsZero = false;
                PartialDeter = -PartialDeter; // Switching rows changes
    // the determinant by a factor −1
            }
        }
    } // void Pivot2
    // - - - - - - - - - - DetermBMatrix - - - - - - - - - - - -
    TcompD DetermBMatrixLoc(TBMatrix& Data1)
    //- Input: Dimen, Data
    //- Output: DetermBMatrix
    //- Function returns the determinant of the Data matrix
    {
        const int Dimen = Data1.get_matrSize( );
        TcompD PartialDeter, Multiplier;
        int ReferenceRow ;
        bool DetEqualsZero ;
        TBMatrix Data(Dimen);
        for (int ii = 0; ii<Dimen; ii++ )
            for (int jj = 0; jj<Dimen; jj++ )
                Data[ii][jj] = Data1[ii][jj];
        DetEqualsZero = false;
        PartialDeter = 1;
        ReferenceRow = −1;//0;
    // Make the matrix upper triangular
        while (!(DetEqualsZero)&&(ReferenceRow < Dimen −1))
        {
            ReferenceRow++;
```

```
// If diagonal element is zero then switch rows
        if (abs(Data[ReferenceRow][ReferenceRow]) < TNNearlyZero)
        Pivot2(ReferenceRow, Data, PartialDeter, DetEqualsZero);
        TcompD tmp00 = Data[0][0],
        tmp01 = Data[0][1],
        tmp10 = Data[1][0],
        tmp11 = Data[1][1];
        if (!DetEqualsZero)
        for (int Row=ReferenceRow+1; Row</*=*/Dimen; Row++)
// Make the ReferenceRow element of this row zero
        if (abs(Data[Row][ReferenceRow]) > TNNearlyZero)
        {
            Multiplier = -Data[Row][ReferenceRow] /
            Data[ReferenceRow][ReferenceRow];
            EROmultAdd1(Multiplier, Data, ReferenceRow,Row);
        }
// Multiply the diagonal Term into PartialDeter
PartialDeter = PartialDeter * Data[ReferenceRow][ReferenceRow];
    }
    if (DetEqualsZero)
return 0.;
    else
return PartialDeter;
} // DetermBMatrix
// - - - - - - - - - - DetermBMatrix - - - - - - - - - - - -
TcompD TBMatrix::determinant( )
//- Input: Dimen, Data
//- Output: DetermBMatrix
//- Function returns the determinant of the Data matrix
{
    int Dimen = get_matrSize( );
    TBMatrix Data(Dimen);
    TcompD Res;
    if (Dimen == 2)
        Res = (_matrix[0][0]*_matrix[1][1] –
_matrix[1][0]*_matrix[0][1]);
    else
    {
        for (int ii = 0; ii<Dimen; ii++ )
```

```
                for (int jj = 0; jj<Dimen; jj++ )
                    Data[ii][jj] = _matrix[ii][jj];
            Res = DeterminantBMatrix(Data);
        }
    //if (get_isCheckedInvOfMatr( ))
        {
        if (abs(Res)<1.e-7)
            {   String s = "determinat <1.e-7.";//, err =
"+FloatToStr(r);
                throw " <- determinat < 1.e - 7.";//s.c_str( );
        }
    }
    return Res;
    } // DetermBMatrix
    // - - - - - - - - - - inverse - - - - - - - - - - - -
    TBMatrix TBMatrix::inverse(int& Error)
    //- Function returns the inverce of the Data matrix
    {
        int Dimen = get_matrSize( );
        TBMatrix Data(Dimen);
        TBMatrix Inv(Dimen);
        for (int ii = 0; ii<Dimen; ii++ )
         for (int jj = 0; jj<Dimen; jj++ )
            Data[ii][jj] = _matrix[ii][jj];
        //Data[ii][jj] = get_Matrix( )[ii][jj];
        Inv = InvertBMatrix(Data, Error);
        // if (get_isCheckedInvOfMatr( ))
        {
        TBMatrix FF = Data * Inv;
        RealType r = FF.get_norm( );
        if (fabs(r - 1)>1.e - 7)
            { //String s = " <- Inverse matrix is bad.";//, err =
"+FloatToStr(r);
                throw " <- Inverse matrix is bad.";//s.c_str( );
            }
        }
    return Inv;
    } // inverse
```

CHAPTER 21

Spontaneous Emission of Atom in Microsphere

21.1 Introduction

In this chapter we consider the program for the calculation of the temporal dynamics of spontaneous emission in metallized microsphere. The physics of this problem is explained in Secs. 12 and 13. In Fig. 21.1 the structure of classes is shown. The basic class is *TQuantField* in which the description of an electromagnetic field as a quantum object is concentrated. This class already has been explained above. According to physics the interaction of field and atom gives rise to an atoms transition. Therefore class *TQuantField* has two derivative classes *TElMagnField* and *TDipolAtom*. Two objects representatives of these classes are placed in the container class *TCoatedMicrosphere*. They are *TElMagnField *_elMagnField* and *TDipolAtom *_dipolAtom*. For the class *TCoatedMicrosphere* an object is used for modeling in the basic program. There are also various auxiliary mathematical classes, which contain the evaluation of Bessel functions, as well as the solving of the complex differential equations for the probability amplitudes and other necessary program codes (see the program codes below).

Before the work begins all objects like field, atom and coated sphere should be created and initialized. Here the developer acts as a supreme force. In Fig. 21.2 the graphical user interface, which was created in Borland C++ Builder (bcb), is shown. In the figure one can write in the TabSheet the parameters most frequently used while calculating. They are the principal ν and spherical m quantum numbers, the ratio of the frequency of

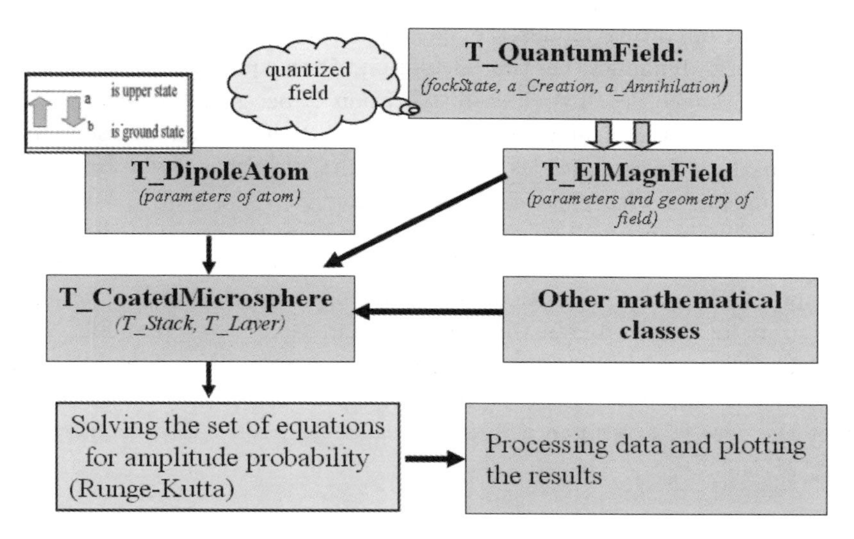

Fig. 21.1 The structure of classes for the calculation of spontaneous emission of two-level atom in microsphere.

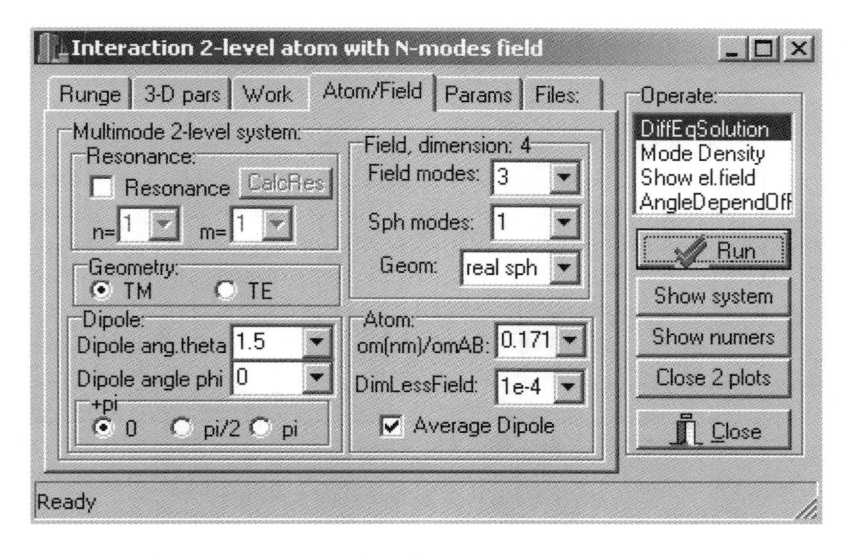

Fig. 21.2 Graphical user interface for the spontaneous emission problem.

a field to a frequency of atoms transition ω/ω_{ab}, the parameter of field, TE or TM modes. These are examined to find out whether are they resonance of the atom transition frequency with some fields mode, or are atomic dipole angles. Other parameters can be changed from other TabSheets. The list of operating modes are specified. The first mode is to research the temporal dynamics, i.e. the calculation of the equations for amplitudes probabilities (see Sec. 13). The initialization of necessary objects appears after clicks on the Run button (see Fig. 21.2) and then the solving of the equations for amplitudes begins. Results of such calculation appear in pop-up forms with plots. In Figs. 21.3 and 21.4 the typical example of such window is shown. The dynamics of the expected value of the excited state of atom, and the fast oscillating curve shows the dynamic polarization of the atom transition. Thus the norm of the wave function of the atom/field system remains constant and equal to 1 (see Fig. 21.3). On the following graphic form the frequency spectrum of dynamic polarization and its autocorrelation function is shown (see Fig. 21.4). In this plot the case of a resonance is shown, thus the Rabi splitting arises (see Fig. 21.4).

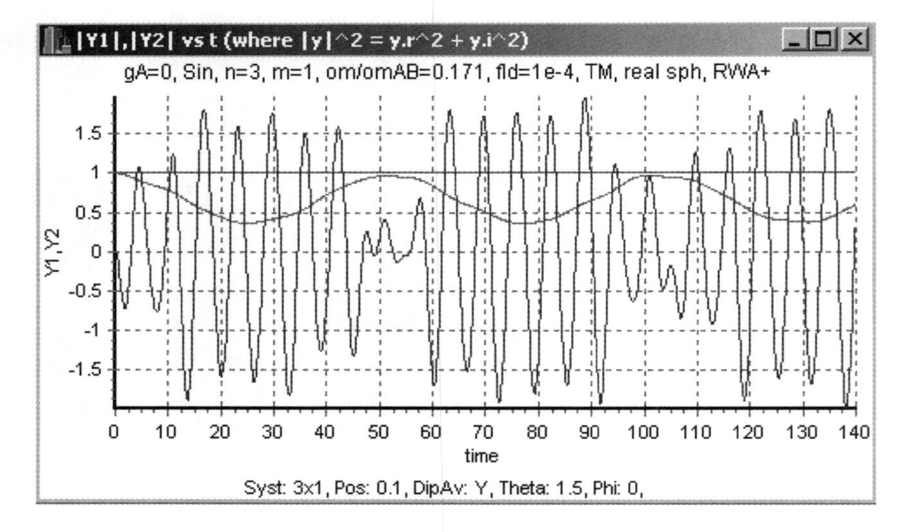

Fig. 21.3 The temporal dynamics of atomic polarization of two-level atoms.

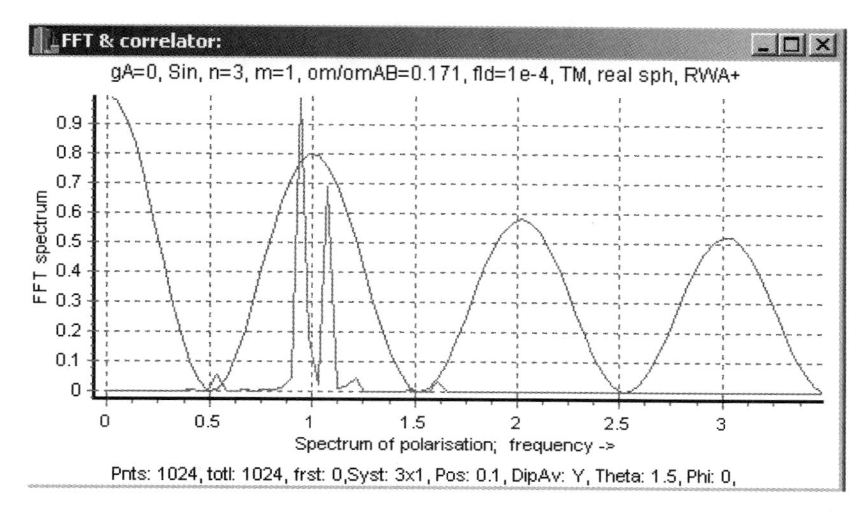

Fig. 21.4 The frequency spectrum of dynamic atomic polarization and its temporal autocorrelation function.

21.2 Code

The program code of this problem can be found on the website, indicated in the introduction to this book.

21.3 Classes

```
typedef std::vector<int> TFockState;
    // - - - - - - - class TQuantField - - - - - - - - - - - -
    class TQuantField
    {
    private:
    TFockState _fockState;
    int _sizeFockState;
    protected:
    public:
    TQuantField( )
    { _sizeFockState = 5;
    _fockState.resize(5);
    for(int ii = 0; ii<_sizeFockState; ii++)
    _fockState[ii] = 0;
    }
```

```
//
double aCreation(int nState)
{
_fockState[nState]++;
return sqrt(_fockState[nState]);
}
//
double aAnnigilation(int nState)
{
if (_fockState[nState] > 0)
_fockState[nState]--;
return sqrt(_fockState[nState]);
}
//
TFockState get_fockState( )
{
return _fockState;
}
//
int get_sizeFockState( )
{
return _fockState.size( );
}
//
void set_sizeFockState(int n)
{
int oldN = _fockState.size( );
_fockState.resize(n);
for(int ii = oldN-1; ii<_fockState.size( ); ii++)
_fockState[ii] = 0;
}
//
virtual String toString( )
{
String s = "{";
for (int ii = 0; ii<get_sizeFockState( ); ii++)
s += IntToStr(get_fockState( )[ii]) + ", ";
return s+"}";
}
```

```
String printPhotonsState( )
{
String s = "{";
for (int ii = 0; ii<get_sizeFockState( ); ii++)
s + = IntToStr(get_fockState( )[ii]) + ", ";
return s+"}";
}
};//class TQuantField
// - - - - - - - class TElMagnField - - - - - - - - - - - -
class TElMagnField: public TQuantField
{
private:
bool _isTMcaseField;
bool _isInhomField;
int _indField; // sin or cos(wt) ?
int _typeGeometryOfCavity;
int _MaxPrincNmbField;
int _MaxSphNmbField;
int _MinPrincNmbField;
int _MinSphNmbField;
String _sInit;
RealType _PulseWidth;
RealType _freqRatio;
RealType _parametrMie;
RealType _freqMin;
RealType _freqMax;
RealType _par1;
RealType _par2;
RealType **_arrayOfFieldRoots; // array of zeros of eigen equation for
field
RealType **_arrayOfSpaceField; // array of el.field in an atom (dipol)
point
RealType **_arrayOfAmplVacuumField; // array of Amplitude of
Vacuum Field
//
public:
TElMagnField( )
{
globalSt::addTo_stringToExchange("TElMagnField -> is
created.");
```

```
    _isTMcaseField = _isInhomField = false;
    _indField = _typeGeometryOfCavity = - 1;
    _MaxPrincNmbField = _MaxSphNmbField = - 1;
    _PulseWidth = _freqRatio = - 1;
    _parametrMie = _freqMin = _freqMax = - 1;
    _sInit = "TElMagnField is initialized";
    }
    virtual ~TElMagnField( )
    {
    globalSt::addTo_stringToExchange("TElMagnField -> is deleted.");
    }
    String get_sInit( ){return _sInit;}
    RealType get_arrayOfFieldRoots(int i, int j)
{return _arrayOfFieldRoots[i][j];}
    // _arrayOfFieldRoots
    void set_arrayOfFieldRoots(RealType x, int i, int j)
{ _arrayOfFieldRoots[i][j] = x;}
    void create_arrayOfFieldRoots( ){_arrayOfFieldRoots =
dmatrix(_MinSphNmbField, _MaxSphNmbField,
_MinPrincNmbField, _MaxPrincNmbField);}
    void free_arrayOfFieldRoots( ){ free_dmatrix(_arrayOfFieldRoots,
_MinSphNmbField, _MaxSphNmbField, _MinPrincNmbField,
_MaxPrincNmbField);}
    // _arrayOfAmplVacuumField
    RealType calcAmplVacuumField(int i, int j);
    RealType get_arrayOfAmplVacuumField(int i, int j){return
_arrayOfAmplVacuumField[i][j];}
    void set_arrayOfAmplVacuumField( )
    {
    int MaxPrincNmb = get_MaxPrincNmbField( );
    int MaxSphNmb = get_MaxSphNmbField( );
    int MinSphNmb = get_MinSphNmbField( );
    int MinPrincNmb = get_MinPrincNmbField( );
    for(int spherNmb = MinSphNmb; spherNmb<=MaxSphNmb;
spherNmb++)// over spherical numbers m
    for(int numberMode = MinPrincNmb; numberMode<=MaxPrincNmb;
numberMode++)// over all radial modes on m = const
    _arrayOfAmplVacuumField[spherNmb][numberMode] = calcAmplVac-
uumField(spherNmb,numberMode);
    }//
```

```
void create_arrayOfAmplVacuumField(){_arrayOfAmplVacuumField =
  dmatrix(_MinSphNmbField, _MaxSphNmbField, _MinPrincNmbField,
_MaxPrincNmbField);}
   void free_arrayOfAmplVacuumField(){
free_dmatrix(_arrayOfAmplVacuumField, _MinSphNmbField,
_MaxSphNmbField, _MinPrincNmbField, _MaxPrincNmbField);}
   // _arrayOfSpaceField
   RealType get_arrayOfSpaceField(int i,
int j){return _arrayOfSpaceField[i][j];}
   void create_arrayOfSpaceField(){_arrayOfSpaceField = dmatrix
(_MinSphNmbField, _MaxSphNmbField, _MinPrincNmbField,
_MaxPrincNmbField);}
   void free_arrayOfSpaceField(){
free_dmatrix(_arrayOfSpaceField, _MinSphNmbField, _MaxSphNmbField,
_MinPrincNmbField, _MaxPrincNmbField);}
   void set_arrayOfSpaceField(RealType x, int i,
int j){ _arrayOfSpaceField[i][j] = x;}
   //
   bool get_isTMcaseField(){return _isTMcaseField;}
   void set_isTMcaseField(bool s){ _isTMcaseField = s;}
   int get_MaxPrincNmbField(){return _MaxPrincNmbField;}
   void set_MaxPrincNmbField(int i){ _MaxPrincNmbField = i;}
   int get_MaxSphNmbField(){return _MaxSphNmbField;}
   void set_MaxSphNmbField(int i){ _MaxSphNmbField = i;}
   int get_MinPrincNmbField(){return _MinPrincNmbField;}
   void set_MinPrincNmbField(int i){ _MinPrincNmbField = i;}
   int get_MinSphNmbField(){return _MinSphNmbField;}
   void set_MinSphNmbField(int i){ _MinSphNmbField = i;}
   int get_indField(){return _indField;}
   void set_indField(int i){ _indField = i;}
   int get_typeGeometryOfCavity(){return _typeGeometryOfCavity;}
   void set_typeGeometryOfCavity(int i){ _typeGeometryOfCavity = i;}
   RealType get_PulseWidth(){return _PulseWidth;}
   void set_PulseWidth(RealType b){ _PulseWidth = b;}
   RealType get_freqMin(){return _freqMin;}
   void set_freqMin(RealType b){ _freqMin = b;}
   RealType get_freqMax(){return _freqMax;}
   void set_freqMax(RealType b){ _freqMax = b;}
```

```
RealType get_freqRatio(){return _freqRatio;}
void set_freqRatio(RealType b){ _freqRatio = b;}
RealType get_par1(){return _par1;}
void set_par1(RealType b){ _par1 = b;}
RealType get_par2(){return _par2;}
void set_par2(RealType b){ _par2 = b;}
RealType get_parametrMie(){return _parametrMie;}
void set_parametrMie(RealType b){ _parametrMie = b;}
bool get_isInhomField(){return _isInhomField;}
void set_isInhomField(bool b){ _isInhomField = b;}
//
TcompD FieldFunction(RealType W, RealType t, int ind);
//
RealType CalcNormalEigenFrequencyOfMode(int numberMode, int
spherNmb, int &iNum);
RealType AverageDipoleField4(RealType posit, int m, int n,RealType
dipoleAngleTheta, RealType dipoleAnglePhi);
RealType ElectrFieldOfHollowSphere3(RealType x, RealType mpar1,
RealType par2);
//
int CalcDegeneracyOfEigValue(int spherNmb, int numberMode);
//
void CalcEigFreqOfHollowSphere();
void SetFreqMaxMin();
//
void setParamsOfField(int MMax, int NMax, int MMin, int NMin)
{
set_MaxPrincNmbField(NMax);
set_MaxSphNmbField(MMax);
set_MinPrincNmbField(NMin);
set_MinSphNmbField(MMin);
create_arrayOfFieldRoots();
create_arrayOfSpaceField();
create_arrayOfAmplVacuumField();
}//
//
virtual String toString()
{
String ends = "\r\n";
```

```
String out = _sInit
+ ends+"_pulseWidth: " + FloatToStrF(get_PulseWidth( ), ffGeneral,
3,3)
+ ends+"_freqRatio: " + FloatToStrF(get_freqRatio( ), ffGeneral, 3,3)
// + "\r\n_parametrMie: " + FloatToStrF(get_parametrMie( ),
ffGeneral, 3,3)
+ ends+"_freqMin: " + FloatToStrF(get_freqMin( ), ffGeneral, 3,3)
+ ends+"_freqMax: " + FloatToStrF(get_freqMax( ), ffGeneral, 3,3)
+ ends+"_isTMcaseField: " + IntToStr(get_isTMcaseField( ))
+ ends+"_indField: " + IntToStr(get_indField( ))
+ ends+"_maxPrincNmbField: " +
IntToStr(get_MaxPrincNmbField( ))
+ ends+"_maxSphNmbField: " + IntToStr(get_MaxSphNmbField( ))
+ ends+"_minPrincNmbField: " + IntToStr(get_MinPrincNmbField( ))
+ ends+"_minSphNmbField: " + IntToStr(get_MinSphNmbField( ))
+ ends+"_typeGeomOfCavity: " +
IntToStr(get_typeGeometryOfCavity( ))
+ ends+"_isInhomField: " + IntToStr(get_isInhomField( ))
+ ends+"_par1: " + FloatToStrF(get_par1( ), ffGeneral, 3,3)
+ ends+"_par2: " + FloatToStrF(get_par2( ), ffGeneral, 3,3)
+ ends+"_eigFreqs:\r\n" ;
RealType freqRatio = get_freqRatio( ); // <- omegaField/omegaAtom
int MaxPrincNmb = get_MaxPrincNmbField( );
int MaxSphNmb = get_MaxSphNmbField( );
int MinSphNmb = get_MinSphNmbField( );
int MinPrincNmb = get_MinPrincNmbField( );
for(int spherNmb = MinSphNmb; spherNmb<=MaxSphNmb;
spherNmb++)// over spherical numbers m
{ out += " _m = "+IntToStr(spherNmb) + "->";
for(int numberMode = MinPrincNmb; numberMode <=
MaxPrincNmb; numberMode++)// over all radial modes on m = const
{
if (get_isInhomField( ))
{
RealType root = get_arrayOfFieldRoots(spherNmb,numberMode);
out += FloatToStrF(freqRatio*root, ffFixed,3,3)+ ", ";
if (globalSt::isShowVainshtein)
{
RealType vain = 0;
```

```
if (get_isTMcaseField()) // Formula Vainshtein
vain = (spherNmb + (2*numberMode-1))*M_PI/2; //TM
else //TE
vain = (spherNmb + (2*numberMode-0))*M_PI/2;
//out += "["+FloatToStrF(fabs(freqRatio*(root-vain)/root),
ffExponent,1,1)+ "], ";
out += "["+FloatToStrF(fabs(freqRatio*(root-vain)/1),
ffExponent,1,1)+ "], ";
}
}
else
{ int itmp;
out += FloatToStrF(CalcNormalEigenFrequencyOfMode
(spherNmb,numberMode, itmp), ffFixed,3,3)+ ", ";
}
}
out += "\r\n";
}
out += "field per photon:\r\n";
for(int spherNmb = MinSphNmb; spherNmb<=MaxSphNmb;
spherNmb++)// over spherical numbers m
{ out +=" _m = "+IntToStr(spherNmb) + "->";
for(int numberMode = MinPrincNmb; numberMode<=MaxPrincNmb;
numberMode++)// over all radial modes on m = const
{
if (get_isInhomField())
out += FloatToStrF(get_arrayOfAmplVacuumField
(spherNmb,numberMode), ffFixed,4,4)+ ", ";
else
{ int itmp;
out += FloatToStrF(CalcNormalEigenFrequencyOfMode
(spherNmb,numberMode, itmp), ffFixed,3,3)+ ", ";
}
}
out +="\r\n";
}
return out + "_photonsState:" + printPhotonsState();
}// toString
}; //class TElMagnField
```

```
// - - - - - - - class TDipolAtom - - - - - - - - - - - -
class TDipolAtom
{
private:
RealType _omAB;
RealType _gammaA;
RealType _vacuumEmissionRate;
RealType _dipolMoment;
RealType _dipoleAngleTheta;
RealType _dipoleAnglePhi;
bool _isAverageDipole;
String _infoModesFreq;
String _sInit;
int _sizeOfHisto;
public:
TDipolAtom( )
{
globalSt::addTo_stringToExchange("TDipolAtom( )-> is created.");
_slst = new TStringList( );
_omAB = _gammaA = _vacuumEmissionRate = _dipolMoment = -1;
_dipoleAngleTheta = _dipoleAnglePhi = -1;
_sInit = "TDipolAtom is initialized";
}
virtual ~TDipolAtom( )
{
globalSt::addTo_stringToExchange("TDipolAtom -> is deleted.");
delete _slst;
}
//
TStringList *_slst;
//
void calc_vacuumEmissionRate(RealType mult, RealType freqRatio,
RealType dimLessField)
{
RealType gam0 = (8.*M_PI/9)*pow(dimLessField,2)
pow(freqRatio,-3);
_vacuumEmissionRate = gam0/mult;
}
```

```
//
bool get_isAverageDipole(){return _isAverageDipole;}
void set_isAverageDipole(bool s){ _isAverageDipole = s;}
int get_sizeOfHisto(){return _sizeOfHisto;}
void set_sizeOfHisto(int i){ _sizeOfHisto = i;}
//
RealType get_vacuumEmissionRate(){return _vacuumEmissionRate;}
RealType get_dipoleAngleTheta(){return _dipoleAngleTheta;}
void set_dipoleAngleTheta(RealType b){ _dipoleAngleTheta = b;}
RealType get_dipoleAnglePhi(){return _dipoleAnglePhi;}
void set_dipoleAnglePhi(RealType b){ _dipoleAnglePhi = b;}
RealType get_dipolMoment(){return _dipolMoment;}
void set_dipolMoment(RealType b){ _dipolMoment = b;}
RealType get_omAB(){return _omAB;}
void set_omAB(RealType b){ _omAB = b;}
RealType get_gammaA(){return _gammaA ;}
void set_gammaA(RealType b){ _gammaA = b;}
//
String get_infoModesFreq(){return _infoModesFreq;}
void inc_infoModesFreq(String s){ _infoModesFreq += s;}
//
void SetOfDiffEqs(RealType t, TcompD XValue[], TcompD XDeriv[],
TcompD params[]);
void calcFreqDistribution(String fNameToSave);
void CalcMinMaxFreq();
void CalcElFieldInAtomPoint();
//
TcompD Polarisation(RealType t, TcompD AmplsProbability[],
TcompD params[]);
//
RealType TotalNorma(RealType t, TcompD AmplsProbability[],
TcompD params[]);
RealType calcFldCoeff(RealType radius, RealType lambdaAB, Real-
Type &emissVelocity);
String get_sInit(){return _sInit;}
```

```
String toString( )
{
String out = _sInit
+"\r\n_OmBA: " + FloatToStrF(get_omAB( ), ffGeneral, 3,3)
+ "\r\n_GamaA: " + FloatToStrF(get_gammaA( ), ffGeneral, 3,3)
+ "\r\n_dipolMoment: " + FloatToStrF(get_dipolMoment( ),
ffGeneral, 3,3)
+ "\r\n_vacuumEmissRate: " +
FloatToStrF(get_vacuumEmissionRate( ), ffGeneral, 3,3)
+ "\r\n_isAverageDipole: " + IntToStr(get_isAverageDipole( ))
+ "\r\n_dipoleAngleTheta: " + FloatToStr(get_dipoleAngleTheta( ))
+ "\r\n_dipoleAnglePhi: " + FloatToStr(get_dipoleAnglePhi( ))
;
return out;
}//
};// class TDipolAtom
// - - - - - - - class TCoatedMicrosphere - - - - - - - - - - - -
class TCoatedMicrosphere
{
private:
TElMagnField *_elMagnField;
TDipolAtom *_dipolAtom;
bool _isRWA;
RealType _rr0PositionAtomInSphere;
RealType _DimLessField;
String _sInit;
protected:
public:
TCoatedMicrosphere( )
{
_elMagnField = new TElMagnField( );
_dipolAtom = new TDipolAtom( );
_sInit = "TCoatedMicrosphere is initialized";
_rr0PositionAtomInSphere = _DimLessField = −1;
}
//
~TCoatedMicrosphere( )
{
get_elMagnField( )->free_arrayOfFieldRoots( );
```

```
get_elMagnField()->free_arrayOfSpaceField();
get_elMagnField()->free_arrayOfAmplVacuumField();
delete _elMagnField;
delete _dipolAtom;
}
//
TElMagnField *get_elMagnField(){return _elMagnField;}
TDipolAtom *get_dipolAtom(){return _dipolAtom;}
String get_sInit(){return _sInit;}
RealType get_DimLessField(){return _DimLessField;}
void set_DimLessField(RealType b){ _DimLessField = b;}
RealType get_rr0PositionAtomInSphere()
{return _rr0PositionAtomInSphere;}
void set_rr0PositionAtomInSphere(RealType b)
{ _rr0PositionAtomInSphere = b;}
bool get_isRWA(){return _isRWA;}
void set_isRWA(bool s){ _isRWA = s;}
void SetOfDiffEqs(RealType t, TcompD XValue[], TcompD XDeriv[],
TcompD params[]);
void CalcMinMaxFreq();
void CalcElFieldInAtomPoint();
void calcFreqDistribution(String fNameToSave);
TcompD Polarisation(RealType t, TcompD AmplsProbability[],
TcompD params[]);
//
RealType TotalNorma(RealType t, TcompD AmplsProbability[],
TcompD params[]);
RealType calcFldCoeff(RealType radius, RealType lambdaAtom,
RealType &emissVelocity);
RealType AverageDipoleFieldScalProduct4(int m, int n);
//
void setParamsOfField(int MMax, int NMax)
{
get_elMagnField()->set_MaxPrincNmbField(NMax);
get_elMagnField()->set_MaxSphNmbField(MMax);
get_elMagnField()->create_arrayOfFieldRoots();
get_elMagnField()->create_arrayOfSpaceField();
}//setParamsOfField
```

```
String toString( )
{
String out = _sInit
+ "\r\n_DimLessField: " + FloatToStrF(get_DimLessField( ),
ffGeneral, 3,3)
+ "\r\n_isRWA: " + IntToStr(get_isRWA( ))
+ "\r\n_position atom: " + FloatToStr
(get_rr0PositionAtomInSphere( )) ;
return out;
}//toString
};// class TCoatedMicrosphere
```

CHAPTER 22

Electromagnetic Oscillations in Layered Microsphere

22.1 Introduction

In this chapter we consider the program for calculation of the electromagnetic properties of the multilayered microsphere. Such a program is more complicate compared to the spontaneous emission program. The physics of this problem is explained in Secs. 1.3, 1.4 and 1.5. In Figs. 22.1, and 22.2 the structure of classes is shown. So far as the program code of this problem

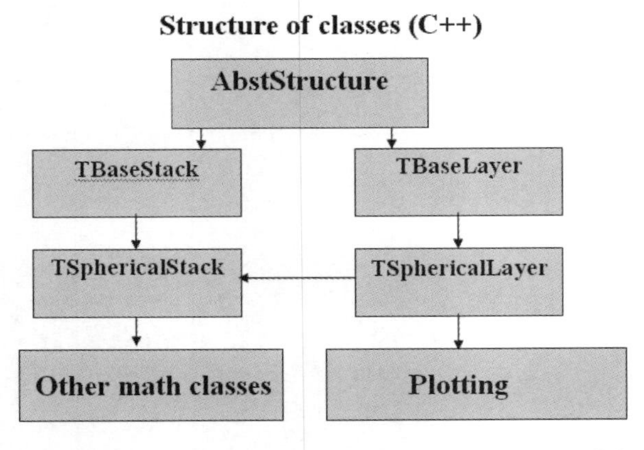

Fig. 22.1 The structure of classes for calculation of properties of electromagnetic eigen oscillations in multilayered microsphere. The general forms.

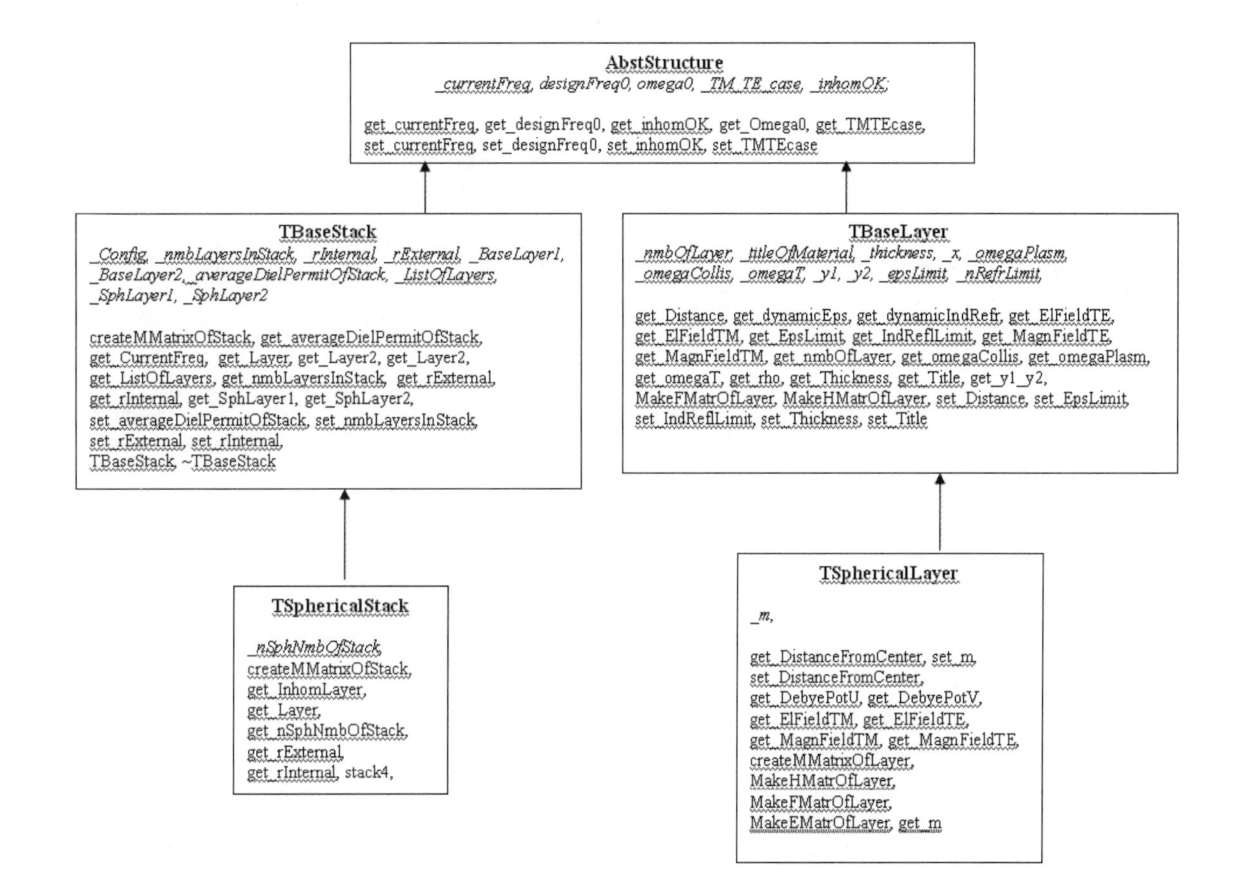

Fig. 22.2 The structure of classes for calculation of properties of electromagnetic eigen oscillations in multilayered microsphere.

is too long, it can be found in the web site, indicated in the Introduction to this book.

In Fig. 22.3 the graphical user interface, which was created in Borland C++ Builder (bcb), is shown. In the TabSheet named "WORK" (see Fig. 22.3) one can put the parameters most frequently used while calculating. They are the number of layers, the structures of layers, the spherical quantum number, number points to plot, *TE* or *TM* modes are examined, etc. Some of the accessible operations are specified. In Figs. 22.4 and 22.5 the typical examples are shown. Figure 22.4 shows the frequency dependence of the reflection and transmittance coefficients, whereas Fig. 22.5 shows the radial distribution of both electric and magnetic optical fields in the microsphere and spherical stack.

In this part we explain some features of the program, which realizes the calculation of the properties of microsphere with a spherical stack deposited on it. The logic of the program is presented in Figs. 22.1 and 22.2, and the program code can be found in the address, which is specified in the Introduction. The complete description of the program would essentially increase the size of this book. However we adhered to the style at which the names of variables and types have sensible meanings. For instance, the basic class of layers refers to TBaseLayer, etc. It makes the structure of the

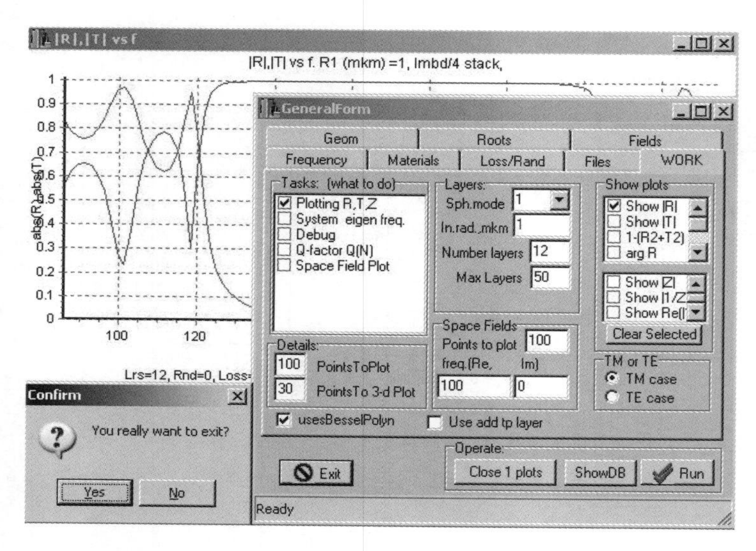

Fig. 22.3 Electromagnetic oscillations in layered microsphere. The graphical user interface.

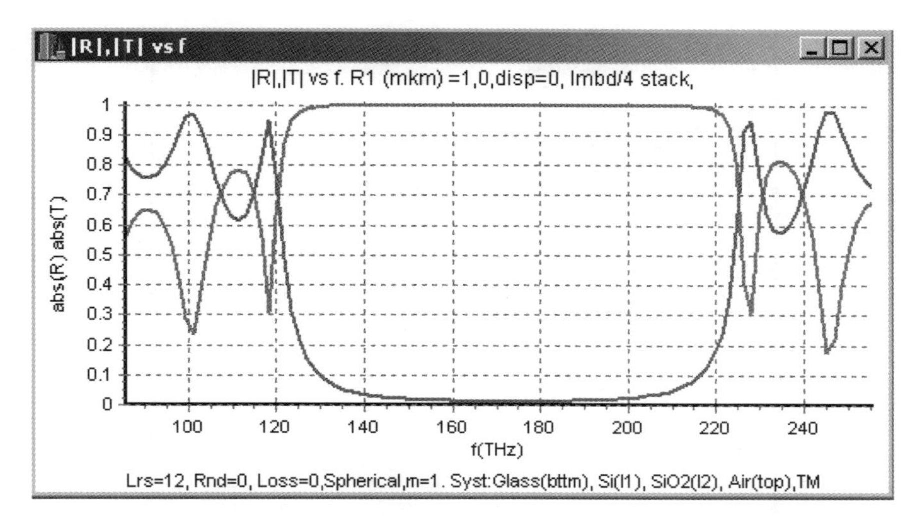

Fig. 22.4 The frequency dependence of the reflection and transmittance coefficients at number of layers 12, spherical quantum number $m = 1$. Other parameters are written in the top and bottom of this plot.

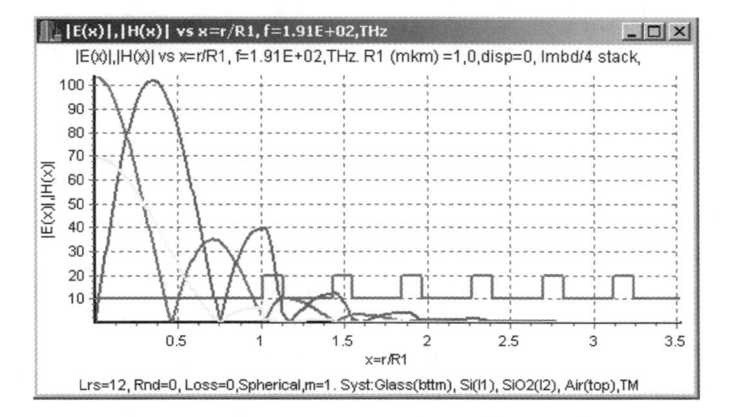

Fig. 22.5 The radial distribution of both electric and magnetic optical fields in microsphere and spherical stack at number of layers 12, spherical quantum number $m = 1$. Other parameters are written in the top and bottom of this plot.

program intuitively clear and the program code becomes rather transparent. We shall touch here only the part of the code describing the formation of a spherical stack on a surface of microsphere.

Whereas both stack and layer contain a lot of various properties it is good idea to create the base class TAbsStruct, in which the general

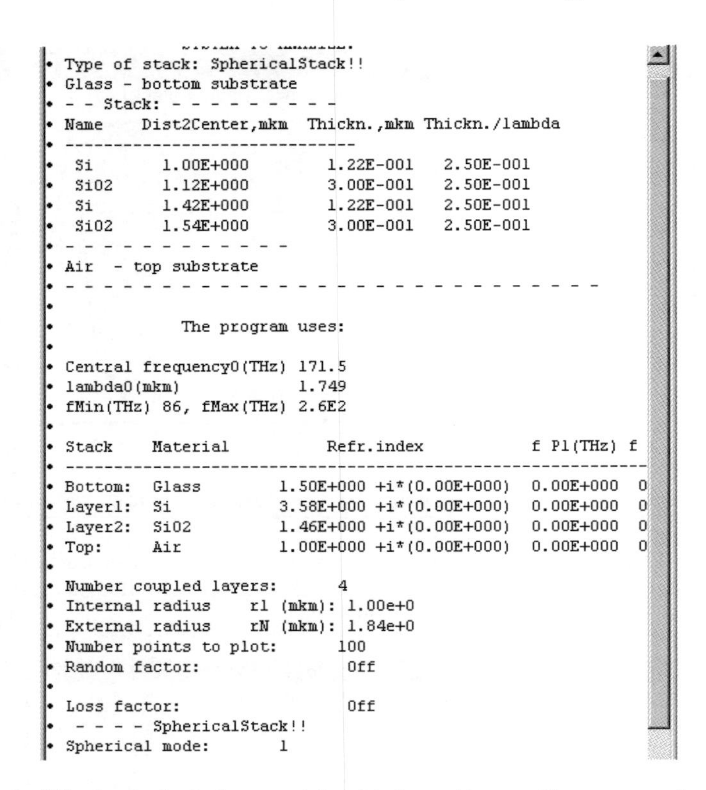

Fig. 22.6 The typical window contained information on the concerned problem.

properties for a stack and a layer, such as the data and methods are concentrated, for instance, current frequency, design frequency, case TE or TM waves, etc. Any individual data can be already described later in special inherited classes. One can see from Fig. 22.1, that class TAbsStruct is base class for two classes TBaseStack and TBaseLayer, which serve as prototypes of a stack and a layer. The class TBaseStack accumulates all base matrix operations for transfer matrixes calculation (see. Sec 1.3–1.5), and class TBaseLayer describes the basic properties necessary for the description of a layer (sizes, material, frequency properties, refraction index, etc.). Clear that both mentioned classes have access to data and methods of class TAbsStruct. It is important, that class TBaseStack contains field TListOfBaseLayers *_ListOfBaseLayers (std:: vector), in which are numerated all layers of this stack.

In the main program there is a variable (type of pointer) TBaseStack * _generalStack2 which corresponds to a considered stack. Formation of a stack is made by the function void TGeneralForm:: BuildNewStack () as follows. Omitting insignificant details, it is possible to say that the new object of a stack _generalStack2 in which necessary number of desired alternating layers will be written is created. The constructor of class TSphericalStack is written below.

```
void TGeneralForm::BuildNewStack()
{
...
_generalStack2 =
    new TSphericalStack(_TheSubstrateMaterials[cBottomSbstr],
      _TheSubstrateMaterials[cTopSbstr],
      _TheStackMaterials[cLayer1],
      _TheStackMaterials[cLayer2],
      _TheStackMaterials[cQuantLayer],
      get_Frequensy0(), get_NLayers(), TETMcase ,
      get_sphNmbOfMode(), usesBesselPolynom,
      "TGeneralForm:: generates SphericalStack.");
......
}
```

After the stack is created, it is possible to have access to all to its public methods and to evaluate calculation for example of such integrated characteristics of whole stack as factors of reflection or transmittance coefficients on the given frequency. Besides, if the materials of internal microsphere and a surrounding environment are known, one can calculate eigenfrequencies, using solver to evaluate the complex roots by Muller method, described in the previous chapters. Certainly the work on the correct arrangement of a stack and layers may be great. Nevertheless it has auxiliary or only technical character and provided that the classes' inheritance hierarchy is well organized such a part is well separated from the physical meaning part. The details of realization of properties of object normally have not to influence in resulting characteristics (the encapsulation property). In this a case, having correctly created object such as TSphericalStack, it is possible to study it from any side, even with that it was not planned at the initial stage.

Further we bring some fragments of the program code which can be useful to an illustration of the aforesaid.

```
typedef std::vector<TBaseLayer *> TListOfBaseLayers ;
```

```
// - - - - constructor of class TSphericalStack- - - - - - - - - - -
TSphericalStack::TSphericalStack(TMaterial BottomMaterial,
   TMaterial TopMaterial,
   TMaterial Layer1Material,
   TMaterial Layer2Material,
   TMaterial QuantLayerMaterial,
   TcompD freq0, int nmb, int TETMcase, int nSph, bool
   usesBesselPolynom, String comment)
   : TBaseStack(BottomMaterial, TopMaterial, Layer1Material,
      Layer2Material,
         QuantLayerMaterial,
         freq0, nmb, TETMcase, usesBesselPolynom, comment +
         "TSphericalStack.")
{ //create a spherical stack
_nSphNmbOfStack = nSph; //spherical harmonic for the stack
TSphericalLayer *newLayer;
newLayer = new TSphericalLayer(BottomMaterial, freq0, TETMcase,
   nSph, 0,
"In TSphericalStack is created a new bottom substrate::");
newLayer->set_typeOfDispersion(globalSt::TypeOfDispers);
newLayer->set_typeOfLayer("TSphericalLayer");
newLayer->set_isMaterialActive(globalSt::isBottSubstrActive);
newLayer->set_isBesselPolynom(usesBesselPolynom);
_ListOfBaseLayers.push_back(newLayer);
RealType rDist = BottomMaterial.rR1; //-dimensional distance from the
centre to a floor of first layer
  for (int ii=1; ii <= nmb; ++ii) //for all layers
   { // separate to the alternative stack:
     if (ii % 2 == 1)
       {
        Layer1Material.rR1 = rDist;
        newLayer = new TSphericalLayer(Layer1Material, freq0,
         TETMcase, nSph, ii,
            "In TSphericalStack is created a new layer::");
        newLayer->set_isMaterialActive(globalSt::isLayer1Active);
        newLayer->set_typeOfLayer("TSphericalLayer");
        _Layer1 = newLayer; // get pointer to 1st layer in the stack
        rDist += Layer1Material.thickness;
```

```
        }
        else
        {
         Layer2Material.rR1 = rDist;
         newLayer = new TSphericalLayer(Layer2Material, freq0,
          TETMcase, nSph, ii,
         "In TSphericalStack is created a new layer::");
         newLayer->set_isMaterialActive(globalSt::isLayer2Active);
         newLayer->set_typeOfLayer("TSphericalLayer");
         _Layer2 = newLayer; // get pointer to 1st layer in the stack
         rDist += Layer2Material.thickness;
        }
       newLayer-¿set_typeOfDispersion(globalSt::TypeOfDispers);
       newLayer-¿set_isBesselPolynom(usesBesselPolynom);
       _ListOfBaseLayers.push_back(newLayer);
       }
      .........
      TopMaterial.rR1 = rDist;
      newLayer = new TSphericalLayer(TopMaterial, freq0, TETMcase,
        nSph, nmb+2,
      "In TSphericalStack is created a new top substrate::");
      ...
      _ListOfBaseLayers.push_back(newLayer); // accumulate current layer
       in the stack
  }

//- - - - constructor of class TSphericalLayer - - - - - -
TSphericalLayer::TSphericalLayer(TMaterial theMaterial, TcompD freq0,
     int TETMcase, int nSph, int nmbOfLayer, String comment)
   : TBaseLayer(theMaterial, freq0, TETMcase, nmbOfLayer,
     comment+" TSphericalLayer: " )
{
_m = nSph; // num. of a sph.mode
_nmbOfThisLayer = nmbOfLayer;
}

// - - - - constructor of class TBaseLayer - - - - - - -
TBaseLayer::TBaseLayer(TMaterial theMaterial, TcompD freq0,
      int TETMcase,
```

```
int nmbOfLayer, String comment)
{
    _TM_TE_case = TETMcase;
    _titleOfMaterial = theMaterial.Name; //a name of a layers material
    _thickness = theMaterial.thickness; //thickness of layer
    _nmbOfThisLayer = nmbOfLayer;
    TcompD nFreq = theMaterial.n_freq;
...
}
```

Appendix A: Calculation of Field's Energy in a Sphere

We assume the volume V with field is spherical with radius a. First we calculate the "transverse" part

$$
\begin{aligned}
&\int \varepsilon_0 \varepsilon (F_\theta^2 + F_\varphi^2)\, dV \\
&= \varepsilon_0 \int \varepsilon \frac{1}{r^2 k_0^4 \varepsilon} \left[\left(\frac{\partial R}{\partial r}\right)^2 \left(\frac{\partial Y}{\partial \theta}\right)^2 + \frac{1}{\sin^2\theta} \left(\frac{\partial R}{\partial r}\right)^2 \left(\frac{\partial Y}{\partial \varphi}\right)^2 \right] r^2\, dr\, dS \\
&= \frac{\varepsilon_0}{k_0^4} \int_0^a \frac{1}{\varepsilon} \left(\frac{\partial R}{\partial r}\right)^2 dr \cdot \int_S \left[\left(\frac{\partial Y}{\partial \theta}\right)^2 + \frac{1}{\sin^2\theta} \left(\frac{\partial Y}{\partial \varphi}\right)^2 \right] dS, \qquad \text{(A.1)}
\end{aligned}
$$

where $k_0 = \omega/c, R(r)$ and $Y(\theta, \varphi)$ are the radial and angular parts of the Debye potential accordingly. In (A.1) through integrating by parts we calculate the radial integral as the following

$$
\begin{aligned}
\int_0^a \frac{1}{\varepsilon} \left(\frac{\partial R}{\partial r}\right)^2 dr &= \int_0^a \left\{ \frac{\partial}{\partial r} \left(\frac{R}{\varepsilon} \frac{\partial R}{\partial r} \right) - R \frac{\partial}{\partial r} \left(\frac{1}{\varepsilon} \frac{\partial R}{\partial r} \right) \right\} dr \\
&= \frac{1}{\varepsilon} \left[R \frac{\partial R}{\partial r} \right]_0^a - \int_0^a R \frac{\partial}{\partial r} \left(\frac{1}{\varepsilon} \frac{\partial R}{\partial r} \right) dr \\
&= \frac{1}{\varepsilon} \left[R \frac{\partial R}{\partial r} \right]_0^a - \int_0^a R \left\{ -k_0^2 R + \frac{m(m+1)}{\varepsilon r^2} R \right\} dr \\
&= \frac{1}{\varepsilon} \left[R \frac{\partial R}{\partial r} \right]_0^a + k_0^2 \int_0^a R^2\, dr - \int_0^a \frac{m(m+1)}{\varepsilon r^2} R^2\, dr, \qquad \text{(A.2)}
\end{aligned}
$$

where $[x]_0^a = x(a) - x(0)$, $[R\frac{\partial R}{\partial r}]_0^a = R(a)\frac{\partial R(a)}{\partial r} - R(0)\frac{\partial R(0)}{\partial r}$. The "longitude" part of electric energy has the following form

$$\begin{aligned}\int_V \varepsilon_0 \varepsilon F_r^2 \, dV &= \int_0^a \varepsilon_0 \varepsilon \frac{1}{k_0^4 \varepsilon} \frac{m^2(m+1)^2}{r^4} R^2 r^2 \, dr \cdot \int_S Y^2 \, dS \\ &= \frac{\varepsilon_0}{k_0^4} \int_0^a \frac{m(m+1)}{\varepsilon r^2} R^2 \, dr \cdot m(m+1) \cdot \int_S Y^2 \, dS.\end{aligned} \tag{A.3}$$

So the full electric part of energy becomes

$$\begin{aligned}\overline{\beta}_e^2 &= \int_V \varepsilon_0 \varepsilon (F_r^2 + F_\theta^2 + F_\varphi^2) \, dV \\ &= \frac{\varepsilon_0}{k_0^4} \int_0^a \frac{m(m+1)}{\varepsilon r^2} R^2 \, dr \cdot m(m+1) \cdot \int_S Y^2 \, dS + \frac{\varepsilon_0}{k_0^4} \frac{1}{\varepsilon} \left[R \frac{\partial R}{\partial r} \right]_0^a \\ &\quad + \frac{\varepsilon_0}{k_0^4} \left\{ k_0^2 \int_0^a R^2 \, dr - \int_0^a \frac{m(m+1)}{\varepsilon r^2} R^2 \, dr \right\} \\ &\quad \cdot \int_S \left[\left(\frac{\partial Y}{\partial \theta} \right)^2 + \frac{1}{\sin^2 \theta} \left(\frac{\partial Y}{\partial \varphi} \right)^2 \right] dS \\ &= \frac{\varepsilon_0}{k_0^4 \varepsilon} \left[R \frac{\partial R}{\partial r} \right]_0^a + \frac{\varepsilon_0}{k_0^4} k_0^2 m(m+1) \cdot \int_0^a R^2 \, dr.\end{aligned} \tag{A.4}$$

The magnetic part of energy $\overline{\beta}_h^2$ one can be calculated by a similar way

$$\begin{aligned}\overline{\beta}_h^2 &= \int_V \mu_0 (H_\theta^2 + H_\varphi^2) \, dV \to \mu_0 \int \frac{c^2 \varepsilon_0^2}{r^2 k_0^2} \left[\frac{1}{\sin^2 \theta} \left(\frac{\partial \Pi}{\partial \varphi} \right)^2 + \left(\frac{\partial \Pi}{\partial \theta} \right)^2 \right] dV \\ &= \frac{\mu_0 c^2 \varepsilon_0^2}{k_0^2} \int r^2 \frac{R^2}{r^2} \, dr \cdot \int_S \left[\frac{1}{\sin^2 \theta} \left(\frac{\partial Y}{\partial \varphi} \right)^2 + \left(\frac{\partial Y}{\partial \theta} \right)^2 \right] dS \\ &= \frac{\mu_0 c^2 \varepsilon_0 \varepsilon_0}{k_0^2} \int R^2 \, dr \cdot G_m^2 = \frac{\varepsilon_0}{k_0^2} G_m^2 \cdot \int_0^a R^2 \, dr \\ &= \frac{\varepsilon_0}{k_0^2} m(m+1) \int_0^a R^2 \, dr,\end{aligned} \tag{A.5}$$

where $G_m^2 = \int_S \left[\frac{1}{\sin^2 \theta} \left(\frac{\partial Y}{\partial \varphi} \right)^2 + \left(\frac{\partial Y}{\partial \theta} \right)^2 \right] dS = m(m+1)$ (see Appendix B). We perform a change variable $k_0 r = y_0$, $k_0 \, dr = dy_0$ in the integral $\int_0^a R^2 \, dr$.

Now $\overline{\beta}_e^2$ and $\overline{\beta}_h^2$ are given by

$$\overline{\beta}_e^2 = \frac{\varepsilon_0}{k_0^4 \varepsilon}\left[R\frac{\partial R}{\partial r}\right]_0^a + \varepsilon_0 \frac{m(m+1)a^3}{\alpha^3}\int_0^\alpha R^2(y_0)\,dy_0, \tag{A.6}$$

$$\overline{\beta}_h^2 = \frac{\varepsilon_0 m(m+1)}{k_0^2 k_0}\int_0^{k_0 a} R^2(y)\,dy = \frac{\varepsilon_0 m(m+1)a^3}{\alpha^3}\int_0^\alpha R^2(y_0)\,dy_0, \tag{A.7}$$

where $\alpha = k_0 a = \omega a/c$. Remember that $R(y_0)$ satisfies the equation

$$\varepsilon\frac{\partial}{\partial y_0}\left[\frac{1}{\varepsilon}\frac{\partial R}{\partial y_0}\right] + \left[\varepsilon - \frac{m(m+1)}{y_0^2}\right] R = 0. \tag{A.8}$$

Equations for the *TE* fields can be derived by the similar way.

Appendix B: Calculation of Surface Integral

Let us calculate

$$G_m^2 = \int_0^{2\pi} d\varphi \int_0^{\pi} d\theta \sin\theta \left[\frac{1}{\sin^2\theta} \left(\frac{\partial Y_m^j}{\partial \varphi} \right)^2 + \left(\frac{\partial Y_m^j}{\partial \theta} \right)^2 \right]. \tag{B.1}$$

We start from formula [Korn & Korn, 1961]

$$\frac{\partial^2 Y}{\partial \theta^2} + \cot\theta \frac{\partial Y}{\partial \theta} + \frac{1}{\sin^2\theta} \frac{\partial^2 Y}{\partial \theta^2} = -m(m+1)Y \tag{B.2}$$

for spherical function $Y = Y_m^j(\theta, \varphi)$. We multiply (B.2) on the right and on the left by Y_m^j and then integrate over a spherical surface $\int_0^{2\pi} d\varphi \int_0^{\pi} d\theta \sin\theta$. On the left we integrate by parts and take into account the periodic boundary conditions for Y_m^j. As a result the left part becomes form (B.1). On the right we take into account the orthogonality conditions for the spherical functions in the form $\int_0^{2\pi} d\varphi \int_0^{\pi} d\theta \sin\theta [Y_m^j(\theta, \varphi)]^2 = 1$. As a result we receive $G_m^2 = m(m+1)$.

Appendix C: Continuity of Tangential Fields

The form of Eq. (1.118) does not change in the case of alternating spherical confocal layered structure with discontinuity of the dielectric permittivity. By integrating (1.114) with respect to radial coordinate r between k and $k+1$ layers the next sum becomes

$$\begin{aligned}
&[R_k(r)R'_k(r)/\varepsilon(r)]_{r_k}^{r_{k+1}} - [R_{k+1}(r)R'_{k+1}(r)/\varepsilon(r)]_{r_{k+1}}^{r_{k+2}} \\
&\quad = [R_k(r_k)R'_k(r_k)/\varepsilon(r_k) - R_k(r_{k+1})R'_k(r_{k+1})/\varepsilon(r_{k+1})] \\
&\qquad - [R_{k+1}(r_{k+2})R'_{k+1}(r_{k+2})/\varepsilon(r_{k+2}) - R_{k+1}(r_{k+1})R'_{k+1}(r_{k+1})/\varepsilon(r_{k+1})] \\
&\quad = R_k(r_k)R'_k(r_k)/\varepsilon(r_k) - R_{k+1}(r_{k+2})R'_{k+1}(r_{k+2})/\varepsilon(r_{k+2}).
\end{aligned} \tag{C.1}$$

(C.1) takes into account that the internal boundary of layers $r = r_1$ and the equality $R_k(r_{k+1})R'_k(r_{k+1})/\varepsilon(r_{k+1}) = R_{k+1}(r_{k+1})R'_{k+1}(r_{k+1})/\varepsilon(r_{k+1})$ obeys the boundary conditions of continuity of the tangential fields. This is valid for all other terms in the internal boundaries. The resulting formula contains only terms from the external boundaries, as it is written in (1.118).

Appendix D: Integral on Bessel Functions

For calculating integral $I_m(\alpha)$ we use the next formula [Korn & Korn, 1961]

$$M_l(\alpha) = \int_0^1 [J_l(\alpha x)]^2 x\, dx = \frac{1}{2}[J_l'(\alpha)]^2 + \frac{1}{2}\left(1 - \frac{l^2}{\alpha^2}\right)[J_l(\alpha)]^2. \quad \text{(D.1)}$$

For $l = m + \frac{1}{2}$ in the TE waves case the eigenfrequency equation is $J_{m+1/2}(\alpha) = 0$, therefore $M_{m+1/2}(\alpha) = (1/2)[J'_{m+1/2}(\alpha)]^2 = \frac{1}{2}[J_{m+3/2}(\alpha)]$. In the TM-case the eigenfrequency equation is $J_{m+1/2}(\alpha) + 2\alpha J'_{m+1/2}(\alpha) = 0$, therefore for TM waves $M_{m+1/2}(\alpha) = (1/2)\{1 - m(m+1)/\alpha^2\}[J_{m+1/2}(\alpha)]^2$.

Appendix E: Surface Integrals for Dipole

Since we take into account the contribution of spherical field modes (emitted photons) we have to integrate $(\vec{\wp}\vec{E}_{nmj})$ in Eq. (13.4) over complete solid angle $d\Omega = \sin\theta\, d\theta\, d\varphi$ in the form

$$\int_0^{2\pi} d\varphi \int_0^{\pi} d\theta \sin\theta \cdot \{\cos(\theta_p - \theta)\varepsilon(z,\theta,\varphi) + \sin(\theta_p - \theta)\varepsilon_\theta(z,\theta,\varphi)\}. \tag{E.1}$$

Substituting the radial and angular parts of ε and ε_θ from (13.6) to (E.1) one can evaluate the first integral in (E.1) in the form

$$\int_0^{2\pi} d\varphi \int_0^{\pi} \sin\theta\cos(\theta_p - \theta)\varepsilon(y,\theta,\varphi)d\theta$$
$$= 2\pi\delta_{j0}A_m(y)\left[\tfrac{2}{3}\cos\theta_p\delta_{m1} - \sin\theta_p\widetilde{A}_m\right]. \tag{E.2}$$

Here we adopt the convention of summation over repeated vector-component indices. Second integrals can be evaluated by nearly the same way

$$\int_0^{2\pi} d\varphi \int_0^{\pi} \sin\theta\cos(\theta_p - \theta)\varepsilon_\theta(y,\theta,\varphi)d\theta$$
$$= 2\pi\delta_{j0}B_m(y)\left[\tfrac{4}{3}\cos\theta_p\delta_{m1} + \sin\theta_p\widetilde{B}_m\right], \tag{E.3}$$

where $A_m(y) = \frac{m(m+1)}{y^2}R_m(y)$, $B_m(y) = \frac{1}{y^2}\frac{dR_m(y)}{dy}$. Functions $\widetilde{A}_m$ and $\widetilde{B}_m$ can be written as $\widetilde{A}_m = \int_{-1}^{1}(1-x^2)^{1/2}P_m(x)\,dx$ and $\widetilde{B}_m = \int_{-1}^{1}x(1-x^2)^{1/2}\frac{dP_m(x)}{dx}\,dx$, where $P_m(x)$ is the Legendre polynomial [Korn & Korn, 1961]. Due to the parity condition for both integrals we have $\widetilde{A}_m = \widetilde{B}_m = 0$ for odd m. For some even m the values are as such: $\widetilde{A}_0 = \widetilde{B}_0 = 0$, $\widetilde{A}_2 = \pi/16$,

$\widetilde{B}_2 = 3\pi/8$, $\widetilde{A}_4 = \pi/128$, $\widetilde{B}_4 = 5\pi/32$, so $\widetilde{A}_m$ and $\widetilde{B}_m$ decrease since m increases. More general forms of $\widetilde{A}_m$ and $\widetilde{B}_m$ can be found in [Gradshteyn & Ryzhik, 2000]. While (E.2) and (E.3) we have used the next relationship: $\int_0^{2\pi} d\varphi\, e^{ij\varphi} = 2\pi\delta_{j0}$, the mutual orthogonality of Legendre polynomials in the form $\int_{-1}^{1} P_m(x)P_n(x)\, dx = \frac{2\delta_{nm}}{2n+1}$ and other corresponding formulas from [Korn & Korn, 1961]. Substituting (E.2) and (E.3) in (E.1), we obtain (13.7) and (13.8).

Appendix F: Some Mathematical Formulas

F.1 Legendre Polynomials

Associated Legendre equation has the form

$$(1-z^2)\frac{d^2w}{dz^2} - 2z\frac{dw}{dz} + \left[j(j+1) - \frac{m^2}{1-z^2}\right]w = 0$$

where $P_j^{(m)}(z)$ is associated Legendre function

$$P_j^{(m)}(z) = w, \quad P_j^{(0)}(z) \equiv P(z)$$

Various formulas

$$P_j^{(x)}(x) = (1-x^2)^{m/2}\frac{d^m}{dx^m}P_j(x), \quad P_j^{(m)}(x) = 0 \quad \text{if } m > j$$

Asymptotic

$$j \to \infty \;\Rightarrow\; P_j^m(\cos\theta)$$

$$= (-1)^m\sqrt{\frac{2}{\pi j \sin\theta}}\cos\left[\left(j+\frac{1}{2}\right)\theta - \frac{\pi}{4} + \frac{m\pi}{2}\right] + \theta(j^{-(3/2)}), \quad 0 < \theta < \pi$$

Orthogonality

$$\int_{-1}^{1} P_j^m(x)P_z^m(x)\,dx = \frac{2}{2j+1}\frac{(j+m)!}{(j-m)!}\delta_{jz}, \quad j, z = 0, 1, 2, \ldots, m \leq j$$

$$\int_0^1 [P_j^m(x)]^2\,dx = \frac{1}{2j+1}\frac{(j+m)!}{(j-m)!}$$

$$\int_0^1 \frac{[P_d^{(m)}(x)]^2}{1-x^2}\,dx = \frac{1}{2m}\frac{(j+m)!}{(j-m)!}$$

F.2 Spherical Functions

Spherical function equation has the form

$$\frac{1}{\sin\theta}\frac{\partial}{\partial\theta}\left(\sin\theta\frac{\partial Y}{\partial\theta}\right)+\frac{1}{\sin^2\theta}\frac{\partial^2 Y}{\partial\varphi^2}+j(j+1)Y$$

$$=\frac{\partial^2 Y}{\partial\theta^2}+\cot\theta\frac{\partial Y}{\partial\theta}+\frac{1}{\sin^2\theta}\frac{\partial^2 Y}{\partial\varphi^2}+j(j+1)Y=0$$

where

$$Y=Y_j^m(\theta,\varphi)=\frac{1}{2}\sqrt{\frac{2j+1}{\pi}\frac{(j-|m|)!}{(j-|m|)!}}P_j^{|m|}(\cos\theta)\,e^{im\varphi},$$

where $j=0,1,2,\ldots,m=0,\pm1,\ldots,\pm j$. For every functions Y_j^m

$$(f,h)=\int_0^{2\pi}d\psi\int_0^{\pi}f^*(\theta,\psi)\cdot h(\theta,\psi)\sin\theta\,d\theta=\begin{cases}0 & \text{if } f\neq h\\ 1 & \text{if } f=h\end{cases}$$

$$\phi(\theta,\psi)=\sum_{j=0}^{\infty}\sum_{m=-j}^{j}\gamma_{j\omega}P_j^{|\omega|}(\cos\theta)\,e^{im\varphi}$$

$$\int_{-1}^{1}P_m(x)P_n(x)\,dx=\frac{2}{2n+1}\delta_{mn},\quad P_0(x)=1,\quad P_m(\pm1)=(\pm)^m$$

$$\int_{-1}^{1}P_n^m(z)P_k^m(z)\,dz=\frac{\left(n+\frac{1}{2}\right)^{-1}(n-m)!}{(n-m)!}\delta_{nk}$$

$$\int_{-1}^{1}P_n^m(z)P_n^k(1-z^2)^{-1}\,dz=\frac{(n+m)!}{(n-m)!}\delta_{mk}$$

$$Y_n(\theta,\varphi)=a_0P_n(\cos\theta)+\sum_{k=1}^{n}(a_k\cos k\varphi+\beta_k\sin k\varphi)P_n^k(\cos\theta),$$

$$\iint_S Y_n(\theta,\psi)P_k(\cos\theta)\,dS=\frac{4\pi R^2}{2n+1}Y_n(\theta_0,\psi_0)\delta_{nk}$$

$$\iint_S Y_\omega(\theta,\varphi)Y_k(\theta,\varphi)\,dS=0,\quad m\neq k$$

$$d^2=z_1^2+z_2^2-2z_1z_2\cos\gamma,\quad x=\cos\gamma,\quad z=\frac{z_2}{z_1}<1$$

$$1>\frac{z_1}{d}=(1-2xz+z^2)^{-(1/2)}=\sum_{n=0}^{\infty}P_n(x)z^n$$

F.3 Bessel Functions

$$\int_0^1 J_m(\mu_i x) J_m(\mu_k x) x\, dx = \frac{1}{2}[J'_m(\mu_i)]^2 \delta_{ik} = J_{m+1}(\mu_i)$$

$$\int_0^1 [J_l(\alpha x)]^2 x\, dx = \frac{1}{2}[J'_l(\alpha)]^2 + \frac{1}{2}\left(1 - \frac{l^2}{\alpha^2}\right)[J_l(\alpha)]^2$$

$$Z_{m+1}(z) = \frac{m}{z} Z_m(z) - Z_m(z)$$

$$\alpha J_\upsilon(x) + \beta x J_{\upsilon'}(x) = 0, \quad \upsilon > -1$$

$$\Rightarrow \int_0^l x J_\upsilon^2\left(\mu \frac{x}{l}\right) dx = \frac{l^2}{2}\left(1 + \frac{d^2 - \beta^2 \upsilon^2}{\beta^2 \mu^2}\right) J_\upsilon^2(\mu)$$

$$J_\upsilon(x) = 0 \;\Rightarrow\; \int_0^l x J_\upsilon\left(\mu_i \frac{x}{l}\right) J_\upsilon\left(\mu_j \frac{x}{l}\right) dx = 0, \quad i \neq j$$

$$\int_0^l x J_\upsilon^2\left(\mu \frac{x}{l}\right) dx = \frac{l^2}{2} J'_\upsilon(\mu)$$

F.4 Spherical Bessel Functions

$$j_n(z) = \sqrt{\frac{\pi}{2z}} J_{n+\frac{1}{2}}(z), \qquad J_{n+\frac{1}{2}}(z) = \sqrt{\frac{2z}{\pi}} j_n(z);$$

$$\begin{aligned}
\int_0^1 J_{n+\frac{1}{2}}(\mu_i x) J_{n+\frac{1}{2}}(\mu_k x) x\, dx &= \frac{xz}{\pi} \int_0^1 j_n(\mu_i x) j_n(\mu_k x) x\, dx \\
&= \frac{2}{\pi} \int_0^1 j_n(\mu_i x) j_n(\mu_k x) x^2\, dx \\
&= \frac{1}{2}\left[\left(j_n \sqrt{\frac{2x}{\pi}}\right)'\right]^2 \delta_{ik} \\
&= \frac{1}{2}[j_{n+1}(\mu_i)]^2 \delta_{ik} \frac{2}{\pi}
\end{aligned}$$

Appendix G: Various Head *.h Files

G.1 File BGNDefg.h

```
#ifndef BGNDef9H
#define BGNDef9H
#include <complex.h>
#include <vector>

typedef Extended RealType;
typedef complex<RealType> TcompD;

const int c501 = 501+1600, c17 = 17;
const int TNArraySize = 200; // Maximum size of vectors }
const double TNNearlyZero = 1E-015;
const TcompD I(0,1); // Imaginary Unit

typedef TcompD TComplVect2a[2];
typedef TcompD TComplVect2a[2];
typedef TcompD TcompDArr10[10];
typedef TcompD TComplVect10[10];
typedef TcompD TComplVect301[301];
typedef TcompD TComplVect100[100];
typedef TcompD TComplVect501[c501];
typedef TcompD TNCvector[TNArraySize];

typedef TcompD TComplMart2x2a[2][2];
typedef TcompD TComplMatrix4x4[4][4];
typedef TcompD TComplMatrix100x15[100][15];

typedef RealType Tarr1x4[4];
typedef RealType TRealVect10[10];
```

```
typedef RealType TRealVect17[17];
typedef RealType TRealVect100[100];

typedef RealType TRealVect301[301];
typedef RealType TRealVect512[512];
typedef RealType TRealVect501[c501];
typedef RealType TNvector[TNArraySize];

typedef RealType TDateAll8x5[8][5];
typedef RealType TDateAll50x4[50][4];
typedef RealType TRealMatrix4x4[4][4];
typedef RealType TDateAll100x4[100][4];
typedef RealType TDateAll500x4[500][4];
typedef RealType TRealMatrix150x17[150][17];
typedef RealType TRealMatrix501x17[c501][17];

typedef int TIntVect10[10];
typedef int TIntVect10[10];
typedef String TStrarr17[17];
typedef bool TBoolVect10[10];

typedef TcompD (*TMyFuncEq1)(TcompD ,TcompD& ); //For Muller5

typedef std::vector<TcompD> TComplArray1;
typedef std::vector<TComplArray1> TComplArray2;
typedef std::vector<RealType> TRealArray1;
typedef std::vector<TRealArray1> TRealArray2;

struct TMaterial
   {RealType rR1; RealType thickness; TcompD n_Limit; TcompD n_freq;
   RealType wp; RealType nu; RealType wT;
   String Name; bool isDispersive; bool isInhom; bool isActive; bool
   isEmpty;};
// - - - - - - - - - - - - - -
const RealType cLight_SI = 3e8; //light velocity m/sec
const RealType e_electr_SI = 1.602e-19;//electron charge, coul
const RealType Me_electr_SI = 9.11e-31; //mass rest electron, kg
const RealType eps0_vacuum_SI= 8.854187817e-12;//eps0 vacuum
const RealType hPlank_SI = 1.05e-34; //Plank const, J/sec
const RealType Rz_SI = 6370e3; // R Earth, m
```

```
const int cHomogenCase = 0;
const int cInomogenODECase = 1;
const int cInomogenSplineCase = 2;

#endif
```

G.2 File BGNutil1.h

```
#ifndef BGNutil1H
#define BGNutil1H
#include "BgnDef9.h"

inline RealType FMAX(RealType a, RealType b)
   {return ( a > b ? a: b);}

inline RealType FMIN(RealType a, RealType b)
   {return ( a > b ? b: a);}

inline RealType IMIN(int a, int b){return ( a > b ? b: a);}

inline RealType SIGN(RealType a, RealType b)
   {return ( (b)  >=  0.0 ? fabs(a) : -fabs(a));}

inline int IMAX(int a, int b){return ( a > b ? a: b);}

void nrerror(String error_text);
void matsub(RealType **a, RealType **b, RealType **c, int nx, int ny);
void free_ivector(int *v, int nl, int nh);
void free_dvector(RealType *v, int nl, int nh);
void free_cvector(TcompD *v, int nl, int nh);
void free_matrix(RealType **m, int nrl,int nrh, int ncl, int nch);
void free_dmatrix(RealType **m, int nrl, int nrh, int ncl, int nch);
void free_cmatrix(TcompD **m, int nrl, int nrh, int ncl, int nch);
void canorm2Matr(RealType **res, TcompD **a, int nx, int ny);
void anorm2Matr(RealType **res, RealType **a, int nx, int ny);
void cGradMatr(TcompD **grad1,TcompD **grad2, TcompD **inp,
   int nx, int ny, RealType hx, RealType hy);


int *ivector(int nl,int nh);

TcompD *cvector(int nl,int nh);
TcompD **cmatrix(int nrl,int nrh,int ncl,int nch);
```

```
RealType *dvector(int nl, int nh);
RealType **dmatrix(int nrl,int nrh,int ncl,int nch);
RealType anorm2(RealType **a, int nx, int ny);
RealType anorm22(RealType **a, int nx, int ny);
RealType canorm2(TcompD **a, int nx, int ny);
RealType canorm22(TcompD **a, int nx, int ny);
RealType summaEnries(RealType **a, int nx, int ny);
RealType MaxNormU(RealType **a, int nx, int ny);

#endif
```

Bibliography

Airy, G. B. (1838). *Trans. Cambridge Philos. Soc.*, 6, 379.

Alhassid, Y. (2000). The statistical theory of quantum dots. *Rev. Mod. Phys.*, 72, 895.

An, K. & Moon, H.-J. (2003). Laser oscillations with pumping-independent ultra-highcavity quality factors in evanescent-wave-coupled-gainmicrosphere dye lasers. *J. Phys. Soc. Jpn.*, 72, 773–776.

Arai, Y., Yano, T., & Shibata, S. (2003). High refractive-index microspheres of optical cavity structure. *Appl. Phys. Lett.*, 82, 3173–3175.

Arciniegas, F. (2002). *C++ XML.* New Riders Publishing.

Artemyev, M. V. & Woggon, U. (2000). Quantum dots in photonic dots. *Appl. Phys. Lett.*, 76, 1353.

Artemyev, M. V., Woggon, U., & Wannemacher, R. (2001a). Photons confined in hollow microspheres. *Appl. Phys. Lett.*, 78, 1032–1034.

Artemyev, M. V., Woggon, U., Wannemacher, R., Jaschinski, H., & Langbeint, W. (2001b). Light trapped in a photonic dot: microspheres act as a cavity for quantum dot emission. *Nano Lett.*, 1, 309–314.

Barash, Yu. S. & Ginzburg, V. L. (1976). On expressions for the energy density and the evolved heat in electrodynamics of dispersive and absorbing environment. *Uspekhi Fizicheskikh Nauk (Phys. -Uspehi)* (pp. Uspekhi Fizicheskikh Nauk. 118, 523–537.)

Barut, A. O. & Dowling, J. P. (1987). Quantum electrodynamics based on self-energy: Spontaneous emission in cavities. *Phys. Rev. A*, 36, 649–654.

Basdevant, J.-L. & Dalibard, J. (2002). *Quantum Mechanics.* Springer.

Beige, A., Pachos, J., & Walther, H. (2002). Spontaneous emission of an atom in front of a mirror. *Phys. Rev. A*, 66, 063801.

Berg, T. W. & Mork, J. (2003). Quantum dot amplifiers with high output power and low noise. *Appl. Phys. Lett.*, 82, 3083–3085.

Berreman, D. W. (1963). Infrared absorption at longitudinal optic frequency in cubic crystal films. *Phys. Rev.*, 130, 2193.

Besset, D. H. (2001). *Object-Oriented Implementation of Numerical Methods: An Introduction with Java and Smalltalk.* Academic Press.

Bimberg, D. *et al.* (2001). Quantum dot laser powers past 10 W. A highly efficient quantum dot laser emits almost 12 W in quasi-continuous-wave mode. *Electron. Lett.*, 38, 883.

Birkl, G., Buchkremer, F., Dumke, R., & Ertmer, W. (2001). Atom optics with microfabricated optical elements. *Opt. Commun.*, 191, 67–81.

Bishop, A. I., Nieminen, T. A., Heckenberg, N. R., & Rubinsztein-Dunlop, H. (2003). Optical application and measurement of torque on microparticles of isotropic non-absorbing material. *Phys. Rev. A* (pp. 033802 (8 pages)).

Bondarev, I., Slepyan, G., & Maksimenko, S. A. (2002). Spontaneous decay of excited atomic states near a carbon nanotube. *Phys. Rev. Lett.*, 89, 5504.

Boone, T. D., Tsukamoto, H., & Woodall, J. M. (2003). Intensity and spatial modulation of spontaneous emission in GaAs by field aperture selecting transport. *Appl. Phys. Lett.*, 82, 3197–3199.

Born, M. & Wolf, E. (1980). *Principles of Optics.* Pergamon, New York.

Bouwmeester, D., Ekert, A., & Aeilinger, A. (2001). *The Physics of Quantum Information: Quantum Cryptography, Quantum Teleportation, Quantum Computation.* Springer-Verlag.

Bozhevolnyi, S. & Erland, J. (2001). Waveguiding in surface plasmon polariton band gap structures. *Phys. Rev. Lett.*, 86, 3008.

Brady, D., Papen, G., & Sipe, J. E. (1993). Spherical distributed dielectric resonators. *Jour. Opt. Soc. Am. B*, 10, 644.

Braginsky, V. & Ilchenko, V. (1987). *Sov. Phys. Dokl.*, 32, 307.

Braginsky, V. B., Gorodetsky, M., & Ilchenko, V. (1989). Quality-factor and nonlinear properties of optical whispering-gallery modes. *Phys. Lett. A*, 137, 393.

Braunstein, D., Khazanov, A. M., Koganov, G. A., & Shuker, R. (1996). Lowering of threshold conditions for nonlinear effects in a microsphere. *Phys. Rev. A*, 53, 3565.

Brun, T. A. & Wang, H. (2000). Coupling nanocrystals to a high-Q silica microsphere: Entanglement in quantum dots via photon exchange. *Phys. Rev. A*, 61, 032307.

Bryllert, T., Borgstrom, M., Wernersson, L.-E., Seifert, W., & Samuelson, L. (2003). Transport through an isolated artificial molecule formed from stacked self-assembled quantum dots. *Appl. Phys. Lett.*, 82, 2655–2657.

Buck, J. R. & Kimble, H. J. (2003). Optimal sizes of dielectric microspheres for cavity QED with strong coupling. *Phys. Rev. A*, 67, 033806.

Bunkin, F. V. & Oraevsky, A. N. (1959). On spontaneous emission of molecule inside cavity. *Izv. Vyssh. Uchebn. Zaved., Radiofiz.* (pp. Radiofiz. 2, 181–186 (1959)(2)).

Burlak, G. (2002). Optical radiation from coated microsphere with active core. *Phys. Lett. A*, 299, 94.

Burlak, G., Koshevaya, S., Sanchez-Mondragon, J., & Grimalsky, V. (2000). Electromagnetic oscillations in a multilayer spherical stack. *Opt. Commun.*, 180, 49–58.

Burlak, G., Koshevaya, S., Hayakawa, M., Gutierrez-D, E., & Grimalsky, V. (2000). Acousto-optic solitons in fibers. *Opt. Rev.*, 7, 323.

Burlak, G., Koshevaya, S., Sanchez-Mondragon, J., & Grimalsky, V. (2001b). Electromagnetic eigen oscillations and fields in a dielectric microsphere with multilayer spherical stack. *Opt. Commun.*, 187, 91.

Burlak, G., Marquez, P. A., & Starostenko, O. (2003). Dynamics of spontaneous emission of two-level atom in microspheres. Direct calculation. *Phys. Lett. A*, 309, 146.

Burlak, G., Zamudio-Lara, A., & Juarez, D. (2001a). Confinement of electromagnetic oscillations in a dielectric microsphere coated by the frequency dispersive multilayers. *Phys. Lett. A*, 289, 99–105.

Burlak, G., Zamudio-Lara, A., Castro-Beltran, H., & Rivera-Lopez, A. (2002). Transmittance and resonance tunneling of the optical fields in the microspherical metal-dielectric structures. *Opt. Commun.*, 206, 27.

Burlak, G. N., Koshevaya, S. V., Mansurova, S. S., & Hayakawa, M. (2002). Four wave acousto-electromagnetic interactions in crystals with a nonlinear electrostriction. *Physica D.*, 166, 197–207.

Cai, M., Painter, O., & Vahala, K. J. (2000). Observation of critical coupling in a fiber taper to a silica-microsphere whispering-gallery mode system. *Phys. Rev. Lett.*, 85, 74–77.

Calvert, C. *et al.* (1998). *Borland C++ Builder 3.* SAMS.

Campillo, A. J., Eversole, J. D., & Lin, H.-B. (1991). Cavity quantum electrodynamic enhancement of stimulated emission in microdroplets. *Phys. Rev. Lett.*, 67, 437–440.

Campillo, A. J., Eversole, J. D., & Lin, H.-B. (1992). *Mod. Phys. Lett.*, 6, 447–457.

Cao, H., Hall, D. B., Torkelson, J. M., & Cao, C.-Q. (2000). Large enhancement of second harmonic generation in polymer films by microcavities. *Appl. Phys. Lett.*, 76, 538–540.

Centini, M. & Sibitia, C. (1999). Dispersive properties of finite, one-dimensional photonic band gap structures: Applications to nonlinear quadratic interactions. *Phys. Rev. E*, 60, 4891–4898.

Chen, G., Chang, R., Hill, S., & Barber, P. (1991). Frequency splitting of degenerate spherical cavity mode: Stimulated Raman scattering spectrum of deformed droplets. 16.

Chen, G., Mazumder, M. M., Chemla, Y., Chang, A. S., & Hill, S. (1993a). Wavelength variation of laser emission along the entire rim of slightly deformed microparticles. 18, 1993–1995.

Chen, G., Serpenguzel, A., Chang, R. K., & Acker, W. P. (1993b). Relative evaporation rates of droplets in a segmented stream determined by droplet cavity fluorescence peak shifts. *Proc. SPIE.* (pp. 200–208).

Chen, G., Swindal, J. C., & Chang, R. K. (1992). Frequency splitting and precession of cavity modes of a droplet deformed by inertial forces. *Proc. SPIE.* 1726, 292–298.

Chen, Y. N. & Brandes, D. S. C. T. (2003). Current detection of superradiance and induced entanglement of double quantum dot excitons. *Phys. Rev. Lett.*, 90, 166802.

Chew, H. (1987). Transition rates of atoms near spherical surfaces. *J. Chem. Phys.*, 87, 1355.

Chew, H. (1988). Radiation and lifetimes of atoms inside dielectric particles. *Phys. Rev. A*, 38, 3410–3416.

Chew, W. (1996). *Waves and Fields in Inhomogeneous Media.* IEEE Press, New York.

Ching, S. C., Lai, H. M., & Young, K. (1987a). *J. Opt. Soc. Am. B*, 4, 1995–2003.

Ching, S. C., Lai, H. M., & Young, K. (1987b). *J. Opt. Soc. Am. B*, 4, 2004–2009.

Chylek, P. (1990). Resonance structure of Mie scattering: Distance between resonances. *J. Opt. Soc. Am. B*, 7.

Chylek, P., Lin, H.-B., & Campillo, J. D. E. (1991). Absorption effects on microdroplet resonant emission structure. *Opt. Lett.*, 16, 1723–1725.

Cohen-Tannoudji, C., Dupont-Roc, J., & Grynberg, G. (1989). *Photons and Atoms.* John Wiley, New York.

Cohen-Tannoudji, C., Dupont-Roc, J., & Grynberg, G. (1998). *Atom-Photon Interactions: Basic Processes and Applications.* Wiley-Interscience.

Convertino, A., Valentini, A., & Cingolani, R. (1999). Organic multilayers as distributed Bragg reflectors. *Appl. Phys. Lett.*, 75, 322.

Corya, D. G., Price, M. D., & Havel, T. F. (1998). Nuclear magnetic resonance spectroscopy: An experimentally accessible paradigm for quantum computing. *Physica D*, 120, 82–101.

Costantini, G., Manzano, C., Songmuang, R., Schmidt, O. G., & Kern, K. (2003). InAs/GaAs(001) quantum dots close to thermodynamic equilibrium. *Appl. Phys. Lett.*, 82, 3194–3196.

Crenshaw, M. E. (2003). Microscopic foundation of macroscopic quantum optics. *Phys. Rev. A*, 67, 033805.

Crooker, S. A., Barrick, T., Hollingsworth, J. A., & Klimov, V. I. (2003). Multiple temperature regimes of radiative decay in CdSe nanocrystal quantum dots: Intrinsic limits to the dark-exciton lifetime. *Appl. Phys. Lett.*, 82, 2793–2795.

Cusack, M. A., Briddon, P. R., North, S. M., Kitchin, M. R., & Jaros, M. (2001). Si/Ge self-assembled quantum dots for infrared applications. *Semicond. Sci. Technol.*, 16, 181–184.

da Silva, M. J., Quivy, A. A., Martini, S., Lamas, T. E., da Silva, E. C. F., & Leite, J. R. (2003). InAs/GaAs quantum dots optically active at 1.5 mkm. *Appl. Phys. Lett.*, 82, 2646–2648.

Datsyuk, V. V. (1992). Some characteristics of resonant electromagnetic modes in adielectric sphere. *Appl. Phys. B*, 54, 184–187.

Datsyuk, V. V. (2001). Optics of microdroplets. *J. Mol. Liq.*, 93, 159–175.

Datsyuk, V. V. (2002a). Gain effects on microsphere resonant emission structures. *J. Opt. Soc. Am. B*, 19, 142.

Datsyuk, V. V. (2002b). Spontaneous radiation of molecules in open cavities. *JETP Lett.*, 75, 368–372.

Datsyuk, V. V. & Izmailov, I. A. (2001). Optics of microdroplets. *Phys.-Uspekhi.*, 44, 1061–1073.

Datsyuk, V. V., Izmailov, I. A., & Kochelap, V. (1993). Anomalous luminescence of disperse media during stimulated emission into whispering gallery modes. *J. Opt. Soc. Am. B*, 10, 1941–1946.

Datsyuk, V. V., Izmailov, I. A., & Kochelap, V. A. (1990). Generation of the whispering-gallery modes in a medium with condensed disperse phase. *Kvant. Electron.*, 38, 56–65.

Debye, P. (1909). *Ann. Phys.*, 30, 57.

Deitel, H. M. & Deitel, P. (2002a). *Java How to Program.* Prentice Hall; 5th edition.

Deitel, H. M. & Deitel, P. J. (2002b). *C++ How to Program.* Prentice Hall; 4th edition.

Deitel, H. M., Dietel, P. J., Listfield, J. A., Nieto, T., Yaeger, C. H., & Zlatkina, M. (2001). *C# How to Program.* Prentice Hall; 1st edition.

Descartes, R. (1639). *Discourse on Method, Optics, Geometry and Meteorology,* Bobbs-Merrill.

Deumie, C., Voarino, P., & Amra, C. (2002). Overcoated microspheres for specific optical powders. *Appl. Opt.*, 41, 3299.

Dung, H. T., Knoll, L., & Welsch, D.-G. (1998). Three-dimensional quantization of the electromagnetic field in dispersive and absorbing inhomogeneous dielectrics. *Phys. Rev. A*, 57, 3931–3942.

Dung, H. T., Knoll, L., & Welsch, D.-G. (2000). Spontaneous decay in the presence of dispersing and absorbing bodies: General theory and application to a spherical cavity. *Phys. Rev. A*, 62, 053804.

Dung, H. T., Knoll, L., & Welsch, D.-G. (2001). Decay of an excited atom near an absorbing microsphere. *Phys. Rev. A*, 64, 013804.

Dung, H. T., Knoll, L., & Welsch, D.-G. (2002). Intermolecular energy transfer in the presence of dispersing and absorbing media. *Phys. Rev. A*, 65, 043813 (13 pages).

Dung, H. T., Knoll, L., & Welsch, D.-G. (2003). Defect band-gap structures for triggering single-photon emission. *Phys. Rev. A*, 67, 021801.

Dung, H. T. & Ujihara, K. (1999). Three-dimensional nonperturbative analysis of spontaneous emission in a Fabry-Perot microcavity. *Phys. Rev. A*, 60, 4067.

Eckel, B. (2002). *Thinking in Java.* Prentice Hall PTR; 3rd edition.

Ed. Hummel, R. E. & Guenther, K. H. (1995). *Handbook of Optical Properties. V.1. Thin Films for Optical Coating.* CRC Press, Boca Raton, Boston, London, New York, Washington, D.C.

Ewing, W. M., Jardetsky, W. S., & Press, E. (1957). *Elastic Waves in Layered Media.* McGraw-Hill, New York.

Fan, X., Lonergan, M. C., & Wang, Y. Z. H. (2001). Enhanced spontaneous emission from semiconductor nanocrystals embedded in whispering gallery optical microcavities. *Phys. Rev. B*, 64, 115310 (5 pages).

Ferdos, F., Wang, S., Wei, Y., Larsson, A., Sadeghi, M., & Zhao, Q. (2002). Influence of a thin GaAs cap layer on structural and optical properties of InAs quantum dots. *Appl. Phys. Lett.*, 81, 1195–1197.

Fiore, A., Oesterle, U., Stanley, R. P., Houdre, R., Lelarge, F., Ilegems, M., & A.O., P. B. (2001). Structural and electrooptical characteristics of quantum dots emitting at 1.3 mkm on Gallium Arsenide. *IEEE J. Quantum Electron.*, 37, 1050.

Fleischhauer, M. (1999). Spontaneous emission and level shifts in absorbing disordered dielectrics and dense atomic gases: A Green's-function approach. *Phys. Rev. A*, 60, 2534–2539.

Fogel, I., Bendickson, J., & Tocci, M. (1998). *Appl. Opt.*, 7, 393.

Fujiwara, H. & Sasaki, K. (1999). Upconversion lasing of a thalium-ion-dopedfluorozirconate glass microsphere. *J. Appl. Phys.*, 86, 2385–2388.

Fuller, K. A. (1993). Scattering of light by coated spheres. 18, 257–259.

Furukawa, H. & Tenjimbayashi, K. (2001). Light propagation in periodic microcavities. *Appl. Phys. Lett.*, 80, 192.

Gacuteerard, J. M., Sermage, B., Legrand, B. G. B., Costard, E., & Thierry-Mieg, V. (1998). Enhanced spontaneous emission by quantum boxes in amonolithic optical microcavity. *Phys. Rev. Lett.*, 81, 1110–1113.

Gaponenko, S. V. (1998). *Optical Properties of Semiconductor Nanocrystals.* Cambridge University Press, Cambridge.

Gaunaurd, G. C. & Uberall, H. (1983). RST analysis of monostatic and bistatic acoustic echoes from an elastic sphere. *J. Acoust. Soc. Am.*, 73, 1.

Geints, Y. E., Zemlyanov, A. A., Kabanov, V. E. Z. A. M., & Pogodaev, V. A. (1999). *Nonlinear Optics of Atmospheric Aerosol.*

Geller, M., Kapteyn, C., Muller-Kirsch, L., Heitz, R., & Bimberg, D. (2003). 450 meV hole localization in GaSb/GaAs quantum dots. *Appl. Phys. Lett.*, 82, 2706–2708.

Gerard, A. (1983). Scattering by spherical elastic layers: Exact solution and interpretation for a scalar field. *J. Acoust. Soc. Am.*, 73, 13.

Ginzburg, V. L. (1989). *Applications of Electrodynamics in Theoretical Physics and Astronomy.* Gordon and Breach, New York.

Glauber, R. J. & Lewenstein, M. (1991). Quantum optics of dielectric media. *Phys. Rev. A*, 43, 467–491.

Gorodetsky, M. L. & Ilchenko, V. S. (1994). High-Q optical whispering-gallery microresonators: Precession approach for spherical mode analysis and emission patterns with prism couplers. *Opt. Commun.*, 113, 133.

Gorodetsky, M. L. & Ilchenko, V. S. (1999). Optical microsphere resonators: optimal coupling to high-Q whispering-gallery modes. *J. Opt. Soc. Am. B*, 16, 147–154.

Gradshteyn, I. & Ryzhik, I. (2000). *Table of Integrals, Series, and Products.* Academic Press, San Diego, San Francisco, and New York.

Gruner, T. & Welsch, D.-G. (1996). Green-function approach to the radiation-field quantization for homogeneous and inhomogeneous Kramers-Kronig dielectrics. *Phys. Rev. A*, 53, 1818–1829.

Hamming, R. (1987). *Numerical Methods for Scientists and Engineers.* Dover Publications; 2nd edition.

Hans, A. B. (1998). *A Guide to Experiments in Quantum Optics.* Wiley-Vch, Weinheim.

Hanselman, D. & Littlefield, B. (2001). *Mastering MATLAB 6: A Comprehensive Tutorial and Reference.* Prentice Hall.

Hayata, K. & Koshiba, M. (1992). Theory of surface-emitting second-harmonic generation from optically trapped microspheres. *Phys. Rev. A*, 46, 6104.

Hill, S. C. & Benner, R. E. (1986). *J. Opt. Soc. Am. B*, 3, 1509–1514.

Hodgson, N. & Weber, H. (1997). *Optical Resonators.* Springer, London.

Hollingworth, J., Butterfield, D., Swart, B., & Allsop, J. (2000). *C++ Builder 5 Developer's Guide.* SAMS; Book and CD-ROM edition.

Hollingworth, J., Gustavson, P., & Cashman, M. (2002). *Borland C++Builder 6 Developer's Guide.* SAMS; 2nd Book and CD-ROM edition.

Hooijer, C., Xiang, L., Klaas, A., G., & Lenstra, D. (2001). Spontaneous emission in multilayer semiconductor structures. *IEEE J. Quantum Electron.*, 37, 1161.

Hours, J., Varoutsis, S., Gallart, M., Bloch, J., Robert-Philip, I., Cavanna, A., Abram, I., Laruelle, F., & Gerard, J. M. (2003). Single photon emission from individual GaAs quantum dots. *Appl. Phys. Lett.*, 82, 2206–2208.

Ilchenko, V. S., Volikov, P. S., Velichansky, V. L., Treussart, F., Lefevre-Seguin, V., Raimond, J. M., & Haroche, S. (1998). Strain-tunable high-Q optical microsphere resonator. *Opt. Commun.*, 145, 86.

Jacak, L., Krasnyj, J., & Bujkiewicz, D. J. L. (2002). Far-infrared laser action from quantum dots created by electric-field focusing. *Phys. Rev. A*, 65, 063813 (12 pages).

Jackson, J. D. (1975). *Classical Electrodynamics.* John Wiley and Sons.

Ji, Y., Chen, G., Tang, N., Wang, Q., Wang, X. G., Shao, J., Chen, X. S., & Lu, W. (2003). Proton-implantation-induced photoluminescence enhancement in self-assembled InAs/GaAs quantum dots. *Appl. Phys. Lett.*, 82, 2802–2804.

Jia, R., Jiang, D.-S., Tan, P.-H., & Sun, B.-Q. (2001). Quantum dots in glass spherical microcavity. *Appl. Phys. Lett.*, 79, 153–155.

Juzeliunas, G. (1997). Spontaneous emission in absorbing dielectrics: A microscopic approach. *Phys. Rev. A*, 55, R4015–R4018.

Kai, L. & Massoli, P. (1994). Scattering of electromagnetic-plane waves by radially inhomogeneous spheres: A finely stratified sphere model. *Appl. Opt.*, 33, 501.

Kalitievski, M. A., Brand, S., Abram, R. A., & Nikolaev, V. V. (2001). Optical eigenmodes of a multilayered spherical microcavity. *J. Mod. Optics.*, 48, 1503.

Kamke, E. (1983). *Differentialgleichungen, Bd.1, Gewhnliche Differentialgleichungen.* Erscheinungsdatum.

Kane, B. E. (1998). A silicon-based nuclear spin quantum computer. *Nature*, 393, 133–137.

Kapale, K. T., Zhu, M. O., & Zubairy, M. S. (2003). Quenching of spontaneous emission through interference of incoherent pump processes. *Phys. Rev. A*, 67, 023804.

Kartashov, Y. V., Vysloukh, V. A., Marti-Panameno, E., Artigas, D., & Torner, L. (2003a). Dispersion-managed conoidal pulse trains. *Phys. Rev. E* (pp. 026613).

Kartashov, Y. V., Vysloukh, V. A., & Torner, L. (2003b). Conoidal wave patterns in quadratic nonlinear media. *Phys. Rev. E* (pp. 066612).

Kerker, M. (1969). *The Scattering of Light, and Other Electro-Magnetic Radiation.* Academic Press.

Khodjasteh, K. & Lidar, D. A. (2003). Quantum computing in the presence of spontaneous emission by a combined dynamical decoupling and quantum-error-correction strategy. *Phys. Rev. A*, (pp. 022322 (6 pages)).

Kien, F. L., Quang, N. H., & Hakuta, K. (2000). Spontaneous emission from an atom inside a dielectric sphere. *Opt. Commun.*, 178, 151.

Kittel, C. (1976). *Introduction to Solid State Physics.* John Wiley and Sons, New York.

Klimov, V., Ducloy, M., & Letokhov, V. (1996). Spontaneous emission rate and level shift of an atom inside a dielectric microsphere. *J. Mod. Optics.*, 43, 549–563.

Klimov, V. & Letokhov, V. (1999). Enhancement and inhibition of spontaneous emission rates in nanobubbles. *Chem. Phys. Lett.*, 301, 441–448.

Klimov, V. V., Ducloy, M., & Letokhov, V. S. (1999). Strong interaction between a two-level atom and the whispering-gallery modes of a dielectric microsphere: Quantum-mechanical consideration. *Phys. Rev. A*, 59, 2491–3014.

Knoll, L., Scheel, S., & Welsch, D.-G. (2000). Coherence and statistics of photons and atoms. *arXiv:quant-ph/0006121.*

Knoll, L., Scheel, S., & Welsch, D.-G. (2001). QED in dispersing and absorbing media. *arXiv:quant-ph/0006121v431May.* (pp. arXiv:quant–ph/0006121 v4 31 May).

Kogelnik, H. & Li, T. (1966). Laser beams and resonators. *Proc. IEEE.*, 54, 1312.

Korn, G. A. & Korn, T. (1961). *Mathematical Handbook for Scientists and Engineers.* McGraw-Hill Book Company.

Kumar, S. A., Nagendra, C. L., Shanbhogue, H. G., & Thutupalli, G. K. M. (1999). Near-infrared bandpass filters from Si/SiO_2 multilayer coatings. *Optical Engineering*, 38, 368.

Laboratories, B. & Inc (2001). TechNote 101 ProActive Microspheres. (www.bangslabs.com).

Lai, H. M., Leung, P. T., & Young, K. (1990a). Limitation on the photon storage lifetime in electromagnetic resonances of highly transparent microdroplets. *Phys. Rev. A*, 41.

Lai, H. M., Leung, P. T., Young, K., Barber, P. W., & Hill, S. C. (1990b). Time-independent perturbation for leaking electromagnetic modes in open systems with application to resonators in microdroplets. *Phys. Rev. A*, 41, 5187–5198.

Lai, H. M., Leung, T., & Young, K. (1988). Electromagnetic decay into a narrow resonance in an optical cavity. *Phys. Rev. A*, 37, 1597.

Lam, C. C., Leung, P. T., & Young, K. (1992). Explicit asymptotic formulas for the positions, withstand strengths of resonances in M. *J. Opt. Soc. Am. B*, 9, 1585–1592.

Landau, L. D. & Lifshits, E. M. (1975). *The Classical Theory of Fields.* Pergamon Press.

Landau, L. D. & Lifshits, E. M. (1977). *Quantum Mechanics (Non-Relativistic Theory), 3rd* edition. Oxford, England: Pergamon Press.

Landau, L. D., Lifshitz, E. M., & Pitaevskii, L. P. (1984). *Electrodynamics of Continuous Media.* Pergamon, Oxford.

Lange, S. & Schweiger, G. (1994). *J. Opt. Soc. Am. B*, 11, 2444–2451.

Lau, H. T. (2003). *A Numerical Library in Java for Scientists and Engineers.* CRC Press.

Lee, Y. & Yamanishi, M. (1995). Theory on the effect of optical dephasing on spontaneous emission in microcavities with dispersive dielectric media. *Phys. Rev. A*, 52, 2312–2318.

Lelong, P. & Lin, S. H. (2002). Polaron model in self-assembled InAs/GaAs quantum dots — A perturbative approach. *Appl. Phys. Lett.*, 81, 1002–1004.

Lewenstein, M., Zakrzewskit, J., & Mossberg, T. W. (1988). Spontaneous emission of atoms coupled to frequency-dependent reservoirs. *Phys. Rev. A*, 38.

Lewis, G. M., Smowton, P. M., Thomson, J. D., Summers, H. D., & Blood, P. (2002). Measurement of true spontaneous emission spectra from the facet of diode laser structures. *Appl. Phys. Lett.*, 80, 1–3.

Li, L.-W., Kang, X.-K., Leong, M.-S., Kooi, P.-S., & Yeo, T.-S. (2001). Electromagnetic dyadic Green's functions for multilayered spheroidal structures. I: formulation. *IEEE Trans. Microwave Theory Tech.*, 49, 532.

Li, L.-W., Kooi, P.-S., Leong, M.-S., & Yee, T.-S. (1994). Electromagnetic dyadic Green's function in spherically multilayered media. *IEEE Trans. Microwave Theory Tech.*, 42, 2302–2310.

Li, Z.-Y., Lin, L.-L., & Zhang, Z.-Q. (2000). Spontaneous emission from photonic crystals: full vectorial calculations. *Phys. Rev. Lett.*, 84, 4341.

Liboff & L., R. (1998). *Introductory Quantum Mechanics.* Addison-Wesley Publishing Company.

Lin, H.-B., Eversole, J. D., Meritt, C., & Campillo, A. (1992). *Phys. Rev. A*, 45, 6756–6760.

Lippman, S. B. (2002). *C# Primer: A Practical Approach.* Pearson Education.

Lorenz, L. V. (1890). *Vidensk. Selsk. Shrifter.*, 6, 1.

Loudon, R. (1994). *The Quantum Theory of Light.* Clarendon Press-Oxford.

Louisell, W. H. (1964). *Radiation and Noise in Quantum Electronics.* McGraw-Hill Book Company.

Love, A. E. (1899). *Proc. Lond. Math. Soc.*, 30, 308.

Lu, W., Pfeiffer, L., & Rimberg, K. W. W. A. J. (2003). Real-time detection of electron tunnelling in a quantum dot. *Nature*, 423, 422–425.

Macovei, M. & Keitel, C. H. (2003). Laser control of collective spontaneous emission. *Phys. Rev. Lett.*, 91, 123601.

Maitland, A. & Dunn, M. (1969). *Laser Physics.* North-Holland Publishing Company.

Makino, T., Andre, R., Gerard, J.-M., Romestain, R., Dang, L. S., Bartels, M., Lischka, K., & Schikora, D. (2003). Single quantum dot spectroscopy of CdSe/ZnSe grown on vicinal GaAs substrates. *Appl. Phys. Lett.*, 82, 2227–2229.

Maksimenko, S. A., Slepyan, G. Y., Ledentsov, N. N., Kaloshaf, V. P., Hoffmann, A., & Bimberg, D. (2000). Light confinement in a quantum dot. *Semicond. Sci. Technol.*, 15, 491–496.

Mandel, L. & Wolf, E. (1995). *Optical Coherence and Quantum Optics.* Cambridge University Press.

Matloob, R. (2001). Canonical theory of electromagnetic field quantization in dielectrics. *Opt. Commun.*, 192, 287.

Meystre, P. & III., M. (1998). *Elements of Quantum Optics.* Springer, London.

Mie, G. (1908). *Ann. Phys.*, 25, 377 (Leipzig).

Miyazaki, H., Miyazaki, H., Ohtaka, K., & Sato, T. (2000). Photonic band in two-dimensional lattices of micrometer-sized spheres mechanically arranged under a scanning electron microscope. *J. Appl. Phys.*, 87, 7152.

Moller, B., Artemyev, M. V., Woggon, U., & Wannemacher, R. (2002). Mode identification in spherical microcavities doped with quantum dots. *Appl. Phys. Lett.*, 80, 3253.

Morgan, M. (1999). *Java 2 For Professional Developers.* SAMS.

Nicolai, J. (1999). *The C++ Standard Library: A Tutorial and Reference.* Addison-Wesley.

Oraevskii, A. N. (1994). Spontaneous emission in a cavity. *Uspekhi Fizicheskikh Nauk (Phys. -Uspehi)*, 164, 415–427.

Oraevsky, A. N. (2002). Waves of whispering gallery. *Kvant. Elektron.*, 32, 377–400.

Ortega, J. M. & Grimshaw, A. S. (1999). *An Introduction to C++ and Numerical Methods.* Oxford University Press, Inc., Oxford, New York.

Ozawa, M. (2002). Conservative quantum computing. *Phys. Rev. Lett.*, 89, 057902.

Pachos, J. K. & Knight, P. L. (2003). Quantum computation with a one-dimensional optical lattice. *Phys. Rev. Lett.*, 91, 107902.

Panofsky, W. & Phillips, M. (1962). *Classical Electricity and Magnetism.* Addison-Wesley Publishing Company, Reading, Massachusetts.

Pekar, V. S. (1975). *Sov. Phys. -JETP.*, 40.

Pendleton, J. D. & Hill, S. C. (1997). Collection of emission from an oscillating dipole inside a sphere: Analytical integration over a circular aperture. *Appl. Opt.*, 36, 8729–8737.

Perelman, A. Y. (1996). Scattering by particles with radially variable refractive indices. *Appl. Opt.*, 35, 5452–5460.

Potter, D. (1973). *Computational Physics.* John Wiley and Sons.

Press, H., Teukovsky, W. A., Vetterling, S. T., & Flannery, B. (2002). *Numerical Recipes in C++.* Cambridge, University Press, Cambridge.

Press, W. H., Flannery, B., Teukolsky, S. A., & Vetterling, W. (1992). *Numerical Recipes in Fortran.* Cambridge University Press.

Press, W. H., Flannery, B., Teukolsky, S. A., & Vetterling, W. (1993). *Numerical Recipes in C: The Art of Scientific Computing.* Cambridge University Press; 2nd edition.

Purcell, E. M. (1946a). Spontaneous emission probabilities at radio frequencies. *Phys. Rev.*, 69, 681.

Rahman, A. & Bryant, G. W. (2002). Spontaneous emission in microcavity electrodynamics. *Phys. Rev. A*, 65, 033817.

Ramo, S., Whinnery, J. R., & Duzer, T. V. (1994). *Fields and Waves in Communication Electronics.* Wiley, New York.

Raussendorf, R., Browne, D. E., & Briegel, H. J. (2003). Measurement-based quantum computation on cluster states. *Phys. Rev. A* (pp. 022312 (32 pages)).

Rayleigh, J. W. L. (1910). *Philos. Mag.*, 20, 1001.

Riley, K. F., Hobson, M. P., & Bence, S. J. (1998). *Mathematical Methods for Physics and Engineering.* Cambridge University Press.

Rogobete, L., Schniepp, H., Sandoghdar, V., & Henkel, C. (2003). Spontaneous emission in nanoscopic dielectric. *Optics Letters*, 28, 1736–1738.

Roll, G., Kaiser, T., & Schweiger, G. (1999). Eigenmodes of spherical dielectric cavities: Coupling of internal and externalrays. *J. Opt. Soc. Am. A*, 16, 882.

Ruppin, R. (2000). Surface polaritons of a left-handed medium. *Phys. Lett. A*, 277, 61.

Ruppin, R. (2002). Electromagnetic energy density in a dispersive and absorptive material. *Phys. Lett. A*, 299, 309–312.

Sakaguchi, S. & Kubo, S. (1999). Transmission characteristics of multilayer films composed of electro-optic and dielectric materials. *Opt. Commun.*, 170, 187–191.

Sanchez-Mondragon, J. J., Narozhny, N., & H.Eberly, J. (1983). Theory of spontaneous-emission line shape in an ideal cavity. *Phys. Rev. Lett.*, 51, 550.

Sasaki, K., Fujiwara, H., & Masuhara, H. (1997). Photon tunneling from an optically manipulated microsphere to a surface by lasing spectral analysis. *Appl. Phys. Lett.*, 70, 2647–2649.

Scalora, M., Bioemer, M. J., Pethel, A. S., Dowiing, J. P., & Manka, C. M. B. A. S. (1998). Transparent, metallo-dielectric, one-dimensional, photonic band-gap structures. *J. Appl. Phys.*, 83, 2377.

Scheel, S., Knoll, L., & Welsch, D.-G. (1998). QED commutation relations for inhomogeneous Kramers-Kronig dielectrics. *Phys. Rev. A*, 58, 700–706.

Scheel, S., Knoll, L., & Welsch, D.-G. (1999a). Quantum local-field corrections and spontaneous decay. *Phys. Rev. A*, 60, 1590–1597.

Scheel, S., Knoll, L., & Welsch, D.-G. (1999b). Spontaneous decay of an excited atom in an absorbing dielectric. *Phys. Rev. A* (pp. 4094–4104).

Schiller, S. (1993). Asymptotic expansion of morphological resonance frequencies in Miescattering. 32, 2181–2185.

Schiller, S. & Byer, R. L. (1991). High-resolution spectroscopy of whispering gallery modes inlarge dielectris spheres. 16, 1138–1140.

Schniepp, H. & Sandoghdar, V. (2002). Spontaneous emission of europium ions embedded in dielectric nanospheres. *Phys. Rev. Lett.*, 89, 257403.

Scully, O. & Zubairy, M. (1996). *Quantum Optics.* Cambridge University Press.

Shifrin, K. S. & Zolotov, I. G. (1993). *Appl. Opt.*, 32, 5397.

Slepyan, G. Y., Maksimenko, S. A., Hoffmann, A., & Bimberg, D. (2002). Quantum optics of a quantum dot: Local-field effects. *Phys. Rev. A*, 66, 063804.

Smith, D. D. & Fuller, K. A. (2002). Photonics bandgaps in Mie scattering by concentrically stratified spheres. *J. Opt. Soc. Am. B*, 19, 2449–2455.

Solimeno, S., Grosignani, B., & DiPorto, P. (1986). *Guiding, Diffraction, and Confinement of Optical Radiation.* Academic Press, Inc, New York.

Songmuang, R., Kiravittaya, S., & Schmidt, O. G. (2003). Formation of lateral quantum dot molecules around self-assembled nanoholes. *Appl. Phys. Lett.*, 82, 2892–2894.

Sorensen, A. S. & Molmer, K. (2003). Measurement induced entanglement and quantum computation with atoms in optical cavities. *Phys. Rev. Lett.*, 91, 097905.

Spillane, S. M., Kippenberg, T. J., & Vahala, K. J. (2002). Ultralow-threshold Raman laser using a spherical dielectric microcavity. *Nature*, 415, 621–623.

Stratton, A. (1941). *Electromagnetic Theory.* McGraw-Hill, New York.

Stratton, J. A. & Chu, L. J. (1939). Diffraction theory of electromagnetic waves. *Phys. Rev.*, 58, 99–107.

Sullivan, K. (1994). Radiation in spherically symmetric structures. I. The coupled-amplitude equations for vector spherical waves. *Phys. Rev. A*, 50, 2701–2707.

Sullivan, K. & Hall, D. (1994). Radiation in spherically symmetric structures. II. Enhancement and inhibition of dipole radiation in a spherical Bragg cavity. *Phys. Rev. A*, 50, 2708–2718.

Sze, S. M. (1969). *Physics of Semiconductor Devices.* Wiley-Interscience, A division of John Wiley and Sons, New York.

Thijssen, J. M. (2001). *Computational Physics.* Cambridge University Press.

Tom, A. (2001). *Inside C#.* Microsoft Press.

Tomas, M. S. & Lenac, Z. (1999). Spontaneous-emission spectrum in an absorbing Fabry-Perot cavity. *Phys. Rev. A*, 60, 2431–2437.

Tsipenyuk, Y. M. (1965). *High-Power Electronics. Issue 4 (In Russian).* Nauka.

Ueha, S. (1998). Ultrasonic actuators using near-field acoustic levitation. *IEEE Intern. Ultrasonics Symp. Sendai. Miyagi. Japan. October* (pp. 5–8).

Ujihara, K. & Dung, H. T. (2002). Two-atom spontaneous emission in a planar microcavity. *Phys. Rev. A*, 66, 053807.

Vahala, K. J. (2003). Optical microcavities. *Nature*, 424, 839.

Vainstein, L. A. (1969). *Open Resonators and Open Waveguides.* Golem Press, Boulder, Colorado.

Vainstein, L. A. (1988). *Electromagnetic Waves.* Radio I Svyaz', Moscow.

van de Hulst, H. (1946). *Optics of Spherical Particles.* Drukkerij J. F. Duwaer.

Vassiliev, V. V., Velichansky, V. L., Ilchenko, V. S., Gorodetsky, M. L., Hollberc, L., & Yarovitsky, A. V. (1998). Narrow-line-width diode laser with a high-Q microsphere resonator. *Opt. Commun.*, 158, 305.

Vernooy, D. W., Ilchenko, V. S., Mabuchi, H., Streed, E. W., & Kimble, H. J. (1998). High-Q measurements of fused-silica microspheresin the near infrared. *Opt. Lett.*, 23, 247–249.

Viana, N. B., Freire, R. T. S., & Mesquita, O. N. (2002). Dynamic light scattering from an optically trapped microsphere. *Phys. Rev. E*, 65, 041921 (11 pages).

Vlasenko, N. A., Pekar, S. I., & Pekar, V. S. (1973). *Sov. Phys. JETP.*, 37, 223.

Vrijen, R., Yablonovitch, E., Wang, K., Jiang, H. W., Balandin, A., Roychowdhury, V., & DiVincenzo, T. M. D. (2000). Electron-spin-resonance transistors for quantum computing in silicon-germanium heterostructures. *Phys. Rev. A* (pp. 012306 (10 pages)).

Vukocvic, J. & Yamamoto, Y. (2003). Photonic crystal microcavities for cavity quantum electrodynamics with a single quantum dot. *Appl. Phys. Lett.*, 82, 2374–2376.

Wang, R. T. & van de Hulst, H. C. (1991). Rainbows: Mie computations and the Airy approximation. *Appl. Opt.*, 30, 106–117.

Wang, T. H., Li, H. W., & Zhou, J. M. (2003). Characteristics of a field-effect transistor with stacked InAs quantum dots. *Appl. Phys. Lett.*, 82, 3092–3094.

Whithem, G. B. (1974). *Linear and Nonlinear Waves.* John Wiley and Sons.

Woggon, U. (1996). *Optical Properties of Semiconductor Quantum Dots.* Springer, Berlin.

Wu, Z. S., Guo, L. X., Ren, K. F., Gouesbet, G., & Grhan, G. (1997). Improved algorithm for electromagnetic scattering of plane waves and shaped beams by multilayered spheres. *Appl. Opt.*, 36, 5188–5198.

Wubs, M., Suttonp, L. G., & Lageudijk, A. (2003). Multipole interaction between atoms and their photonic environment. *Phys. Rev. A*, 68, 013822 (16 pages).

Wylie, J. M. & Sipe, J. (1984). Quantum electrodynamics near an interface. *Phys. Rev. A*, 30, 1185–1193.

Xu, Y., Lee, R. K., & Yariv, A. (2000a). Finite-difference time-domain analysis of spontaneous emission in a microdisk cavity. *Phys. Rev. A*, 61, 033808.

Xu, Y., Lee, R. K., & Yariv, A. (2000b). Quantum analysis and the classical analysis of spontaneous emission in a microcavity. *Phys. Rev. A*, 61, 033807.

Yablonovich, E. (1987). Inhibited spontaneous emission in solid-state physics and electronics. *Phys. Rev. Lett.*, 58, 2059.

Yamasaki, T., Sumioka, K., & Tsutsui, T. (2000). Organic light-emitting device with an ordered monolayer of silica microspheres as a scattering medium. *Appl. Phys. Lett.*, 76, 1243.

Yang, D. (2000). *C++ and Object-Oriented Numeric Computing for Scientists and Engineers.* Springer-Verlag, New York.

Yang, W. (2003). Improved recursive algorithm for light scattering by a multilayered sphere. *Appl. Opt.*, 42, 1710–1720.

Yeh, P. (1988). *Optical Waves in Layered Media.* John Wiley and Sons, New York.

Yi, S.-S. & Stafsudd, O. M. (1999). Observation of lossless radiative modes of a dielectric sphere. *J. Appl. Phys.*, 86, 3694–3698.

Yokoyama, H. & Ujihara, K. (1995). *Spontaneous Emission and Laser Oscillation in Microcavities.* CRC Press.

Young, M. J. (1998). *Mastering Visual C++ 6.* Sybex; Book and CD-ROM edition.

Zeldovich, Y. B. (1961). *Sov. Phys. JETP.*, 12, 542.

Zhang, W., Chang, C. T., & Sheng, P. (2001). Multiple scattering theory and its application to photonic band gap systems consisting of coated spheres. *Phys. Rev. E*, 8, 203–208.

Zhang, W., Lei, X. Y., & Wang, Z. L. (2000). Robust photonic band gap from tunable scatterers. *Phys. Rev. Lett.*, 84, 2853.

Zhu, Y., Gauthier, D. J., Morin, S. E., Wu, Q., Carmichael, H. J., & Mossberg, T. W. (1990). Vacuum rabi splitting as a feature of linear-dispersion theory: Analysis and experimental observations. *Phys. Rev. Lett.*, 64, 2499.

Index

3D-confined photon states, 55

action, 32
active kernel, 121
active microsphere, 121
address-of operator, 222
amplitude, 7
angular momentum index, 66
angular part, 16
annihilations, 164, 260
anomalous resonances, 81
array, 223
arrays of silica microspheres, 53
asymptotic formula, 190
average Hamiltonian, 36
averaged Lagrangian, 35
azimuthal quantum numbers, 200

backscattered light, 54
backward, 41
backward spherical, 21
base class, 218, 243
Bessel function, 59
Borland *C++*, 219
Borland *C++* Builder, 220
boundary condition, 41

C, 220
C#, 217, 220, 221
C++, 217, 219, 220, 221, 250
C++ operators, 222
catch-block, 250
class hierarchy, 240
classical field, 157
coated microsphere, 124, 129
coated sphere, 84
coaxial jets, 54
coefficient of losses, 95, 97
coherent interaction, 51
coherent state, 208
collapse, 193
Combobox, 259
complex conjugate, 6
complex roots, 87, 270
complex vectorial, 293
conditional statements, 225
conducting sphere, 68
confinement of electromagnetic
oscillations, 113
constructor, 240
Coulomb gauge, 8, 147
coupling, 50
creation, 164
critical coupling, 53

damped oscillator, 146
Debye potential, 13, 21, 24, 187, 349
decoherence time, 201
degeneracy, 10, 180
density of the electric current, 4
derived class, 218, 241–243
destructor, 240
dielectric and magnetic
permittivities, 4

dielectric permittivity, 102
dielectric sphere, 102
dipole approximation, 10, 158
dipole moment, 149
dipole orientation, 187
Dirac notations, 157
dispersive stack, 118
dissipative force, 31
distributed Bragg reflector, 95
Do-while Loops, 230
dynamic dipole moment, 176
dynamic polarization, 150, 193, 197
dynamical electrical polarization, 192
dynamical tunneling, 108

eigenfrequencies, 26, 27, 87, 89, 92, 98, 106, 109, 117, 124
eigenfrequencies equation, 28, 29, 86, 105, 124
Einstein coefficient, 63
elastic scattering, 58
electric $\vec{E}$ and magnetic, 3
electric charge, 4
electromagnetic waves, 6
energy in a layered microsphere, 36
energy of the spherical system, 25
enhancement in the spontaneous emission, 55
equidistant resonances, 103
Euler–Maclaurin, 183
Euler-Lagrange equation, 33, 35
event, 259
exception, 250
excited state, 185
exciton resonance, 123
exponential part, 20

fast Fourier transformation, 154
Feynman's quantum-mechanical computer, 51
field energy, 23
field per photon, 27, 162, 187
fields' quantization, 162
flux of energy, 23
form, 259
FORTRAN, 217, 219
forward, 21, 41
frequency, 5
frequency dispersion, 30, 36
frequency of collision, 103
frequency resonances, 82
function declarations and definitions, 230
function prototypes, 240
function types, 236

gain factor, 123
general solution, 132
graphic user interface (GUI), 219, 249, 258, 259
Green tensor, 62
Green's-function technique, 14

Hamiltonian of field, 34
Hankel spherical functions, 39
harmonic generation, 129
header file, 240
Heisenberg picture, 146
Helmholtz equation, 13
higher-order peaks, 197
homogeneous case, 19

impedance, 41, 43, 45
increments and decrements, 225
influence of sphericity, 47
inherit, 246
inheritance, 246
inhomogeneous equations, 8
inhomogeneous waves, 21
initial conditions, 181
initialization, 224
interaction picture, 168, 170, 171, 175
interaction representation, 169
interference, 10, 203
interference pattern, 102, 111
intermediate case, 187
inverse populations, 32
inverted population, 122
ionic crystal, 114
iteration statements for loops, 227

Java, 217, 221
Jaynes-Cummings model, 186

Lagrangian, 32–34
Laplace operator, 58
lasing emission, 52
leakage, 87
leakage energy, 203
light absorption, 72
loaded cavity, 79
Lorentz–Mie, 57
Lorenz gauge, 8
low-threshold microlasers, 55

magnetic and electric fields, 21
material dispersion, 119, 120
material losses, 45, 87, 88
matrix element, 147
matrix form, 158
matrix operations, 293
Maxwell equations, 3
measurements of the quality factor, 54
metal-coated spheres, 81
metallized microsphere, 28, 41
metallized sphere, 94
metallo-dielectric spherical stack, 204
metallo-dielectric structure, 101
Microsoft, 219
microsphere, 143
Mie coefficient, 60
Mie scattering function, 60
mode density, 82
motion equations, 187
multilayered system, 87
multiple perturbations, 77

nanoclusters, 121
narrow transparency gap, 48
non-resonant background, 196
non-uniform case, 130
non-uniform layers, 131, 134
normalization, 66
not-uniform light-absorbent inclusions, 72
numerical methods, 184, 215, 220
numerical simulation, 216

object is, 240
object-oriented approach, 221
object-oriented technology, 215, 219
one-photon state, 175
open cavity, 76
operator functions, 248
operator op, 248
operators, 164, 248
operators of creations, 260
optical feedback, 53
"optimal" sphere size, 51
optical gain, 122
optical pumping, 53
optical radiation, 127
or field per photon, 167
ordinary differential equations (ODE), 284
organic–inorganic hybrid microspheres, 54
overcoated microspheres, 51
overloading, 247

parameter (or formal argument), 236
particular solution, 132
passing functions as argument, 237
periodicity, 43
phase velocity, 5, 6
photonic band gaps, 81
photonic dot, 50
plane wave, 6
plasma frequency, 32, 103
Poisson equation, 134
population, 153
potentials, 13
potentials of field, 7
Poynting vector, 23
private, 240
private level, 238
probability amplitudes, 146, 148
protected, 238, 240
public, 238, 240

Q factor, 88, 99, 121, 125, 127
quality factor, 69
quality factor degradation, 72
quantization, 157
quantization procedures, 26
quantized, 260
quantized electromagnetic, 218
quantum dots, 121
quantum interference, 209
M mirrors, 51
quarter-wave case, 43, 85
quarter-wave layers, 83
Quarter-wave stack, 95
quasi-stochastic, 194
quasi-stochastic dynamics, 197

Rabi effects, 211
Rabi frequency, 154, 156, 159
Rabi oscillations, 193
Rabi splitting, 55, 197, 211
radial, 200
radial distribution, 84, 91–93, 99, 108, 109, 119
radial fields distribution, 29
radial structure, 128, 190
radiating losses, 184
radiative losses, 80
random deviation, 46, 89
random factor, 46
rapid application development, 255
recurrence algorithm, 130
recursive functions, 234
recursive relations, 39
reflectance, 106
reflection coefficient, 41, 43, 45, 95, 135
reflections, 21
refraction index, 6, 21, 41, 45
regime of generation, 123
Relational expressions, 229
resonance frequency, 55
resonant passbands, 107
retarded case, 9
Riccati–Bessel function, 59
rotating-wave approximation, 178, 186
Runge-Kutta method, 284
RWA (rotative-wave approximation), 160

saturation, 88
scalar potential, 8, 33
scalars, 13
Schrödinger equation, 157
Schrödinger picture, 168, 171, 175
second harmonic generation, 81
semiconductor nanocrystals, 50
separation variables method, 15
shoot method, 134
skin-depth, 102
Sommerfeld's radiation conditions, 41
spatial distribution, 98
spherical, 200
spherical Bessel functions, 19
spherical cavity, 65
spherical coordinates, 13
spherical dispersive stack, 115
spherical function, 24, 352
spherical geometry, 23
spherical Hankel function, 234
spherical harmonic, 59
spherical layers, 84
spherical microcavity, 50
spherical stack, 43
spheroidal dielectric microresonators, 52
spliting, 156
spontaneous emission, 50, 62, 82, 129, 141, 182, 184, 185, 324
stack, 39
steepness, 45
structures, 224
superposition, 7, 21
Switch Statements, 226

TComboBox, 259
TE case, 16, 19
TE wave, 10–12, 16, 18, 22
TEdit, 259
TeeChart, 220
tensor Levi-Civita, 33
The Variational Principle, 32

threshold of generation, 128
TLabel, 259
TM case, 18, 21
TM wave, 10–12, 14, 15
TMemo, 259
TRadioGroup, 259
transfer matrix, 39
transfer matrix method, 37, 131
transmission peaks, 210
transmissions, 21
transmittance, 95, 106
transmittance coefficient, 41
triple photon states, 201
try-block, 250
TType type, 222
tuning, 52
tunneling, 106
two-level, 185
two-level atom, 143, 145
typedef., 236

upconversion lasing, 52

vacuum field, 166
vacuum state, 175, 185
variable, 222, 240
variations, 32, 35
vectorial, 8
vectorial $\vec{A}_M$, 13
vectorial and scalar potentials, 16
vectorial potential, 33
Virtual functions, 243

wave equation, 5
wavelength, 6
waves number, 5
Weisskopf-Wigner approach, 186
While Loops, 229
whispering gallery mode, 82
whispering gallery modes, 50, 184
Whitham's average variational principle, 34
Wigner function, 207, 208
Wronskian, 42